SURVEYING BUILDINGS

by
Malcolm Hollis
BSc(Est.Man.) FRICS FSVA FIAS ACIArb
Chartered Building Surveyor

and

Charles Gibson
MA(Oxon)
of Lincoln's Inn
Barrister

R.I.C.S. BOOKS

12 Great George Street, London SW1P 3AD, England

No responsibility for loss occasioned to any person acting or refraining from action as a result of any material in this publication can be accepted by the authors, publisher or the Royal Institution of Chartered Surveyors. The views expressed and the conclusions drawn in this book are those of the authors.

First published July 1983
Second edition published June 1986
Third edition published January 1991

Photographs
The photographs on pages 152 to 164 are reproduced by permission of Blue Circle Enterprises, Decorative Products Division.
The photographs on pages 71, 72, 211 and 218 to 227 are Crown Copyright and are reproduced by permission of the Building Research Establishment, DOE.
All the remaining photographs are the work of Malcolm Hollis.

RICS Books
12 Great George Street,
London SW1P 3AD, England.

©*RICS 1990*. Copyright in all or part of this publication rests with the RICS, and save by prior consent of the RICS, no part or parts of this publication shall be reproduced in any form or by any means electronic, mechanical, photocopying, recording or otherwise, now known or to be devised.

ISBN 0 85406 464 8

Design and Production by Martin Ash Graphics, Ticehurst, Sussex.
Printed by Otter Print Limited,
Tunbridge Wells, Kent.

SURVEYING BUILDINGS

Contents

What is a Survey? — Page 3
Why a survey may be required
Taking instructions
What does the Client want to know?
Extent of the inspection
Fees

The Contract — 5
When and how it is made
The contractural relationship
Misrepresentation
The contents of the contract in:
    Common Law
    Statute
    Exemption Clauses
Misrepresentation Act 1967
Unfair Contract Terms Act 1977
Supply of Goods and Services Act 1982
The drafting of the contract
Confirmation of instructions
Standard letters for confirming instructions, making appointment to inspect
The appointment to inspect, consents that are required

Checklist in preparing for a survey — 14

The Visit and the Report — 15
The report, cost of repairs, instructions, contents of the report, conclusions, pro-formas, standard clauses, examples of standard clauses

Additional considerations for a survey of flats and apartments — 23
Lifts, leases, caretakers' accommodation, scope of building to be inspected, services

Additional considerations for a survey of commercial property — 26
Access, scope of the inspection and report, services, future maintenance liability, statutory contracts, contents of the report, cost in use, vandalism

The duties of the surveyor, their performance and their breach — 29
The principle of duty, the application of the principle of duty. The surveyor's neighbour in Law, Yianni v Edwin Evans & Sons. Performance and breach of duty, the standard required, with examples of the failure to carry out instructions, inadequate knowledge or experience, inadequate inspection and report. Protection against liability, to the Client, to third parties, to the building owner

The consequences of breach of duty — 37
The principle of compensation, contractural fees, causation and remoteness. The position of the surveyor where claims are made by Clients who purchase the surveyed property, damages, the examination of the trend of damages settled in court decisions. The position where Clients do not purchase the property. Heads of damage common to all claims, contributory negligence

Surveying Equipment — 45
The RICS list of recommended equipment required to carry out a property survey:
    Detailed analysis of equipment, ladders, platforms to gain access
    Diagnosis of damp, carbide meters, electrical resistance meters, laboratory testing
    Measurement of humidity
    Leak detection, ultrasonic detection, sprays
    Concrete reinforcement location and testing for corrosion
    Ultrasonic testing of materials
    Laser measurement
    Sonic measurement
    Photogrammetry
    Infra red thermography
    Infra red photography
    Movement monitoring
    Video
    Electrostatic meters
    Light measurement
    Stud sensor
    Manometer
    Sound measurement
    Scopes, rigid and flexible probes with hot or cold light source
    Metal detection
    Radon detectors
    Electrical supply testing
    Level meter
    Impulse radar
    On site tests: rapid test method for cement content of mortar and chemical analysis of plaster
    Sundry: ventometer, crack gauge, identification and face mask
    Sick building syndrome: water-borne bacteria, drinking water supply, breeding grounds for bacteria and sundry sources of contamination
    How healthy is the house?

The inspection of prefabricated reinforced concrete dwellings — 69
The identification of the building type, research into the common failures of each type, common problems, the method of inspection, the identification of the necessary repairs, typical appearance of the building types most commonly encountered.

	Page
Roofs	73

Flat roof failures, asphalt, roofing felts, parapets, edges and verges, metal flat roofs, pitched roofs, tiles, slates, roof shape, thatch, timber, shingles, asbestos cement, lichen, patio surfaces

Walls	93

Brickwork, types of pointing, mathematical tiles, render, harling, pebbledash, balconies, tile hanging, flashings and fillets, cavity brickwork, wall ties, wall tie failure, cavity insulation, brick defects

Gutters and downpipes	105
Drainage	110
Stone	111

Origin, location, causes of decay, method of cleaning, frost action, weathering, conservation, repairs, repointing, defective stone, redressing, salts. Tables of stone quality and resistance

Concrete	117

Types of concrete, cement, high alumina cement, concrete failures, shrinkage, system building, joint failure

Sealants	123

Performance and characteristics of most sealants, common failures and a guide to recognition

Plastics	129

Performance and characteristics of most plastics and a guide to recognition

Openings	133

Windows, lintels, sills, brick arches, putty, damp penetration, sliding sash windows, sash cords, window locks, burglars, metal and wooden casements, sealants, ventilation

Condensation	139

How it occurs, diagnosis, ventilation, insulation

Timber framed property	141

The inspection of modern timber framed property

Corrosion of metals	143

Relationship and vulnerability of various metals

Asbestos	147

The type of asbestos, the building materials that contain asbestos, the risks and precautions

Frost action	150
Decorations	151

Indications of failure in external paint finishes on various materials

	Page
Cracks in buildings	165

Soils, reaction of various types, characteristics. Subsidence, extent of cracks, categories of crack, repairs required. Clay soils

Radiation in soils	176
Trees	177

Their effect on buildings, root action, safety, estimating the age of a tree, heave in clay where trees have been removed. List of trees in order of risk of property damage, drains, tree failures, Dutch elm disease, branches, fungus on trees

Identification of trees: Crack willow, Lombardy poplar, Oak, Elm, Horse chestnut, Sycamore, Lime, Beech, Plane, Birch.
How to estimate the height of a tree

The importance of accurate observation	194
Timber decay caused by insect attack	195

Various treatments summarised, descriptions of commonly encountered harmless insects. Detailed descriptions of destructive insects: house longhorn beetle, pinhole borers, powder post beetle, furniture beetle, death watch beetle, wharf borer, wood wasp, wood boring weevils, bark borer, bostrich beetle, gribble, shipworm

Timber decay caused by fungus	211

Fungal attack and wood decay, detailed descriptions of harmful fungi: dry rot (*Serpula lacrymans*), wet rot (*Coniophora puteana*), mine fungus (*Fibrioporia vaillantii*), *Paxillus panuoides*, *Donkioporia expansa*, stag's horn fungus (*Lentinus lepideus*), *Daedalia quercina*, slime moulds (*Myxomycetes*), *Phellinus contiguus*, *Poria placenta*. Harmless fungi, mould, problems with woodblock floors

Electrical services	229

Illustrations of obvious failure, information about cables and their age

Appendix 1: Concrete tube testing	231
Appendix 2: testing laboratories	231
Appendix 3: directory of equipment suppliers	232
Bibliography and suggestions for further reading	236
Index	237

Foreword

Surveying buildings is an art. Verifying the cause of a building failure is a science. The surveyor's work involves a combination of both the art and the science.

When I was at school, I was faced with many textbooks which left me with an ambition. I could never understand why I found Ian Fleming's James Bond books capable of being read in one sitting whereas the reading of a single page of a geography textbook was a marathon achievement. My ambition was to try to write a book in which photographs and text appeared on the same page, where the text had some interest and life in it even if it failed to achieve some of the more steamy passages of a Bond novel, and where the overall relationship of text and pictures was pleasing to the eye.

When Lord Denning quoted the opening paragraph of this book in an address at The Royal Institution of Chartered Surveyors in 1984, I felt that I had progressed some way towards achieving that ambition.

No book can be a substitute for practical experience, but I hope that the original and new illustrations which pepper the pages of this third edition will help anyone who is interested in discovering the cause of building failures. If I have achieved this, much of the credit for that success must go to everyone who has contributed to its creation. I particularly wish to thank the staff of RICS Books for all their assistance in producing each edition of this book.

Charles Gibson's friendship as well as his contribution to the legal section of the book, and the correction of the English wot I wrote, has been invaluable. The creation of new friendships with people who share the same enthusiasm for buildings has made the hard work worthwhile.

I wish to acknowledge my debt to the late Professor Oxley for his assistance on the section on dampness. Mike Bailey, both thatcher and surveyor, helped me revise the section on thatching, Geoff Hayes of Blue Circle has continued to add to my photographic knowledge, and Geoff Webb of Jointmasters Limited produced many of the photographs of sealant failures.

Many colleagues and acquaintances have assisted by guiding my attention to interesting failures in buildings up and down the country, and the support of my colleagues at Baxter Payne and Lepper and the University of Reading must be acknowledged.

In my introduction to this edition of Surveying Buildings I refer to the junior surveyors receiving knowledge and experience from the senior members of the profession. I would like to thank Morris Pinto for introducing me to the subject and for passing on to me some of his knowledge.

Professor Malcolm Hollis
University of Reading
November 1990

Introduction

Seven years ago I wrote the introduction to the first edition of this book. In it I pointed out that there was no examination which qualified a person as being competent to inspect and advise upon the failures in a building. There are now over 20 degree and diploma courses in building surveying in universities and polytechnics in the United Kingdom.

What is a surveyor, and what is expected of him? The subject of surveying buildings has moved away from the practical technician into the academic sphere, and come under close scrutiny.

No longer are surveyors just finding defects in the properties they survey. Nowadays they are expected to locate both existing and future problems in a building and its surroundings. They have to anticipate problems in existing buildings, in buildings that are in the course of construction, and buildings when they are no more than a few lines on a piece of paper.

Buildings have become much more complex. A building is more than just the envelope which keeps out the elements, it is now a composite collection of services to create and control the internal climate. The building can be a complex external structure built to comply with economic or site conditions, or designed for the speed of erection required by the accountants who now run the world. The building also influences people who pass by, and casts its influence on other buildings in the neighbourhood.

However, the influence of a building on its surroundings has not been considered in sufficient detail. The windy squares of the 1960's townscapes resulted from inadequate thought being given to the relationship of one building to another. The collapse of a cooling tower is an example of the damage that can be caused by the careless location of buildings. Also consider the poor positioning of some new houses which resulted in the acceleration of air passing between them, which in turn causes water penetration.

The deaths of passers-by who had inhaled minute droplets of contaminated water, and the adverse effect that building components have on the health of the occupiers have made surveyors generally more conscious of the health hazards of buildings.

Surveyors also have to recognise the wider implications for the environment of some building components. For example, the construction industry is a significant user of Chlorofluorocarbons (CFCs), of which about 3,000 tonnes are produced each year in Britain for rigid polyurethane and extruded polystyrene. The rapid growth of concern for the environment has taken the surveyor from the back office and placed him in the public gaze. His responsibilities have expanded as the problems with buildings and their surroundings have increased.

The recognition of a building failure, or the anticipation of a failure occurring, is an art. Most people can see a problem but do not understand the consequences of the failure. The surveyor is a trained observer who can not only note and advise upon the failures that have occurred, but also warn of future problems that may develop within the property.

Surveyors nowadays will have learned their profession within academic establishments, while their predecessors had skills handed down to them by senior members of the profession. It is essential, however, that the profession maintains the correct balance between technical skill and academic knowledge for future generations of surveyors.

What is a Survey?

A survey is the inspection and the investigation of the construction and services of a property which enables the surveyor to advise his client upon the condition of that property. The extent of the inspection must be sufficient to enable him to advise upon the future problems that may occur with the various components of the building. He must also be in a position to advise his client of any alterations that need to be carried out to the property before it will meet the requirements of current statutes and legislation. A survey should however always be part of a property valuation. A valuation of a property for its open market value is outside the concept of a structural survey.

A survey may be required for a large number of reasons. These will include:

1. **The sale of a property**
   (a) A report required by a prospective purchaser who needs to know the condition of his proposed dream house, factory, office, shop, investment etc.
   (b) A report required by a vendor who wishes to disclose to his purchaser the defects or to be prepared for any arguments that are put forward to reduce the price. This is often called for if work has been undertaken which is not available for inspection such as underpinning, alterations, damp-proofing, dry rot eradication etc.

2. **Company sales or changes in ownership**
   The proposed investigation of the assets of a company and their condition prior to the sale or acquisition of a company.

3. **Valuation**
   Valuation purposes. No valuer should ever produce a market valuation of a property without a thorough knowledge of the condition of the buildings involved.

4. **Repair cost budgeting**
   The evaluation of property so that a budget can be produced of annual repairs or maintenance costs.

5. **Pre-emptive maintenance**
   The report requested by the client who is intelligent enough to realise the savings that are made if one systematically replaces building components prior to damage. (The surveying profession should raise a statue to this client when he is born.)

6. **Repair of failures**
   The report which is supposed to tell the client why it fell off and what he should do about it.

The report which is submitted following such a survey should satisfy the agreed instructions which form the contract between client and surveyor but it is questionable whether the extent of the inspection should vary.

Taking instructions

Nearly everyone who carries out surveys is aware of the situation where instructions are received by telephone and where the property must be inspected so quickly that it does not enable the formal confirmation of those instructions to have taken place before the inspection occurs. In many cases the client does not have adequate time to respond to any error in the letter confirming the instructions.

Even the most efficient person is unlikely to be able to prepare, write, send and have his letter confirming the instructions received by the client within 24 to 36 hours of the initial request to perform the duty, let alone have a signed copy of his letter returned to complete the contract with his client.

Where the request is to report on part of the building or a specific element of the services, it is essential that the client should be given ample opportunity to receive a letter and to make his observations upon the instructions which have been confirmed. The letter should be drafted and read over on the telephone to the client prior to the inspection being made if there is not sufficient time for the the post office to transport one's epistle prior to the date of the inspection.

The taking of instructions relating to the carrying out of inspections and surveys is important. In many busy practices this work is carried out by assistants and secretarial staff. It is essential that they have a full knowledge of the work which is required in carrying out an inspection and anybody who is going to fulfil this function must have accompanied a surveyor on a number of occasions. The client's initial contact with an office must give him a feeling of confidence. The surveyor or his staff can achieve this by setting out the work which is carried out by the surveyor and those items of the inspection which they do not carry out, or for which specialists are retained.

It is not always practical for the surveyor to speak to the client prior to the carrying out of a survey, even though such contact can be very valuable. As most clients seem reluctant to bring their building into the surveyor's office, the surveyor spends his working day visiting properties in order to report on their condition. This absence from his office desk does not endear the surveyor to his clients as he is rarely available on the telephone. The message dealing with the instructions should be clear and precise and should answer the following questions.

1. **What does the client want to know?**
   From this it will be determined whether the surveyor has to carry out –
   (a) a structural survey;
   (b) an appraisal relating to proposed alterations of the premises;
   (c) an inspection so that he can advise certain alterations if work can be carried out;
   (d) an inspection so that he can advise if certain remedial work should be carried out;
   (e) an evaluation of the cost of rebuilding the property for insurance purposes;

WHAT IS A SURVEY?

(f) a limited inspection so that he can comment on a specific problem or a specific part of a building.

2. The date by which the information is required

If an exchange of contracts has been set for a specific date, it is essential that the surveyor has that information so that he can arrange his inspection and report accordingly.

3. The approximate size, shape and age of the building which is to be inspected

This is important as it gives most surveyors a fighting chance of being able to assess the approximate amount of time that his inspection may take. It is practical on a September evening to consider leaving the office at 3 p.m. to do an inspection of a one room flat, but it is not possible to complete an inspection when it turns out to be a 17th century mansion. Most people are aware that the sun sets earlier in winter than in the summer, when it often never rises, and completing inspections of empty buildings after 3.30 p.m. in the winter months is very difficult.

4. Does the client intend to carry out any alterations if he proceeds with the purchase?

This should give information relating to any alterations which the client has already decided to carry out. There is no point in producing a detailed report on all the bathroom installations and defects of the plumbing fittings, if the client had made up his mind to replumb the property with bathrooms in different locations.

5. Does the client require a detailed report on boundaries, outbuildings and the grounds of the property?

In many cases the client knows that the woodshed has no door, the garage fell down last week, and the fences don't keep out the local vandals and is not interested in being told this when his concern is only for a report on the main buildings.

6. Does the client require an estimate of the rebuilding costs of the property?

Is this needed for a proposed insurance company or by a mortgage company? It will be necessary to know if the client is registered for VAT (see page 7).

7. Details of lease, or other special information

If the property is leasehold then the following information will be needed. A copy of the lease should be provided so that comments may be made in relation to the liability that the prospective purchaser would have. If the property is a flat the client may incur a liability for part of the repainting costs for the complete building as opposed to the repairing costs for the individual unit that he is proposing to acquire.

The fee

Confirm the approximate fees. It is important that there should be some estimate of the fees quoted to prospective clients at the time that they make their initial call. It is not easy to provide a schedule of fee charges as the charges will relate to the amount of time that the surveyor will take to carry out an inspection and prepare the necessary report.

An example of an instruction sheet which should be completed by the surveyor or his staff during their conversation with the client is shown on page 9. The queries and questions that need to be discussed are set out, and the standard advice that has to be given is printed at the bottom so that it may be read out to the client over the phone. This ensures that the client is aware of the extent of the work that will be carried out by the surveyor and any limitation of the work which would normally be referred to in the final report. This information will be duplicated on the letter confirming the instructions but it is important that it is raised at the inception of the instruction, so that the letter of confirmation is a true confirmation of the conversation and not the introduction of new clauses and conditions within the contract.

It is important to confirm these instructions even if the letter will arrive after the inspection has been completed.

Framed building in the Sultanate of Oman. Traditional timber frame construction is reflected in modern concrete design.

When and how it is made

The surveyor owes duties at law to clients and to other people. He needs to know what those duties are, how they arise and how he may best equip himself to fulfil them.

The contractual relationship

The time at which, and the terms on which, parties have entered into a contractual relationship can provide material for intricate argument among surveyors and lawyers. The surveyor should be aware of certain broad principles and try to avoid certain common misunderstandings.

To many, the word "contract" means a written document. They fondly believe that they are safe from the snares and pitfalls of a contractual relationship so long as they do not commit themselves to writing. When they do commit themselves in writing they believe that nothing that has been said before can bind or otherwise affect them, unless it is repeated in the document that they put forward or sign.

The formation of a contract does not depend on writing. It is created by an offer and acceptance between parties who intend to enter into a legally binding relationship. Many people, having a sense of the formality of the relationship, seek to embody it in writing. But many contracts are made only partly, some not at all, in writing. A contract may be founded on words spoken on one occasion, or on conversations spread over a period. It may be implied from conduct. If people act in a way that would lead an observer to the conclusion that they intend there to be a contract between them, then a contract will be implied from their behaviour.

No binding contract, however, exists when there is no intention to create legal relations or when a party has not furnished valuable consideration for the promise that he seeks to enforce. This valuable consideration can vary from a peppercorn to large sums of money.

Where the surveyor does intend to enter into a contractual relationship he should take care to avoid entering into an oral contract. If he enters into an oral contract, and later seeks to "confirm" the contract by introducing for the first time written terms designed to protect his position, he will be too late. He will be bound by a contract free from those terms. His endeavour should be to conclude with a potential client a written contract on terms that are clear and that will be upheld by the courts, if it is ever subjected to their scrutiny.

The surveyor's negotiations before he commits himself to an inspection will often be wholly oral and will rarely be wholly written. He is likely to be telephoned in the first instance. The potential client may ask him, to take the obvious example, to carry out a structural survey on a house that he hopes, subject to survey, to buy for a particular price. It is open to the surveyor to conclude a contract before the end of the conversation; but such a contract will be open to debate as to its terms. The surveyor should make it clear that if he is engaged the engagement will be on the terms of a letter that he will post to the client. By doing so he avoids the pitfall of contracting in haste and repenting at leisure.

Misrepresentation

Care must be taken to avoid making any assertion which the surveyor does not intend to incorporate into the contract but on which the prospective client is likely to rely in entering into the contract. It is inappropriate in this book to make any substantial reference to the law of misrepresentation. Suffice it to say that the surveyor who makes a significant misrepresentation is likely to face a claim, and he may be seriously inhibited by statute if he seeks to protect himself against claims by means of a contractual term.[1]

The contents of the contract: the common law approach

The common law has largely left it to the parties to conclude the terms on which they contract; and it rarely interferes with the relationship. As an exception to that general rule, it implies terms where the contract is silent but which the parties would regard as really too obvious to merit express inclusion. Common law can in certain cases override a contract and render it void. The surveyor must be careful that he is not entering into a contract with a minor, for the minor will not be bound by it. Nor should he contract with anyone whose mental capacity to understand what he is doing he suspects, which may rule out quite a few existing clients. Where a contract offends against public policy, for example by being founded on illegality, it will not be enforced.

In accordance with the general rule, there was at common law no reason why the surveyor or other professional man should not stipulate terms so favourable to himself as to absolve him from liability for even the most egregious incompetence. If his exemption clause was tightly and clearly drawn, the common law, strive though it generally did to defeat unmeritorious defences, would uphold it. Reputable professional men were inhibited not so much by legal considerations as by care for their professional reputation.

In the absence of any clause to a contrary effect, there was and is implied at common law a term in the contract between the professional man and his client that **the work should be carried out with the skill and care reasonably to be expected of a competent man exercising the particular calling and professing the particular skill in question.** There is nothing peculiar to surveyors in this principle. It applies, with exceptions that are not relevant in the context of this book, to all professional or otherwise skilled men. The nature of the principle will be more fully discussed later on. For the present, its existence must be borne in mind, and the contract between the surveyor and his client should be so

THE CONTRACT

worded that the client cannot have expectations of the surveyor that the surveyor cannot reasonably hope to fulfil.

Brief reference has already been made to the law of misrepresentation and, by way of a footnote, to the Misrepresentation Act 1967.

The principal statute to be considered at this stage is the Unfair Contract Terms Act 1977. This statute is somewhat misleadingly named. It prevents reliance on certain contractual provisions without requiring proof of unfairness. It does not provide a general power to intervene in a contract on the basis that it contains "unfair terms". Its impact is not confined to the relations between the parties to a particular contract, nor, indeed, is it confined to contracts.[2]

At this stage we are concerned with the Act only insofar as it affects the drafting of a written contract between the surveyor and his client.

The provisions with which we are concerned apply only to "business liability", that is liability for breach of obligations or duties arising from things done by a person in the course of a business or from the occupation of premises used for the business purposes of the occupier.[3] All the

This render-faced cottage is a timber-framed building with brick infill between the framework. This type of construction is quite common in agricultural buildings. There is a risk of deterioration in the timbers because of the absence of a damp proof course at the head of the plinth. Wall thicknesses tend to be about 100mm (4") but can be as little as 50mm (2") as can be seen in the illustration of the wall below the window.

When a surveyor is confronted with any building which is render-faced, he must ask himself whether the render has been applied after the building was constructed. Although some properties were designed to be faced in pebbledash, render or similar facings, many have them applied later to conceal the original construction which may have become defective. Faulty brickwork, extensive fracturing to the exterior of the building or decaying timber frame may be hidden behind a plaster face. In this case, an inspection of this building failed to reveal the timber framework with brick infill which was the main construction. The timber framework was in such poor condition, being subject to both wet and dry rot at various parts, that the building had to be condemned and has now been demolished. The timber framework had been supported on a brick plinth which did not have any form of damp proof course. This resulted in damp penetration to the timber framework and the resulting deterioration.

The very small wall thickness should have drawn the surveyor's attention to the probability of the construction not being conventional brickwork. Instead of being 225mm (9") thick, the walls were only 100mm (4") reducing to approximately 50mm (2") below the window. Wherever there is a render-face to a building, extreme care must be taken in determining the construction which has been used, and in deciding why the render was applied to the property.

liability of a surveyor to his client will be a "business liability", for "business" includes a profession.[4]

It is pointless for a surveyor to include in the contract any term purporting to exclude or restrict his liability for death or personal injury resulting from negligence.[5] Any such liability, provided it is a "business liability", cannot be so excluded or restricted.[6] Negligence in this context is widely defined so as to include breach of any contractual obligation to take reasonable care or to exercise reasonable skill in the performance of the contract.[7] This provision frustrates not only an attempt to exclude or restrict liability for breach of an admitted duty to take care, but also an attempt to negate the existence of such a duty.[8]

In the case of other loss or damage "business liability" for negligence cannot be excluded or restricted by reference to a contract term except insofar as the term satisfies "the requirement of reasonableness".[9] Some, though not much, help is given by the Act as to the ambit of that requirement. The very use of the word "reasonableness" suggests that each case is to depend substantially on its own facts. The requirement is that the term in question shall have been a fair and reasonable one to be included, having regard to the circumstances which were, or ought reasonably to have been, known to or in the contemplation of parties when the contract was made.[10] The only further help given by the Act as to the cases with which we are concerned relates to terms purporting to restrict liability to a specified sum of money. In such cases regard shall be had in particular to the resources which the party relying on the term could expect to be available to him for the purpose of meeting the liability should it arise, and how far it was open to him to cover himself by insurance.[11]

For many years surveyors were kept in suspense as to the extent to which their relations with persons with whom they contracted and with other persons affected by their work would be governed by the Act. Now, as we shall see below,[12] the House of Lords in *Smith* v *Eric S. Bush* and *Harris* v *Wyre Forest D.C.*[13] has made the position clear so far as ordinary domestic house purchases are concerned. In such cases, the requirement of reasonableness is not satisfied by a term purporting to exclude or restrict liability for negligence. The reasons given by the House were these: the surveyor is a professional man offering his services for reward; he is paid, directly or indirectly, by the purchaser for those services; he knows that some 90% of purchasers rely on a valuation made for mortgage purposes; such purchasers are under pressure so as to rely because many could not afford to commission reports of their own or because when a prospective sale goes off money will be wasted; he knows that a failure on his part may be disastrous to the purchaser; by contrast he may protect himself by insurance; and the task which he undertakes is not a difficult one if it is undertaken with reasonable skill and care. However, different considerations are likely to apply in the case of industrial property, large blocks of flats and very expensive houses: the purchaser's expectations and the insurance position are likely to be quite different.

In every case, it is for those claiming that a contract term satisfies the requirement of reasonableness to show that it does.[14]

The provisions referred to above will cover most liability that the surveyor will incur. It is, however, possible for a surveyor to incur liability other than for negligence, as broadly defined in the Act. For example, he may incur liability for the breach of an express term that a survey should be carried out by a particular member of his firm. The surveyor may also incur liability for breach of an express or implied term as to the period within which the survey is to be carried out. A further provision must therefore be mentioned.[15]

Section 3 of the Unfair Contract Terms Act 1977 is limited in its application in a way described below. Where it applies, it is very similar to that relating to liability for negligence and described above. The party subject to it cannot by reference to any contract term exclude or restrict any liability of his in respect of breach of contract except insofar as the term satisfies the requirement of reasonableness.[16] Nor, with the same exception, can he claim to be entitled to render a contractual performance substantially different from that which was reasonably expected of him or, in respect of the whole or part of his contractual obligation to render no performance at all.[17]

This provision applies only as between contracting parties (in contrast to that referred to above), and only where one of them "deals as consumer" or on the other's written standard terms of business.[18] In many contracts into which a surveyor enters the other party will "deal as consumer"; in many others he will not. It will depend on whether the other party makes the contract in the course of a business or holds himself out as doing so. He will "deal as consumer" only if he does neither of these things. In all cases with which we are concerned, the other relevant test of the client's status as consumer will be satisfied; for the surveyor will make the contract in the course of a business.[19] The burden of proof will be on the surveyor to prove that the client does not deal as consumer.[20]

The provision will also apply where the client deals on the surveyor's written standard terms of business. The ambit of this deceptively simple expression has not been the subject of any reported case. It is likely that the surveyor will wish on every occasion to use a formula, and it is highly desirable that he should do so. But in a particular case it may be agreed that while the bulk of his "written standard terms of business" should apply some part should not, or that all the terms should apply but should be augmented by some further important term. In such a case it might be argued that the provision will not apply. The only guidance that can be given at the present is that the courts are unlikely to find that the provision has been successfully evaded in the absence of some substantial departure from the particular written standard terms of business.

The House of Lords has now shown that the field within which exemption clauses may be deployed is limited. Within that limited field the surveyor will be concerned as to the fate of exemption clauses if he includes them. There is no point in including a provision which may deter the prudent potential client from engaging the surveyor who puts it forward. Nor, for at least two reasons, is it commercially sensible to include a provision likely to be attacked on the basis of the Act. First, the professional man will be anxious as to his reputation if one of his contractual terms is impeached by a "consumer"[19] as not satisfying "the requirement of reasonableness". The very fact of the litigation is likely to embarrass him. Secondly, it is predictable that the courts, at least in the cases involving "consumers"[21] will not be satisfied with less than compelling and cogent evidence of reasonableness; though where there is equality of bargaining power and financial prosperity much of the sting of the argument against terms protecting the surveyor is removed. For all that, it is likely that terms put forward by surveyors, solicitors or other professionals will be scrutinised as carefully as those relied on by any other party. Very high standards of probity and reasonableness are expected of all practitioners.

Brief mention should be made of the Supply of Goods and Services Act 1982, Part II of which relates to services. This is intended to be no more than an interim codification of the existing law; and so far as surveyors are concerned it probably satisfies the description. By Part II there are implied into all contracts for the supply of services, where supplier is acting in the course of a business (including a profession), terms that the supplier will carry out the service with reasonable care and skill and within a reasonable time (unless the time for performance is provided for by the contract or the course of dealing between the parties).

1. Misrepresentation Act 1967.s.3. as amended by Unfair Contract Terms Act 1977.s.8(1).
2. Save in Scotland: Unfair Contract Terms Act, 1977,s.15(1).
3. *Ibid.*,s.1(3); and see s.16(1) in relation to Scotland.
4. *Ibid.*,ss.14 and 25(1).
5. In respect of Scotland, the expression "breach of duty" is used.
6. *Ibid.*,ss.2(1) and 16(1).
7. *Ibid.*,s.1(1)
8. *Smith* v *Eric S. Bush; Harris* v. *Wyre Forest D.C.* [1989] 2 All ER 514
9. *Ibid.*,s.2(2). In Scotland, the same effect is achieved by the words "if it was not fair and reasonable to incorporate the term in the contract":s.16(1)(b).
10. *Ibid.*,ss.11(1) and 24(1).
11. *Ibid.*,ss.11(4) and 24(3).
12. See p.31 below.
13. See note 8 above.
14. *Ibid.*,ss.11(5) and 24(4).
15. *Ibid.*,s.3. The Scottish equivalent is s.17.
16. See note 9 above.
17. *Ibid.*,s.3(2) and cf.s.17(1)
18. *Ibid.*,s.3(1) and cf.s17(1); for Scotland the expression "standard form contract" is used.
19. *Ibid.*,s.12(1), and cf.s.25(1).
20. *Ibid.*s.12(3) and 25(1).
21. The inverted commas denote the technical sense in which the word is used in the Act; not the writer's lack of enthusiasm for it.

THE CONTRACT

The drafting of the contract

The surveyor, in drafting the terms on which he proposes to contract, should concentrate on informing the prospective client of the ambit of the work that is to be done. He should spell out the limitations of the work. This will entail describing precisely those parts of the building to be surveyed to which the surveyor will not gain access. The prospective client, on reading such a description, is likely to ask himself what the value of a report based on a limited inspection will be. His paramount requirement is for a report on which he may confidently rely in making a major business decision – possibly more significant than any business decision that he has previously made. The form of contract must, therefore, be helpful to him, it is not enough for it to protect the surveyor if at the same time it deters the client. The contract should contain a provision that the surveyor will include in his report an assessment of the risk of there being some material and undesirable feature in the uninspected parts of the building and advice as to the course to be taken to investigate such risk further.

It is important that the client should be told as precisely as possible what the exercise is going to cost him. There is, of course, nothing in law to prevent the surveyor from quoting a fixed price for his services or, indeed, for his services as well as those of contractors such as builders, plumbers and electricians. But such a course may not be satisfactory to the surveyor. Particular among his skills employed on an inspection are the assessment of the need for further inspection by himself, the need for certain trades to be engaged and the extent of each of those needs. The best that he can reasonably do in some cases is to inform the client as fully as he can of the likely cost, and to make clear the areas of envisaged work that are not included in any figure that he mentions.

Confirmation of Instructions

It is important that instructions should be confirmed immediately. The confirmation must relate to the instructions received and should refer to whether they came by telephone, the name of the person who received the call and the nature of the instructions that were received. They should, if possible, also give the name of the surveyor who will be carrying out the inspection.

The letter confirming instructions should set out the date upon which the survey probably will be made and the date by which the report probably will be completed. An exact date should not be specified, because otherwise it will become part of the contract. Any variation in the dates upon which an inspection is made could cause complications. A client may feel the surveyor did not satisfactorily comply with the contract, especially when he comes to pay the fees.

Instructions must set out the full record of the inspection which is to be carried out by the surveyor. This should include reference to the extent of the proposed inspection of the building, outbuildings, fences and grounds. It should also point out that where fitted carpets have been installed, these will not be lifted and that it will not be possible to move heavy or fitted furniture to carry out an inspection of the structure concealed by these items. As this information should have been imparted to the client over the telephone, it is possible that he may have requested an inspection to be made of the areas where the carpet is fitted. If this is the case, it is important that the necessary consent is received in writing from the vendor or his accredited agents to enable the raising of the carpet. This point is covered in the suggested letter to the vendor about making an appointment to inspect the building.

The letter must set out the insurance exclusion clause where this is required by the professional indemnity insurance policy of the surveyor. Many companies now no longer require the inclusion of such a clause. It is important to confirm that no inspection will be made of any part of the property which is unavailable for inspection, or where the opening up to gain access will cause damage to the surfaces and decorative finishes.

There are many complaints about the work carried out by surveyors which result from the impression the client has gained from the usual collection of ill-informed advisers as to the scope of work carried out by the surveyor. Many people believe that the surveyor will advise upon the planning position relating to adjoining properties and the prospects or otherwise of any road alterations or revisions in the area. Local searches of the registers of the Local Authority are not part of the work of the surveyor in carrying out structural inspections of a residential property. This work would involve the personal inspection of the registers situated at the town hall or headquarters office of the Local Authority. This may be some distance from the property and such buildings are usually located in positions which are difficult to get to and which call for a five mile hike after parking. The extra cost of carrying out these inspections can be disproportionately high for a structural inspection of a small residential property.

In certain circumstances where inspections of commercial property are to be made the situation is different. It is probable that for surveys of hotels, office buildings, light industrial premises and other forms of commercial property it will be necessary to carry out investigations of current legislative and other controls. These may include the Town and Country Planning Acts, building regulations, certificates relating to the means of escape in case of fire, and the Health and Safety at Work Act, together with any other temporary licences which should be maintained or obtained, and other legislation which is relevant to the property.

The confirmation of instructions should also refer to those tests which will be carried out for the client by the surveyor; it should stipulate where such tests are to be carried out by specialists or other contractors and include the name of the contractor who is to be used. Specialist contractors are usually engaged for the testing and report on drains, the electrical installation, heating and ventilation equipment, lift equipment, fire alarm systems, specialist telecommunications and audio visual systems and, in the case of multi-storey property, the window cleaning arrangements.

This letter should also confirm any instructions to assess, for fire insurance purposes, the cost of rebuilding the property. Such a valuation should be set out clearly to reveal the actual building cost involved, the professional fees, the allowance for inflation and the liability for Value Added Tax (VAT). The VAT position of the client should be recorded in such a valuation so that the surveyor knows whether he should include VAT on the professional fees element of the valuation.

The insurance company will not settle in full a claim which includes VAT on the professional fees, if the client is registered for VAT and can offset the VAT liability in his own trading or business. If the VAT is included in the valuation the client will be paying a premium to insure the VAT on the fee part of the cost when he will never be paid the money by the parsimonious insurance company; so why allow your client to donate money to the insurance company?

A form of confirmation of instructions with paragraphs relating to the various alternatives which could be prepared by a firm or practitioner and preprinted is set out below. The information which is needed to complete the letter could either be typed in or written in longhand. This will naturally speed up the process of issuing this letter. It is important that a copy of the letter sent to the client be kept.

The surveyor's satisfactory performance of a contract in the form set out in this letter should keep him free from liability in litigation, and with luck from litigation at all. The letter does not contain material designed to enable the surveyor to escape or mitigate the normal consequences of his carelessness. It is undesirable that surveyors as professional people should seek to escape such consequences; and we have seen that the Unfair Contract Terms Act would make most attempts by them to do so pointless. As to limitation of liability, the surveyor should rather keep his insurance under review and refrain from accepting work likely to involve him in any uninsured risk. Alternatively, where a particular and unusual transaction is in question, the surveyor would do well to consult his solicitor with a view to a limitation of liability being introduced into the contract. Provided that the client is clearly told that the limit corresponds with the limit to which the surveyor is insured, the term may satisfy the requirement of reasonableness; for the client's need to consider obtaining insurance in respect of any top slice of loss will have to be made plain to him.

INSTRUCTION SHEET

Instruction Sheet for Building Inspection

Job Ref: ....

01 Client: ....
02 Client Contact: ....
03 Address: ....
....

Tel: ....
Fax: ....
Home: ....

04 Property Address: ....
....
....
....
Access to Property: ....
....
....
....

Job Type:
05 Mortgage Valuation
06 House Buyers Report
07 Flat Buyers Report
08 Structural Survey
09 Diagnostic Inspection
10 Valuation
11 Rebuilding Cost Valuation

Type of Building
(ie approx. size of building, number of rooms, floors, etc.)

12 House
13 Flat
14 Office
15 Shop
16 Office
17 Other ....

Location of Building
(How to find it if in isolated position, etc.)

Construction
Brick, solid or cavity walls, concrete frame, steel frame, timber frame, roof, thatch, slate, tile, flat.

18 Price being paid for building £ ....

19 Agreed Fee   £ ....
21 *Quantum meruit*   £ ....per hour
22 Anticipated extra costs £ ....

20 Time allocated: ....hrs

23 Date of Instructions: ....
24 Date Confirmation of Instructions sent: ....
25 Date allocated for inspection: ....

Notes for survey instructions
26 What does client want survey for? (purchase, taking on lease, company records, takeover, investigation of complaint, litigation, other.)
27 Date when report required: ....
28 Does client intend to carry out any alterations?
29 Is the property leasehold or freehold?
30 Are tests required to electrics, drains, services?
31 Client advised of standard limitation of inspection.
32 Special requirements of client.

33 Copies of report to be sent to:
   Solicitors

34 Previous contact with client
35 References
36 Fee to be paid prior to collection/sending of report

Instructions taken by: ....
Instructions received from: ....

Instruction Sheet

The information entered onto this form can be transferred to a computer or word processor. A simple program will then enable the machine to print a confirmation letter addressed to the client. The letter will confirm the level of work, any special instructions and the fee. It will also state the date when the inspection will be made, and the date when the report will be ready. The letter will set out the agreed method of payment. The machine will also print the account at the same time, so that it can be sent with the confirmation if payment is required in advance.

The responses to the 'prompt' questions will enable the machine to select the appropriate paragraphs to adequately describe the extent of the varying inspections that may have been requested.

THE CONTRACT

Standard letter to confirm instructions

The Surveyor must select those phrases that apply to the specific instructions and the property he is to inspect.

Dear

Re: Address of property to be surveyed

We wish to confirm the terms upon which we will inspect the above-mentioned property, as discussed during your telephone conversation with our office on .... (or meeting with, letter from, etc.)

Please read through this letter very carefully. If you wish to amend these instructions we must have the information before the inspection is carried out.

Mr/Miss/Mrs .... of this firm will inspect the property on or around the .... (date). If you do not wish the inspection to proceed you must advise us by phoning this office on 000 000 before .... (time and date)

We note that you have advised us that the property is a detached/semi-detached house/flat/etc. which comprises .... bedrooms, ... living rooms, ... bathrooms, ... kitchens, ... garages, ... and .... acres of ground.

The survey report will state the opinion of the surveyor as to the defects that are present and which can be found in the building at the time of the inspection, or may occur within the next one or two years if no action is taken. The opinion will be based upon the information that the surveyor will be able to obtain following his/her inspection of all parts of the building that can be seen without damaging the property, its decoration or contents.

A survey report is not a guarantee that all defects that are present, or may occur in the future, will be discovered by this limited inspection.

The surveyor will report on those aspects of the condition of the property which are considered likely to materially affect the market value of the property, and which the surveyor believes would influence the decision of a prospective purchaser of the property (or prospective leaseholder, etc.)

The surveyor's inspection, as set out below, may have to be amended if the vendor/occupier/owner refuses or is unable to provide access to a part of the building, or forbids any test or operation.

The surveyor will take a ladder of at least 10 feet in length (3 metres). He/she will inspect all parts that can be safely reached with this ladder, or can be seen from a vantage point within the property or its grounds.

The surveyor will use a moisture meter to locate unacceptable levels of dampness in all accessible parts of the property.

The surveyor will carry a torch, and will inspect within all ducts and traps, provided no damage will be caused to the building or its decorations in opening such access traps.

Carpets will be lifted where they are loose and there is no risk of damage being caused to the carpet, or the property in so doing. Underfelt will be lifted to inspect the floor surface if it is not fixed to the floor. All the surfaces of timber floors which are carpeted or covered in lino, or similar sheet materials, cannot be inspected. The surveyor will give his/her opinion of the condition of the floor surfaces based upon the information he/she is able to gain from the limited inspection that he/she is able to carry out.

The floors will be tested for their level, and the surveyor will advise if there is any indication of settlement due to the floors having a fall greater than 1:100.

We will advise you of the repairs that may be necessary in the future, and the common faults which occur in the property of the type that you are proposing to acquire/lease/etc.

We will not inspect or comment upon the condition of any outbuildings that may be present.
Or
Outbuildings, such as sheds, greenhouses, huts, car ports or garages not joined to the house will be inspected only briefly, and the report will only comment on those defects that are sufficiently major to affect the value of the complete property.
Or
Together with a valuation of the open market value of the property.
Or
Together with a valuation of the approximate cost of reinstatement of the building. You are advised that this assessment should be annually reassessed to ensure that your insurance replacement cost is always maintained at the correct level.

Or
To inspect only ....
(a specified part of the property) and advise you in relation to the condition of this item. Such an inspection will not cover any other parts of the property nor will we advise on the effect that other sections of the property may have upon the specific items we are required to inspect.

We propose to make such an examination of the structure as is practicable at the time of our inspection with a view to advising you of the principle failures in the construction of the property and advise you as to its apparent condition. It must be understood that our examination will not cover those parts of the building that are concealed, inaccessible or unexposed.
Or
Our examination will not cover those parts of the building that are concealed, inaccessible with the use of such equipment as our surveyor normally carries, or unexposed. Unless you specifically so instruct us and the vendor previously consents, we shall not move any heavy furniture, raise any fitted floor coverings or remove any floorboards.
Or
We acknowledge your instructions to move the following specified items of furniture and will arrange for the necessary assistance to be provided.
Or
We note your instructions to move the following specified items of furniture and thank you for your help by providing the requisite assistance in repositioning the units.
Or
We will be unable to raise fitted floor coverings or remove floorboards.
Or
We note your instructions to raise the fitted floor covering in the following areas and have written to the vendors requesting their specific consent to these proposals. In the event of our not receiving their consent, we will be unable to carry out this work because of the risk of damage to the floor coverings.
Or
We note your instructions to raise the floor coverings in the following specified areas and we have arranged for a carpet contractor to call at the premises on ....
(date) at .... to set aside these coverings and to reinstate them after our detailed inspection. The approximate cost of this specialist is £............... which will become part of our fee relating to the inspection of the property.
Or
We note that the invoice for the carpet fitter is to be sent to you for your payment.
Or
We note that you are instructing the carpet contractor yourself and will be responsible for all costs so incurred.

In our report we shall advise you of the extent of the inspection that we carried out, and any major limitations to reporting that were encountered.

If we believe that there is a risk of a defect or future failure in the areas that we have been unable to inspect, we will recommend the further steps that should be taken by you before you exchange contracts.

Roof surfaces will be inspected from ground level or from available vantage points within the demised premises. The surveyor will carry with him a 10 foot (3 metre) ladder. This will mean that he may be unable to inspect the back of gutters or eaves where access cannot be obtained by use of a ladder as referred to above or from available vantage points.

The interior of accessible roof voids will be inspected.

We will advise you as to the effectiveness of ventilation and insulation and comment upon any weakness in relation to sound transmission between the exterior and interior or between adjoining properties.

The exposed elements of all walls and brickwork will be inspected externally and internally as far as is practicable, but it will not be possible to carry out any inspection of the foundations.

The building services will be visually inspected and tested by normal operation. In the event of the termination of gas and electrical supplies, it will not be possible for us to comment upon the physical operation of those fittings which operate in conjunction with these services.

We will comment on the fences, paths, outbuildings and grounds.

A visual inspection will be made of the manholes and drainage connections where the cover is intact and able to be removed.

Or

We shall engage builders to test the drainage systems electricians/plumbers/heating engineers to test the respective services under our supervision. In our report we shall inform you if any of these specialists have been unable to test to our satisfaction and we shall advise you what, if any, further steps can and should be taken.

Or

We refer to your specific instructions in connection with the electrical services and would advise you that we have instructed Messrs ....
to carry out a test which will include:
(a) ring circuit continuity;
(b) protective conductor continuity;
(c) measurement of earth electrode resistance;
(d) measurement of insulation resistance
(e) check of protection by boundaries and enclosures;
(f) measurement of the insulation of non-conducting floors and walls;
(g) verification of polarity;
(h) measurement of earth fault loop impedance;
(i) ....

The contractor's report will be sent to you without comment and their invoice will be in the region of £............................... which will be passed directly to you for your payment.

Or

We note your instructions to arrange for a water test of the drainage system and have written to the vendors to obtain their consent to this action.

As soon as we have received their confirmation, we will instruct Messrs ....to carry out this inspection.

We will be in attendance to check the test and will report to you as soon as possible thereafter.

The attendance of the contractors will cost in the region of £....................... which will be added to our inspection fee. If we are unable to carry out this inspection at the same time as our survey visit, we will have to charge the extra fee of £....................... for our attendance. Every effort will be made to ensure that this inspection can take place at the same time, but in the event that we are unable to obtain the vendor's consent speedily, this may not be possible.

Or

We note your instructions to arrange for specialist heating and ventilation engineers to carry out an inspection of the building and would advise you that we have instructed Messrs .... to carry out this work.

Their report will be sent directly to you. Their invoice for these tests, which will be in the region of £............................... will also be forwarded directly to you for your payment. We will make no comment upon their findings within our report.

Or

We will require a copy of the lease of the property before we carry out our inspection. We will also require copies of any licence, or other documents which may influence the use or control of the buildings or the site.

Or

The fee for the inspection and report upon this property is £.....................

This fee is based upon your description of the property which has been repeated in the third paragraph of this letter. We reserve the right to charge a different fee if the property is not as it has been described to us.

An invoice for this £.................... is enclosed and is to be paid in full before the release of the survey report.

Or

An invoice for £.................... will be

Are all the out-buildings to be inspected and reported upon?

enclosed with the report and is to be paid within 14 days. Interest at 3% above the base lending rate of Barclays Bank will be charged for each complete period of 24 hours up to the time payment is received if payment is received after 14 days from the date of posting the report.

Flats

The inspection of the flat will include the interior of the common parts serving the flat, and the interior of the subject flat.

The exterior of the building will be inspected where access can be gained.

Defects on the exterior of the building will be reported only where they affect the subject flat.

In the absence of a lease we will assume that the subject flat will be liable to pay a proportion of the charge for the repair and maintenance of the building that will be equivalent to the proportion that the flat is to the complete building in which it is situated.

If there is a caretaker's flat we will use our best endeavours to carry out a limited inspection of its interior. This flat will not be surveyed, but we will assess the possible expenditure that is immediately required to provide essential amenities and repairs to this part of the property.

If you have any queries relating to this confirmation of instruction or if you wish to vary the information contained within this letter, we would be grateful if you would contact our office before the date on which the inspection is expected to be carried out.

We hope the details that are set out within this letter accurately record our understanding of the instructions which we have received from you.

We should be grateful if you would confirm your instructions to us by signing and returning the enclosed copy of this letter.

Yours sincerely/faithfully

.................................(signature of surveyor)

....(signature of client)

THE CONTRACT

The appointment to inspect the property

In order to carry out a detailed inspection of the property which will enable the surveyor to advise the client, it is necessary to make arrangements with the vendor or his agent to gain access. In many cases the initial contact is made by telephone. As with the initial communication with the client, one's relationship with the vendor is equally important. It also requires an equivalent degree of care because there is a legal relationship which should be set out in writing in order to avoid the usual forms of misunderstanding which end up being expensively determined in Court.

In making the appointment, it is important that the surveyor or his representative should make clear the amount of time that an inspection is likely to take. Many people selling their property are inundated with various professionals wishing to walk through their prized dwelling. They frequently mistake the ministrations of the Building Society's valuer with those of a structural surveyor and are somewhat surprised to find that the inspection is to take more than 5 or 10 minutes.

It is courtesy to confirm to the vendor the name of the surveyor who is to call, as well as to give a verbal picture of his appearance. Many vendors become somewhat nervous of strange people wandering through their property and are only too grateful that a kind voice on the telephone has given them some indication as to the age, distinguishing features and criminal tendencies of the practitioner.

During the course of making the arrangements for the inspection, it is worthwhile trying to find out if the vendor has been in the property for a long time and to obtain further details about the type and age of the property which is to be inspected.

There may be a number of documents that may help the surveyor in preparing his report and it will help if they are available. As the vendor or occupant is rarely psychic it helps if he has been given prior notice. The following documents or information will be useful:

1. Any lease, licence or any other documents relating to the property, particularly if it is not freehold.
2. Any copies of recent service charge demands for the property if it is a flat or a property where the expenditure on the exterior is shared between the occupants.
3. Any copies of records of recent gas and electricity bills (so that the surveyor may advise the client on the running costs of the equipment within the property or note any disproportionate expenditure which may indicate a concealed failure).
4. Any plans showing the property as it now is or any recent alterations which have been carried out. (These can save hours of measuring if one has to do a valuation of the property for rebuilding purposes. Information relating to recent alterations may also mean that further instructions have to be sought from the client to check approvals or to advise the client to request his solicitor to make the requisite searches.)
5. Any guarantees which may relate to repairs, underpinning, timber treatment or rodent infestation. (Specialist guarantees are rather like life itself; they are fine so long as you don't take them seriously.)
6. The vendors should be asked if they intend to remove any items of equipment of furniture from the property. Both vendor and client should have agreed a list of items to be removed prior to the survey.
7. Information relating to the neighbours and whether there are any disputes over boundaries, trees or similar matters is also useful. Information relating to whether the property has been flooded may also be of value.

All this information provides valuable background material for the surveyor during the course of his inspection and could save him a considerable amount of time. He is also saved the long wait, standing on one leg, while the vendor tries to locate those guarantees, plans, drawings or other important letters, which they know they have filed in a safe place, if only they could remember where that safe place was.

Occupant's or vendor's consent

When a surveyor is instructed to carry out an inspection of a property it is usually where a sale is proposed. If this is the case access is usually granted by the vendor or his agent. In other cases the client may be the occupier, or no sale may be intended because the report is required for finance reasons. The references to the vendor may have to be adjusted in these varying circumstances.

Contractual problems often occur when inspections are to be carried out which may cause damage to the property. In some properties, carpets will have been laid and their edges secured. It is necessary to obtain the consent of the occupant or the vendor to the raising of these carpets, particularly where the surveyor does not have a carpet layer's skills for returning the floor covering to a position which will not inadvertently arrest the passage of the occupier at a future date.

If the surveyor considers it necessary to carry out an inspection of inaccessible sections of timber or solid flooring, he will have to obtain the vendor's consent to the drilling of any necessary holes or to the raising of floor boarding or other surfaces where damage is likely to occur. In the absence of such consent, the surveyor has a licence to carry out an inspection which will not cause damage to the property.

Certain other tests that the surveyor may wish to carry out may also require the vendor's consent. His permission should be obtained where contractors are to carry out tests on the premises. It can be embarrassing if your friendly electrician happens to have had a blazing row with the vendor and his arrival at the premises is met with an "over my dead body is he coming into this building" reaction.

The activities of specialist contractors or of their specialist tests may also cause damage. A water test of the drainage may cause fractures and failures if the drainage beneath the property is in a vulnerable condition. Such tests should be avoided in Central London without a careful examination of the drains. In many properties over one hundred years old failures may be caused by the test particularly if the drains are very deep. On one occasion a vendor expected the surveyor to pay for the reinstatement of drains which had failed during the course of a water test.

It is reasonable that the vendor or occupant should be made fully aware of the risk involved in a water test on drainage which may be in a delicate condition. He should also be advised that sound drainage will not be damaged by a water test.

If the electricity board is to carry out a wiring test for the surveyor, the occupant's or vendor's consent should be obtained. It will be an embarrassment to the surveyor if the electricity board arrive to find that various areas of bare wiring around a switchboard are sufficient reason for terminating the supply at that moment. The surveyor will not be responsible for the activities of the electricity board in carrying out their statutory function, but he may be vulnerable to liability or sharp words for having allowed the electricity board on to the premises without first having obtained the vendor's consent.

There are occasions when one needs to make such an appointment by writing to a vendor, because it is not possible to speak to him on the telephone. In any event, the request for information and the extent of the inspection should be confirmed, and a letter which covers the various queries that may be raised is set out below.

Standard letter to confirm appointment

The Surveyor must select those phrases that apply to the specific instructions and the property he is to inspect.

Dear

**Re: Appointment to view a property
(Address)
(Survey of Property)**

We wish to advise you that we have been instructed by .... to carry out an inspection of your property in order that we may advise (him) (her) (them) as to the condition of the premises, grounds, fences, outbuildings, etc. In order to carry out this inspection we would like to call at your premises on .... (date) at .... (time).

We expect, from the information that we have been given as to the extent of your premises, that we will require access to the property for a continuous period of .... hours.

We have been instructed by our client to arrange for the attendance of various contractors to carry out the inspection of specific sections of the property, and we list below the names of the contractors and the functions that they will perform. We would be grateful if you would advise us if you have any objection to any of these contractors entering your premises:

....
....
....

We hope that we will be able to arrange for their inspections to take place during the course of our visit to the property, but sometimes this is not possible and if necessary further arrangements will be made to find a mutually convenient appointment.

If the contractors are unable to call at the time of our visit, we will not be in attendance during their inspection and would be grateful if you would advise as to whether you object to their entering the premises.

We have been instructed to carry out a water test of the premises. We must advise you that in the event of the drains being in poor condition damage may occur to them as a result of the pressures which are exerted during the course of this test.

We have been requested to carry out an inspection of the underside of the floor surfaces. The surveyor will endeavour to reinstate the boards to their former location, but some damage may occur to the tongues of the boarding if it is tongued and grooved, or to the surface of the boards themselves.

We have been requested to carry out an examination of the floor surfaces which are beneath fitted carpets.

Or

We have been requested to examine all floor surfaces which may be beneath fitted carpets. It is possible that some damage may be caused to the edges of carpets where they are secured and we would be grateful if you could advise us if you know of the method of carpet fixing.

Or

We have been advised by you that carpets within your property are fixed and have arranged for Messrs........................ Carpet Contractors, to call and lift these carpets and reinstate them following our inspection. We would be grateful if you would confirm whether there are any objections to this firm being in attendance at your property during the course of our inspection.

We would be grateful if you would initial the relevant paragraphs on the attached copy of this letter as an indication of your acceptance that those stated procedures may be carried out, and that you will not hold this firm or its contractors liable for any damage that may result from those procedures, unless such damage results from the negligence of any member or employee of this firm or of its contractors.

It will be of great assistance if you could advise the surveyor of any matters which relate to the following queries.

1. Have any structural alterations been carried out to the property? If alterations have been carried out are there any drawings, specifications or other documents relating to the work that was undertaken?
2. Have any major repairs been carried out?
3. Do you hold any guarantees in respect of any of the following:
   Timber treatment
   Eradication of fungal decay
   Underpinning
   Repairs to the structure of the property
   Roof repairs
   Installation of heating or electrical equipment?
4. Do you know whether the property has ever been flooded?
5. Are you aware of any disputes with adjoining owners in relation to the boundaries?
6. Are there any trees within the boundary to the property?
   (a) If yes, please advise if there has been any pruning carried out to the trees.
   (b) Do you know if the trees are protected by the Local Authority?
   (c) Are you aware of any mushroom growth or appearance of toadstools around the base of any of the trees during the summer months or the growth of any mushroom from the side of any tree at any time?
7. Have you a list of items of furniture, fittings or fixtures that you intend to remove from the house, grounds or outbuildings?

We would be grateful if you could arrange for all sections of the property to be available for inspection and that all keys for window locks, security gates and doors to cupboards and rooms are to hand, so that our surveyor may make a thorough inspection of the premises.

We note that the property is leasehold, and we would be grateful to have sight of a copy of the lease, if you hold this document.

We would be grateful to have sight of recent gas and electricity bills relating to the building so that we can comment to our client upon the efficiency of the existing services.

We would like to inspect service charge accounts, or accounts relating to annual expenditure in relation to the property where the maintenance of the structure or various sections of the exterior is shared with adjoining properties.

We hope we may have your assistance as this will enable our surveyor carry out a detailed examination of your property and advise our clients as to its condition.

Yours (sincerely) (faithfully)

...............................(signature of surveyor)

Checklist in preparing for a survey

Preparation – Instructions
- Confirmation of instructions.
- What is clients name?
- Do you know the client?
- Can the client pay the bill?
- Do you need the money in advance?
- What does the client want the inspection for? Purchase, repair, diagnosis of the causes of a failure, cost budgeting, valuation etc?
- What is the full address?
- Can you find it?
- What is the size of the property?
- What type of property is it: Office, Shop etc?
- How long with the survey take?
- How much will you charge?
- How easy is access to the property?
- How soon do they need the report?
- Is the building listed, or in a conservation area?
- Are there any fire certificates, or other consents that are relevant, that you will need to see?
- Are there any guarantees? (Might be good for a laugh)
- Has the client seen the property?
- Does he know if it is suitable for his purposes?
- Do the vendors know how long you will be there?
- Do you have to arrange any specialist tests?
- Have you sent a confirmation of instructions letter?
- When are you going to do the job?

Preparation – Leasehold properties
- Obtain certified copy of the lease.
- Check colour on demised plan in lease.
- Obtain copy of management accounts.
- Is there a caretaker's flat to inspect?
- Can you gain access to the rear, roof etc?
- Are there any special dates for redecoration, or major repairs having to be carried out referred to within the lease?
- Is the property on an estate where the properties must always be maintained to a specific, high standard of repair?

Preparation – the visit
- Have you got all the necessary equipment?
- When was the damp meter last calibrated?
- Are all the batteries adequately charged?
- Have you a list of the special instructions of the client?
- Have you sufficient identification for the benefit of the occupant?

Preparation – the adjoining properties
- Are there any failures in similar properties in the neighbourhood?
- Are there any major works being carried out to a similar property in the area?
- Have changes been made to similar properties in the area?
- What is the condition of adjoining properties, particularly if they are attached?

Preparation – the grounds
- What type of ground is the property built upon?
- Is there any history of settlement in the area?
- Is there a clay soil? If so how deep would you expect the foundations to be? Are they vulnerable to movement?
- In clay soil areas are there any trees that are too close to the building?
- How wet is the ground? Are there any water courses in the area? Is there any history of flooding? How well is it drained?
- What type of foundations would you expect the building to have? Are they likely to be adequate?

Preparation – the grounds
- Boundaries, can you identify them, or locate them?
- Are there any outhouses, sheds? Have you inspected them? If not say that you haven't.
- What is the condition of any fence, wall, hedge or other boundary demarkation? Are there any special reasons for boundaries to be in very good condition? (ie for animals to be kept on the premises?)

Preparation – outside influences
- Noise, of aircraft, neighbours, particularly in flats.
- Noise from local buildings at the time of your inspection (ie factories, schools, or traffic).

The Inspection – exterior
- Record the weather at the time of your inspection, and any limitations in what you can see of the exterior.
- Roof. Type and appearance. Is client likely to have seen it? Do you need to include a sketch or photograph?
- Roof covering. Is the pitch correct? Is the surface the original? If it has been changed, has someone thought through all the construction changes that may have been needed?
- What is the material of the roof covering? How has it lasted in other properties? Are there any disadvantages of this material?
- How is it fixed? If you can't see, point out how it could be and the deficiencies of nail sickness, peg tiles, failed tile nibs, poor laps, ease or difficulty of repair.
- If condensation occurs below the roof surface what will happen?
- Is the roof ventilated? Is any ventilation that may be present adequate?
- Are there any openings through the roof surface? How have they been sealed? Can water penetrate?
- What happens in driving rain? What happens in snow?

Chimneys
- Are there any defective flashings? What is the material of the flashing? What life span and what condition?
- What is the security and condition of pots? Are they special, do they have to be replaced in the same design?
- What is the condition of the flue lining? What are the special needs of the flues, the boiler, the wood fire, the coal fire, ventilation etc?
- What regulations may apply to the location of the flue, and the fuel used?

Rainwater disposal
- Is the capacity of the system adequate? Is it effective? Are there any signs of leakage or failure?
- What materials have been used? Are the materials compatible? Are they prone to corrosion? Can you see all the surfaces to be able to record their condition?
- Is there a history of failure with the disposal system that has been used, or the materials used? Are the materials still available in the event that repairs are needed?

Eaves detail
- Parapet walls, condition of the coping, is there a damp proof course below and does it work?
- If water penetrates into the wall, will it do any damage?
- Security of the parapet. Is it straight, will it fall over, are the bricks sound?
- Eaves. Will water run off into a gutter, or behind it? Are there any stains on the wall to give you a clue?

Walls
- Construction. Solid or cavity? Components of the wall and the implications of the construction.
- Are there any signs of cracks? If so how serious is it? How big are the cracks? How big are they if you add all the cracks together?
- Are they in a particular direction? Is it cavity tie failure, chemical attack, or movement in the foundations or beams or supports above the ground?
- Are the walls straight or are there any bulges, if so how far has the wall bulged? How much disturbance is there to the window jambs or door jambs?
- What is the condition of the windows? What is the condition of the putty, the sill details, the materials, is water trapped anywhere? If it is trapped, how much damage can it do?
- How weather-tight is the design? How good is the seal to head or side?

Damp proof course
- Can you locate it? If so is it clear of the ground, is it bridged? Where are the weep holes around the dpc?
- Are there any air bricks? Should there be any? Are they clear? Is there adequate ventilation?
- Will water, vermin or other undesirables enter through any opening to the outside of the building?

Loft space
- Can you get into it? What is the condition of the timbers? Is there diagonal bracing? Has it been strengthened if a heavier surface has been provided?
- What is the support given to water tanks? Have they covers? Are the overflow pipes supported? Are the service pipes insulated?
- Is the roof insulated? Is it ventilated? What is the deterioration that the colder surface may cause?

Condensation
- What is the condition of the ceiling surface? What is the construction of the ceiling? Is there any wiring in the insulation? What is the condition of the wiring? What type of wire is used?
- Are there any animals etc in the roof void: bats, bees, wasps, pigeons, rats or mice, or squirrels? What damage have they, or could they cause?

Interior
- Damp checks – have you tested under windows, to all walls to exterior, to internal partitions to ground floor, to skirtings etc?
- Soundness of surfaces – is plaster sound, is it loose, has it pulled from backing, is it lath and plaster or plasterboard?
- Are there any cracks? What has caused them? Are they serious?
- Wooden surfaces – are they free from beetle or fungal decay? Can you be sure? Are the symptoms there?
- Condensation? If so why, and what do you do to eliminate it?

This is not a complete list for the inspection of a property, but it aims to set the mind of the surveyor so that he will be asking the question that the building demands that he should ask if he is to be able to report upon the failures that may be present now or may occur in the future.

The surveyor's report should be a clear and precise document in which the surveyor should identify the defects within the property and the cost of carrying out such repairs are necessary. It is important that the report should be a readable document which leaves the client a clear picture as to the condition of the premises, the likely effect of the failure to carry out the repairs that have been referred to, and the cost of any repairs that are required.

Costings

The references to the cost of repairs in the report should be qualified as being a rough estimate. The quotations that the client may obtain if he buys the property will vary considerably, depending upon the amount of work that the contractor has at the same time and the difficulty of obtaining contractors who will carry out small items of work.

Conditions

The report should include a restatement of the instructions which have been given by the client and which are set out within the letter confirming instructions.

The report should restate any exclusion or limitation clauses or other caveats which have been disclosed at the time of the original instructions and were confirmed within the letter of confirmation.

The report should follow a regular pattern so that the client may retrace the steps of the surveyor and identify the failures to which he refers at various stages in the report.

Centre: Be very careful in the inspection of property during the course of construction or renovation. Be sure to state the extent of work that has been completed and report on the condition of the property at the time of your inspection.

Bottom: If your view of the property is obscured state this in your report. Faults may be only too apparent when foliage is removed at a later date.

Overleaf: Picturesque stone warehouses on the River Dart at Totnes, Devon. Such buildings may have developed many problems over the years due to their proximity to water.

The Visit and the Report

THE VISIT AND THE REPORT

Generalities

The property should be clearly identified. The surveyor must check the address placed upon the report to ensure that it is the same address as on the confirmation of instructions. It is not unknown for a report of a totally different property, sharing only minor components of the correct address, to have been prepared.

The conditions of the weather at the time of the inspection should be referred to within the report. Roof surfaces, gutters, downpipes and other pipework may reveal failures in wet weather which are not evident in dry conditions. Frost, snow, mist and fog may impair the surveyor's inspection.

When you arrive to carry out your survey do check the address.

Weather conditions at the time of your inspection should be reported. Leaks under doors are only apparent during heavy rain. Snow could hide an important defect. The cinema below was photographed at mid-day in August 1981!

Content

The practice note prepared by the Royal Institution of Chartered Surveyors called "Structural Surveys of Residential Property" (published by Surveyors Publications) gives the following specific and valuable advice about the content of a report:

"In respect of each section of the construction or services, the narrative should:

(a) State the discovered facts.
(b) Describe the elements which could not be identified and indicate why.
(c) Include recommendations for further enquiry or investigation.
(d) Describe defects and disadvantages in relation to contemporary standards and to those applicable to the period of construction.
(e) Make recommendations for immediate repair or improvement and advise on any long term implications."

Conclusions

The report should clearly set out the conclusions that the surveyor has reached relating to the property. He may, during those conclusions, make references to

17

THE VISIT AND THE REPORT

similar properties in the area in order that the client does not gain the impression that the previous fifteen pages of the report have only reinforced his impression that he was going to buy the worst property in that town.

Pro formas

It is not possible to produce a sufficiently flexible pro forma report. Buildings vary considerably and the modern passion for word processing and computers tries to impose a strict regime with considerable limitations to sundry haphazard collections of bricks and mortar which rarely suit a standard format.

There are exceptions to this statement. Many surveyors inspect properties which are virtually identical, such as those on some large estates. There is a considerable risk in the use of a standard framework because the surveyor may feel he has completed his inspection when he has filled in his form, and he may fail to carry on to investigate the cause of a problem.

Standard paragraphs/clauses

There are certain paragraphs which surveyors tend to trot out regularly. These may cover specific advice or guidance for certain types of property. Here is an example:

"Lath and plaster construction is usually formed by the application of a lime plaster on to the face of willow laths which are fixed to timber joists. The gap between the willow laths enables the plaster to pass between them to form a key to secure the main surface to the woodwork. Over a period of years fractures occur in the plaster where it passes between the laths. The ceiling plaster surface can become separated from the lath construction. When this happens there is a risk of collapse which can cause damage to the furniture and fittings within the room and the risk of injury to anyone who may be in the area when failure occurs. The normal useful life of this type of construction varies between 40 and 70 years. The addition of central heating to a property tends to reduce this anticipated life span."

There is no harm in having a schedule of such paragraphs so that when a report is written or dictated reference may be made to them. This means that the long suffering lady or passionate word processor who types out the final document may simply copy the specific wording set out in the reference material.

This enables the surveyor to attain a uniform standard of report and also to make sure that there are no minor variations in the information contained within these paragraphs.

A section of standard clauses and paragraphs has been incorporated within an appendix to this chapter. They should not be regarded as being a complete list. They are intended mainly as a guide as to the type of paragraphs which are duplicated in many reports.

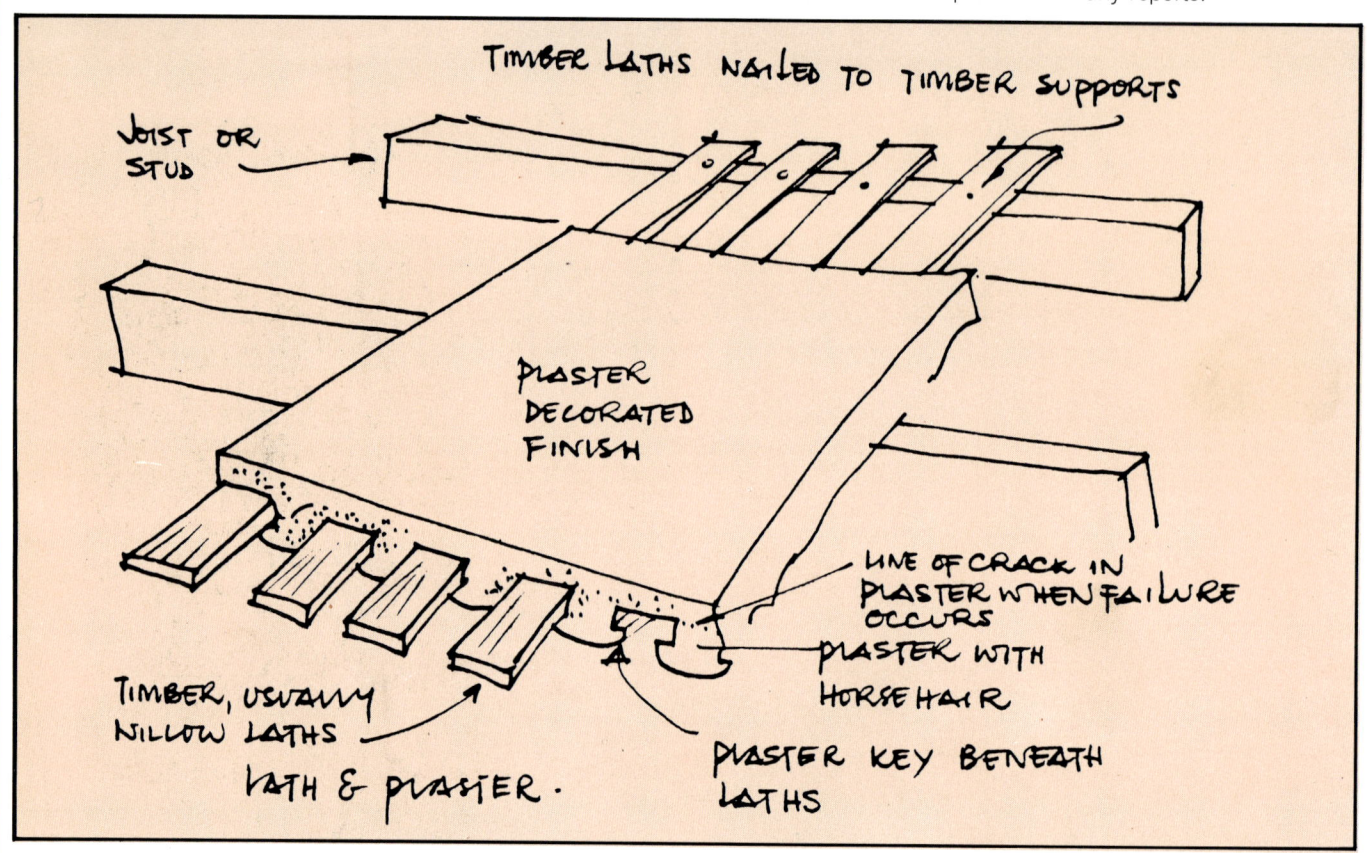

Words used as illustrations

Many surveyors have their own vocabulary of words describing the various parts of a building. The variations in word usage in various parts of the country can be surprising. To some surveyors the underside of an arch may be a SOFFIT, to others a REVEAL.

In order not to confuse the client, let alone other surveyors, a sketch of a property with the various technical words connected to their location can make life easier.

Several examples of these are shown here. These are particularly useful for areas of a building which are unlikely to be examined by the client. The surface of a roof with a central valley which one had to inspect with the aid of a 30' ladder and accomplished acrobatics is unlikely to be visited by your client. Your word picture of this roof may leave no clear image upon a client of exactly what goes on above his head!

The use of photographs greatly aids the appearance of a report. The report is more attractive and easier to read and understand. With today's 24 hour service in developing and printing, 10 photographs can be included in a report for a small amount of money.

The introduction of video and film as a method of presenting a report is discussed in a later chapter, which also outlines the cost of the equipment required.

THE VISIT AND THE REPORT

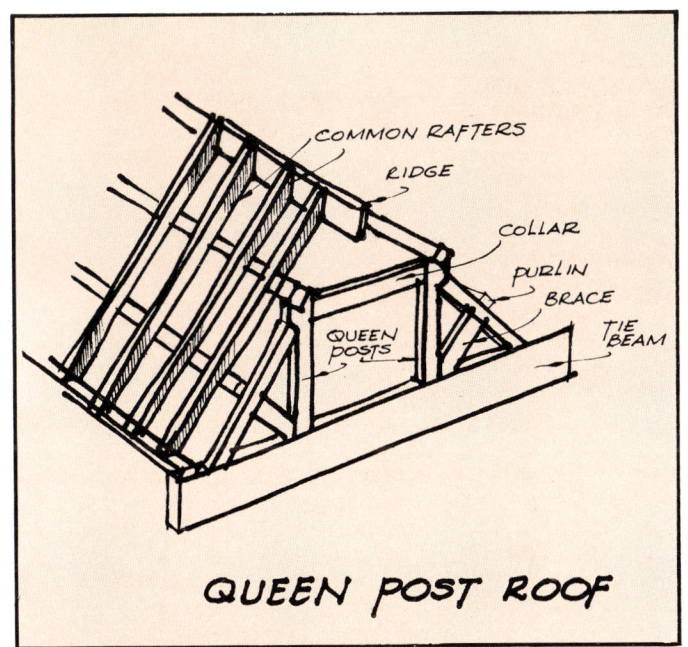

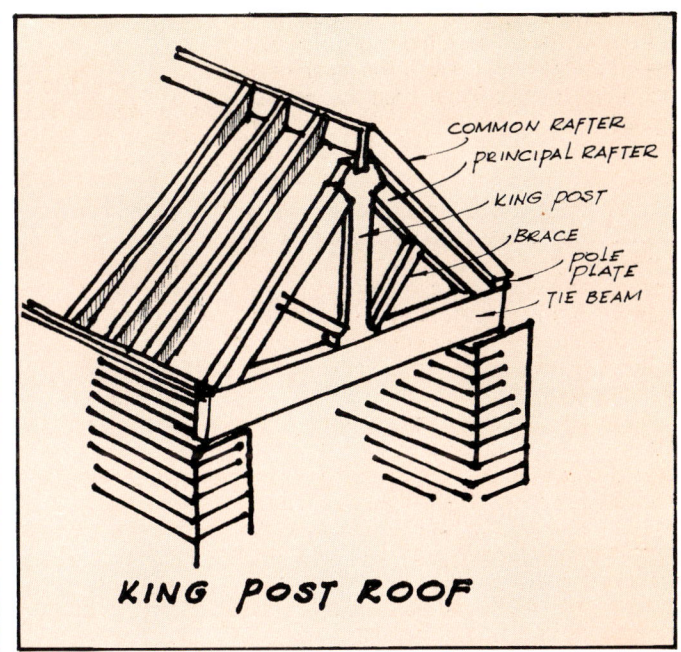

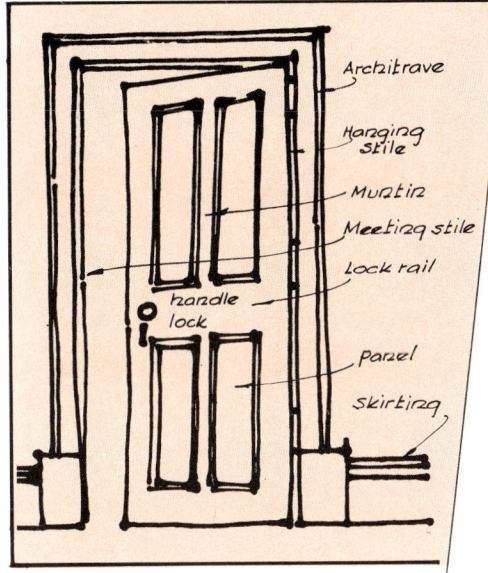

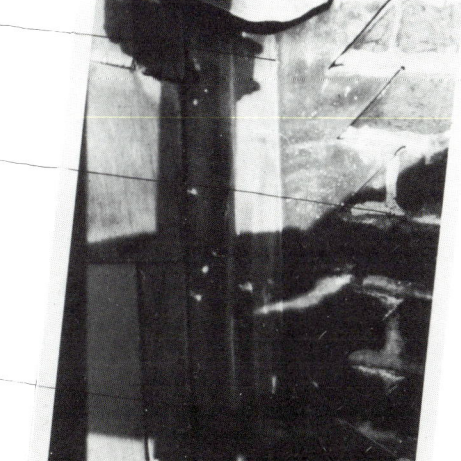

Shown here is a useful example of how photographs can be used in an informal way to clarify a report for a client. (This is a reduced facsimile of an actual A4 sized page).

THE VISIT AND THE REPORT

Examples of some Standard Clauses

Assumptions
The surveyor will assume the following in preparing this report (and Valuation):

No high alumina cement concrete or calcium chloride additive or other potentially damaging material was used in the construction of the property or has since been incorporated.

The property is not subject to any onerous or unusual restrictions, encumbrance or outgoings and that good title can be shown.

The value of the property is not affected by any matters which would be revealed by a Local Search, Replies to Enquiries before Contract, or by a Statutory Notice, and that neither the property, nor its condition, use or proposed use is or will be unlawful.

Report for Addressee Only
The reports shall be for the benefit of the addressee only. We accept no liability to any other party who may seek to rely upon the whole, or any part of this report.

The Main Structure
THE MAIN ROOF
   Where practicable the interior of the roof space has been examined but inspection of the roof coverings has been made from ground level only and without the use of long ladders.

MAIN ROOF COVERING
   It was not possible to inspect the underside of the main roof covering. This is laid on top of roofing felt (close boarding, etc.) which conceals this part of the surface of the tile/slate. The absence of ventilation to the underside of the roof covering can reduce its life expectation, but without examination we are unable to determine the level of ventilation that has been provided.

SUBSIDIARY ROOF AND COVERINGS
   There is no access to the roof structure of the bay roof and therefore no attempt has been made to examine the supporting structure of such feature other than to inspect the underside of ceilings directly beneath it where these are accessible.

DRAINAGE
   There are no visible access manholes to the drainage system. It was not possible to decide whether a separate surface water system or soakaways are provided for the dispersal of rainwater below ground level.

CHIMNEY FLUES
   The fireplaces have been removed in some rooms within the property. If any fireplace was to be restored it is essential that the interior of the flue should be tested and checked through its full height. It is common to find the flues damaged, and dangerous, and the corresponding reinstatement of the flues can be very expensive. The installation of a gas fired coal or log effect fire requires an efficient flue for the full height.

   The existing flue has been inspected where visible. There are indications that the original lining has failed. If the fireplace is to be restored to use it will be necessary to repair the flue lining through the full height of the chimney, to prevent the escape of fumes into upper rooms or the roof space. There is a risk of fire being caused by the use of defective flues.

THE MAIN WALLS
   No examination has been made of any foundations to the building because to do so requires extensive excavation. We cannot confirm the stability of the walls from their support but have drawn what conclusions we can from the surface evidence available at the time of our inspection.

BRICKWORK
   With solid walling of a thickness of about 9″ there is a tendency for condensation to form on the inner face of the external walls in exposed locations.

CAVITY WALLS
   The external walls have been built in cavity construction. This involves two walls of brick separated by a gap of about 2 inches (50 millimetres). The gap between the brickwork is joined together with ties. These ties are often metal, and are known to be vulnerable to corrosion. If the tie fails there is a risk of partial collapse of part of the cavity wall, together with distortion being caused in the roof or internal floors.

   The inspection we have carried out does not include the checking of the existing ties within the wall thickness. When corrosion takes place it can cause horizontal cracks every four, five or six courses of brickwork in the mortar joint, and sometimes this is accompanied by bulging in the outer leaf of the brick wall.

   We have found no indication of failures to suggest that there is an existing defect in the wall ties, but must advise you that our experience is that properties such as this are known to have developed tie failure and remedial work has had to be carried out.

THE MAIN DAMP PROOF COURSE
   It was not possible to inspect the whole of the damp proof course, although sometimes the edge of a bituminous damp course can be seen in modern property. Very often the expected position was covered with exterior rendering and was not visible.

   The failure of the damp proof course is usually discovered by finding unacceptable levels of dampness within the interior of the building. Where high damp readings have been referred to it is essential that a full investigation be carried out before exchange of contracts.

   A failure in a damp course will involve expensive repairs, not only in installing a replacement damp course, but also in replacing plaster within the building, replacing plumbing, electrics and furniture and decorations that may be disturbed in carrying out the work.

   The failure in part or the whole of a damp proof course will also mean that any timbers that may be in the area of any of the damp walls will be vulnerable to fungal decay. This can be very expensive to eradicate and it is essential that the timbers be opened up and checked so that the risk of decay can be quantified, and estimates obtained for any remedial work that may have to be carried out.

   It is essential that this further inspection and investigation is carried out before the exchange of contracts/signing of the lease in connection with this property.

RISK OF DRY ROT
   During the course of our inspection we noted several areas of damp timber, or damp close to or in contact with timber surfaces. These circumstances are ideal for the propagation of fungal decay. Dry rot is the most serious of these decays and is very expensive to eradicate. The rot will spread by up to 1 metre a year, and the eradication includes the removal, not only of the infected timber but also those timbers with 2 feet (600 millimetres) of the last sign of the outbreak. It is essential that the areas which are at risk should be opened up before exchange of contracts so that the extent of such decay as may have already occurred may be determined.

SUB-FLOOR VENTILATION
   The timber floor to the ground floor must be well ventilated beneath to prevent the timbers becoming damp, and being vulnerable to fungal decay. The ventilation is provided by air bricks. These must be kept clear at all times.

   We have not been able to check below the floor surfaces and cannot state that there are no obstructions that might cause future problems.

OPENINGS
   The lintol over door and window openings in older buildings is often timber. It is concealed within the wall and no inspection can be carried out. We are aware of past failures in gutters and rainwater downpipes. These failures will have allowed the walls to become damp

and may have lead to a deterioration of these timbers.

Such timber failures are usually discovered because they result in disturbance to the brickwork to the exterior of the building.

We believe that there is a risk of such a defect being present within this building. We recommend that further investigation be carried out before exchange of contracts to acquire this property.

FOUNDATIONS

We cannot advise you as to the depth and size of the foundations provided to this property. To obtain information relating to the foundations would require the excavation of trial holes around the base of the main walls. In the absence of any visible signs of structural failure or evidence of any past repair work having been carried out we can see no necessity to carry out a detailed examination of the foundations.

Interior

ROOF VOID: CHIMNEY STACKS

We note that the chimney stacks, where they passed within the roof void, are without a render face and there is a risk of sparks escaping through defective pointing. We believe this to be a remote risk, but in view of the condition of the brickwork and pointing we would recommend the chimney surface to be rendered if the flues are to be used.

ROOF VOID: CHIMNEYS

The construction of the roof and supports provided to it is adequate and the timbers are of satisfactory size for their particular purpose. There is no visible evidence of any movement and the loads of the roof are transferred down to proper load bearing partitions.

ROOF VOID: TIMBERS

A general inspection of the roof timbers was made for any evidence of wood-worm or similar defects, and as far as we could determine the timbers are without major failure. We have not examined the surface of every length of timber because some parts of the roof construction are inaccessible.

Our examination of the timbers within the roof void was impeded by the storage of a large amount of material within the void.

The dimensions of the ceiling joists indicate that this roof void is not strong enough to support such a load without causing distortion to the roof support, and possible damage to the building.

ROOF VOID: ROOFING FELT

Due to the lack of any roofing felt usually provided under the roof covering to give a degree of thermal insulation and water tightness, it is possible that during times of heavy and prolonged driving rain some water penetration could take place through the roof covering. At the time of our inspection there was no evidence that this had occurred.

CEILINGS

The underside of ceilings have been tested by applying light pressure to see if there is any loss of key in lath and plaster construction. Lath and plaster construction is usually formed by the application of a lime plaster on to the face of willow laths which are fixed to timber floor or ceiling joists. The gap between the willow laths enables the plaster to pass between them to form a key to secure the main surface to the woodwork. Over a period of years fractures occur in the plaster where it passes between the laths. The ceiling plaster surface can become separated from the lath construction. When this happens there is a risk of collapse which can cause damage to the furniture and fittings within the room and the risk of injury to anyone who may be in the area when failure occurs. The normal useful life of this type of construction varies between 40 and 70 years. The addition of central heating to a property tends to cause failures and may vary this anticipated life span.

FLOORS

We have inspected the surface of floors where possible. We note that the floor surfaces were carpeted in the rooms to the ground/first/second floors. We have not raised any floor boards because to so do would damage the decorations of the property. Our comments are based only on those parts that we have inspected.

INTERNAL PARTITIONS

Most internal partition walls are rendered and decorated on both sides and without the removal of plaster or coverings their construction or internal condition cannot be determined.

STAIR TIMBERS

The underside of the stair timbers were examined from within the (meter) (store) cupboard formed beneath, and as far as we have been able to determine the staircase is in a satisfactory condition. At the time of our inspection the cupboard contained a considerable quantity of stored items which were not completely removed for a full and detailed inspection.

STAIR TIMBERS

We were unable to examine the underside of the stair timbers owing to a plasterboard/asbestos/plaster/other lined surface which precluded the inspection and we cannot comment further.

DECORATIONS

During the course of our inspection the property was fully furnished with the owner/tenant etc. in occupation. When the property is vacated and pictures and furnishings and other objects have been removed, the decorations will present a more marked appearance than at the time of our inspection. We have not commented in detail upon the decorations and assume you are aware of their condition.

Services

WATER SUPPLIES/PRESSURE

Adequate water pressure was obtained from the various sanitary fittings when these were tested over a short period of time, and we therefore consider it reasonable to conclude that this system is satisfactory although we obviously cannot advise upon the internal condition of the service piping.

CENTRAL HEATING

The central heating system was working satisfactorily at the time of our inspection and the radiator surface area appears satisfactory to fulfil normal domestic heating requirements, but this could only be confirmed by a specialist heating engineer's inspection.

ELECTRICAL

In view of the specialist's advice that a new electrical installation is required, we would draw your attention to the fact that this work may cause damage to internal wall decoration and plaster surfaces. Even where the existing wiring is arranged in metal conduits it is not always possible to draw new wires through the conduit without causing damage. It is probable that the existing system will require an extension which will necessitate the chasing in of wires into the plaster surface.

Sundry

HEALTH

There is a risk that you or your family may be allergic to certain elements within this or other properties. We are unable to advise on individual health risks, but must draw your attention to the following:

Lead paint

The lead paint remaining to the windows and doors within the property must not be rubbed down during the redecoration without the appropriate preventative measures being taken.

The Artex ceilings within the .... rooms contain asbestos. These must not be rubbed down at the time of redecoration without the correct preventative measures being taken.

Asbestos lagging to heating pipes maybe/is present. This must be removed as set down by the current Health and Safety regulations. This is an expensive process involving the creation of airtight corridors between the exterior and the source of the asbestos.

Timber treatment

Recommendations have been made in connection with treatment to eradicate insect attack and/or fungal decay. The chemicals used can affect people who have breathing problems, or are pregnant. It is wise to make provision for the property to be vacated during and for one week after treatment work.

ASPECT

The property has an approximate .... aspect for the front and this will mean that the front/back of the house/bungalow/flat/building will derive the maximum/no benefit of natural sunlight. The back/front will have reduced enjoyment.

THE VISIT AND THE REPORT

Additional considerations for a survey of flats and apartments

The Lease
One of the main problems for the surveyor when he is instructed to carry out an inspection of a flat or apartment is how to balance his advice on the defects of the individual unit with the problems or failures that occur in the main building.

Service Charge
An examination of the lease will almost certainly reveal that the owner of the flat which is to be inspected has a responsibility to pay a proportion of the outgoings on the main buildings. If the mansion block is very large this proportion may be less than 1%. Many of the flats in London's Dolphin Square which has some 600 to 700 flats within the block have a service charge responsibility which is to 2 or 3 places of decimals of a percentage point. In order to advise upon the liability for the service charge in a building of this type it will not be possible to carry out a detailed inspection of the complete building.

On other occasions the mansion itself may be part of an Estate of similar buildings and the service contributions be related to the Estate rather than each individual building.

The surveyor must set out in detail the liability that the client will face if he proceeds with the acquisition of the flat. He should also comment upon the past history of expenditure on the buildings.

The surveyor should advise upon the general trends of future service charge expenditure. He may with his practised eye be able to advise upon the probability of capital expenditure on lifts, roofs, external decoration and similar items in the years to come.

Where the examination of the lease reveals that the unit has a service charge which is in excess of 5% of the annual maintenance and repair costs of a building, an inspection and advice upon the complete property will be necessary.

The surveyor should also draw the attention of the client to any specific clauses in the lease which may interfere with his quiet enjoyment of the flat. Many leases contain clauses requiring the flat to be carpeted throughout and it is wise to comment upon this. The cost of carpeting these days can be a major embarrassment if one has not mentioned it, and the client or his wife has ideas of rolling acres of parquet floor rattling through to the neighbour's flat below.

With a modern desire to alter all new property acquisitions into an individual shape, the clauses which control the alteration or amendment of the layout of a flat should be drawn to the attention of the client. Specific advice as to the approvals that he would require would be based upon the surveyors knowledge of the client's proposal for the accommodation if he proceeds with the purchase.

The surveyor should also comment in detail upon the landlord or lessors conventants. There are a number of rogue leases which are still in existence whereby the lessee or tenant has not covenanted to carry out the repairs to window frames, external plaster, external decorations, drains, walls, roofs, etc only to find that his delight in avoiding these obligations is short lived when he finds that the landlord or lessor is not obligated to carry out these works.

A lease should be a balanced equation whereby the responsibilities for the complete building fall upon one party or the other.

Maintenance Charges
Where possible the surveyor should be shown the accounts for the building over the previous three or four years. This is particularly important where the buildings are of such a size that it is not possible to carry out an inspection of all parts of an Estate. The surveyor's ability as company accountant may be stretched fully in unravelling some of the intricacies of the service charge accounts for buildings of this type. It may be necessary for him to seek specialist advice. In some cases the accounts are easy to understand. If the accounts are not available for the past few years, it may be an indication of poor management of the properties. Six months is an adequate time to prepare annual accounts for expenditure.

The Inspection
The major expenditure in the running and maintenance of mansion blocks of flats fall into a number of categories. (1) Roof, (2) External Decorations, (3) Lift Maintenance, (4) Communal heating, and hot water, (5) staff and staff accommodation, (6) internal decorations.

It can be seen that a number of these items relate to the building itself and arrangements should be made so that access can be obtained to the concealed parts of the roof, the lift motor room, the communal boiler and central heating plant so that one may report upon their condition.

Lift
Victorian mansion blocks of flats have lifts which are examples of the craftmanship of a bygone age. The intricate mouldings and polished woodwork of the lift cage are matched only by the ornate mouldings of the wheels, machinery, and the 'out of order' notice stored in the motor room!

Whilst the examination of lift equipment is a specialist engineer's job there are a number of items which the surveyor can comment upon. Because many of these lifts are surrounded by the main staircase the protection that the cables and mechanisms is given can be examined. If one can place one's finger through the bars and touch any of the equipment and machinery the protection is inadequate as may your fingers be if you carry out this test too literally.

The cables should be lightly greased and there should be no indications of any irregularities over the surface such as minor hairlike coverings which are an indication of the wires having failed and individual strands broken away.

The lift should be smooth in operation and should stop accurately to each level. There should be no springing or jerking when the car stops. This may indicate slackness in the cables.

The lift should respond easily to the touch of the requisite call buttons and it should not be necessary to press any button twice.

The alarm system within the lift cage should be tested, after consultation with the caretaker. The absence of an alarm button should be noted within one's report.

An examination of the electrics during the operation of the lift may reveal any arcing between the cut-outs on the board.

From the inspection that one is able to make it would be possible to comment upon the likely remaining life of the lift mechanism and advise as to the replacement cost of lift and shaft or other sections of the equipment.

The engineers reports should be obtained from the managing agents and may be included in the report.

Communal Heating System

If the building has a communal heating system which provides central heating and hot water for the flats and the common parts it is probable that this is going to be a major cost in the service charge. In the majority of cases communal heating is uneconomic and the cost can considerably exceed the individual heating bills of each of the flats in a similar buiding where they have independent heating systems. The surveyor should comment upon this within his report.

The examination of the heating system will reveal a number of items which should be commented upon.

The fuel which is used should be reported on, as must the current cost in use of the material and the efficiency of the boiler equipment. If the building is heated by oil, comments will be made about the savings which may be affected if the system can be converted to gas, and the cost of converting the system to burn a different fuel.

The surveyor will be looking for the mixing of pipe materials such as galvanised pipework and copper cylinders, or galvanised cold water storage tanks. He would stress the risks which occur where such mixtures take place because of the likely corrosion which will take place due to electrolitic reaction.

The condition of pipework and any leaks around the valves and connections would be reported on, particularly if it is felt that they may be indicative of the condition of the pipework in concealed parts of the property.

Caretaker's Accommodation

Most blocks of mansion flats will carry at least one caretaker's flat. Sometimes in larger blocks or estates there is a full set of staff accommodation. When one is making one's inspection arrangements, access to the staff accommodation should be included. The service charge will often include the rentalising of this accommodation and the maintenance and repair of the interior of these staff flats. As it has a direct bearing on the expenditure which the client will face their inspection will enable the surveyor to advise upon the future maintenance charges relating to these flats.

Means of Escape in Case of Fire

Buildings of this type and complexity may incorporate specialised escape systems. These may be the connection between the staircases at various parts of the building over the roofs or by external fire escapes. The condition of the metalwork in a fire escape must be reported upon in detail. The buildings compliance with the statutory requirements relating to fire escapes must be stated. Checks with the Local Authority will have to be made to see if the building has the relavant fire certificates. The conditions imposed upon the owners of the building by statute, or by a fire certificate or its attached appendix will be reported and the buildings compliance commented upon. The cost of compliance with any outstanding items must be included.

When inspecting the interior of the flat the fire standard of doors to the bedrooms, reception rooms and the entrance to the flat should be noted so that they can be incorpoated in one's report under the heading of fire escapes.

In the examination of a multi-storey building which has a secondary means of escape, make sure that you have examined the condition of the adjoining premises. The means of escape on the adjoining premises must also be in good order and the access onto the adjoining premises be safe and secure. Make sure that, in your report, you point out that there is a risk, if the adjoining premises were demolished, or the existing licence for means of escape was terminated.

In many cases it may be virtually impossible to provide an alternative means of escape other than that which is arranged over an adjoining building. If it became necessary to provide a suitable alternative, the cost could be considerable. Most clients who are faced with large bills for which they are not prepared, tend to cast their nets for an alternative signature on the cheque.

Always give careful consideration whether means of escape is over or possibly into an adjoining building and point out that it is probably by licence only which should be investigated.

In checking the fire escape, make a note of how safely it is fixed to the building. The condition of the treads is also important. There is little advantage in escaping from a fire if the fire escape itself is in a lethal condition.

THE VISIT AND THE REPORT

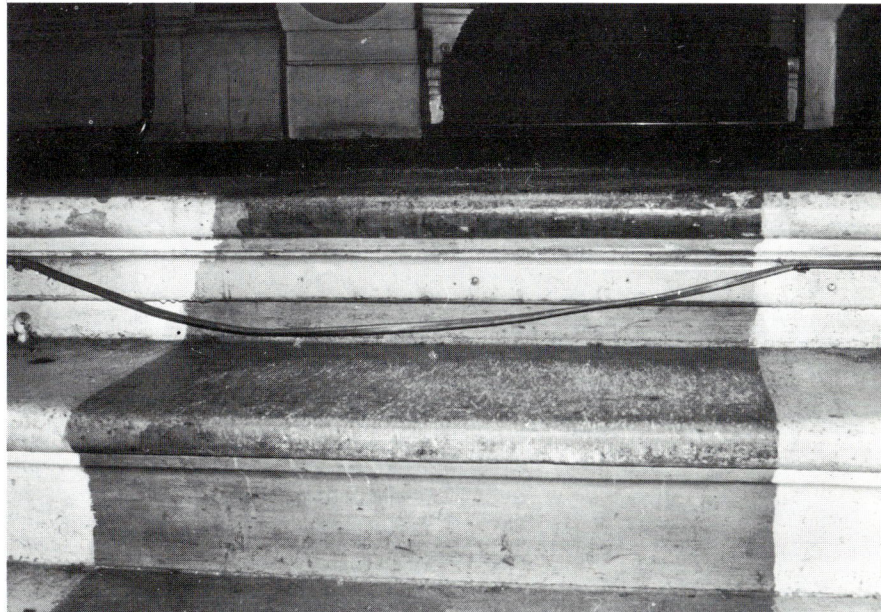

For inspections of property in multiple occupation the report should include comments on failures in communal areas. The poorly fitting carpet and trailing wire are evidence of both lack of care and potential danger. The cost of replacing carpets and repairing failures could be substantial. That cost, in most cases, would be passed on to lessees through the service charge.

Below: A concrete frame in the course of construction. How is the external panelling going to be fixed? Will the securing clamps last as long as the rest of the building? How expensive will the failure be?
Below right: A curtain wall of glass and metal. The high heat loss through external walls can mean that they now have to be examined in order to determine the 'cost in use' of the building.

25

Additional considerations for a survey of commercial property

When one is carrying out an inspection of a larger and more complicated building it is probable that this will have a commercial aspect to its use. It may be a factory, office, shopping complex, warehousing complex or similar building.

The inspection of the building will not materially differ from those which are carried out to residential or other smaller properties. It will still be necessary to determine the condition of the building and to analyse the full extent of any defects. The main variation between the residential and the commercial survey will be in the report which is submitted.

The report on commercial property must relate to the total suitability of the property for the purpose which the client has in mind. It is also a report that will be considered by people who may not have inspected or know anything about the buildings. If the clients intend to invest money in the company which owns the buildings or wish to acquire the buildings for investment purposes their decision will be based upon the report that the survey provides. In the event of the property being acquired for investment purposes the suitability of the property for the existing occupants is important. The long term investment future of the property will be affected if the buildings may soon fall vacant and be difficult to relet on the current market. The surveyor's report will include the information that he has been able to find out about the building or buildings, but it will also cover a much wider area of investigation. The effect of current legislative and other controls will be taken into account, the condition of all the services which are required to the property, the condition of the building and the cost in use of the complex will all be reported on in considerable detail.

Even though this form of report differs in some respects from that of residential property it is important that the same basic rules are followed. If the property is leasehold it is important that a copy of the existing or proposed lease should be obtained prior to the inspection in order that one may report upon the repairing covenants, and determine the liability for any of the repairs recommended.

The covenants of the lease will indicate whether the landlord's consent was required for any improvements or variations which have occurred to the property during the period of the lease. This may assist the valuation members of the fraternity to decide whether the alterations can be rentalised in a future review.

Access to commercial property often causes some problems. This country does not provide an identity card. For those people whose job it is to call on many different properties and worm their way in without a decent identity card this is a problem. All too frequently, one is welcomed into a building with open arms without anyone having checked one's credentials or one's right to be in the building. Nearly every surveyor that one meets has his own story of other people's carelessness in allowing him into bank vaults, top security laboratories, army barracks or police stations without having adequately checked his credentials and allowing him to complete a full inspection before realising that he was in the wrong place. Even with recent terrorist activity the tendency is for people in this country to accept the word of the individual who appears before them and never to consider that it is not an insult to check or ask for his credentials and identity.

The inspection of the buildings will enable a report to be prepared upon their condition. It is probable that there are some areas of the structure of the property which may need specialist testing, for example a concrete frame containing high alumina cement or where reinforcement may have deteriorated as a result of sulphate or chloride attack. The approval of the occupants of the building will have to be sought before any analysis takes place. The removing of samples will cause some damage to the decorations of the property as well as one or two enlarged mouseholes within the concrete frame. You will see from the letter of confirmation of instructions and the letter to the vendor/occupier that these items are covered.

The service installations of the property will form a separate part of the report. This will deal with the statutory services to the property and will record the condition of the equipment, as well as commenting upon its future life. The aim of the comments is to determine the future maintenance liability for each of the aspects of the services within the property.

The services may be listed as follows:
electrical services
hot and cold water plumbing installation
heating services
air circulation or air conditioning facilities
lifts
drainage services
window cleaning equipment
fire alarm system
intruder alarm system
telephone system
escalators
hoists
moving platforms

These services are expensive to install and expensive to maintain and will form a substantial part of the investigation into the total suitability of the property for the client.

The report should deal with the current supplies by the statutory authorities and the flexibility for any extension to those supplies. Comments should be made as to the possibility of the extension of the size of the main service for gas or electricity and as to the flexibility that the telephone service has for providing extra lines. There are limitations to the information that can be obtained, and care must be taken to ensure that the dubious credit for the information is placed upon the party who provided it. Although the gas board may have advised that they are able to provide a 2" main to a particular building, that information may become out of date if there is a sudden increased demand for services and supplies. If you have a client who is likely to acquire the building purely on the basis that such an extension could take place, it is important that confirmation from the statutory authority is obtained in such a way that the advice can be relied upon. If changes are taking place within a reasonably short period of time your client should be in a position to obtain compensation from the authority for any losses he may have suffered.

Top: A concrete framed office block enclosed with calcium silicate brickwork which has shrunk resulting in cracks in the brick face of the building.
Above: Window panels in a modern office block. Are the sealants going to have the correct life span? How are those panels fixed below the projecting bays?

Legislative and other controls

The surveyor must have a working knowledge of the legislation which has an effect on the use, layout and construction of property. This knowledge must be maintained and it is important that the surveyor keeps himself well informed about the continuing amendments to current legislation. One difficulty occurs where a particular minister has the right to make an order in council under a particular statute. This can change the rules almost overnight. Building publications and journals together with *The Times, Guardian* and *Telegraph* newspapers usually keep the surveyor informed. The surveyor has a duty to keep his technical knowledge up to date at all times.

The legislation and other matters which will affect his report will include the following:

1. Town and Country Planning Acts

These Acts control the use to which a building can be put and it is important that they are checked during the course of preparing the report.

It can be embarrassing if one has produced a detailed viability study and report on an hotel complex only to find that it only has a residential user. Under this heading the surveyor will comment on the future planning prospects for sections of land surrounding the building and the possibility of being able to extend the existing buildings within the confines of a site or on to adjoining land.

2. Building Regulations (or the London Building Acts)

These regulations deal with the requirement to comply with various constructional rules or regulations. It will be necessary to comment on any part of the building which does not comply with these regulations; Difficulties may occur with the regulations dealing with thermal conductivity and heat loss for buildings. The required levels have been raised.

3. Fire Precautions Act (or means of escape in case of fire)

There is a requirement for certificates to have been issued for certain types of property such as hotels, and these certificates should be inspected. The existing arrangements must be checked to ensure that they comply with the original plans attached to the relevant consent or certificate.

There is also a need to have provided adequate fire fighting equipment. The schedules of the equipment that should be provided should be listed as well as the condition of the existing equipment reported on. Part of this report may overlap with the services installation report referred to earlier.

The case history of errors within valuations highlights the importance of knowing the means of escape requirements, and reporting upon the cost of complying with the appropriate regulations.

4. Offices, Shops and Railway Premises Act

One of the major provisions within this Act relates to toilet accommodation within the building. The number of cubicles required will vary depending on the number of people who are regularly employed within the property. This may have an effect on the extent of the services which will have to be provided in any extension to the building.

5. Health and Safety at Work Act

It is an interesting point that if you followed the letter of this Act to its logical conclusion surveyors who carry out inspections of property should not be operating from a ladder unless they have somebody standing at the base. As most surveyors always claim that they are in two places at once they probably comply with these requirements.

In multi-storey property safe access to the exterior of the building must be provided for the cleaning of the windows (if the windows are to be cleaned). The likely cost involved in complying with the terms and conditions of this Act must be referred to in the report.

6. Licence Controls

It will also be necessary to list any temporary licences which may have been obtained and to refer to the problems of renewals upon the termination of such licences. Controls are often exerted over the discharge of certain trade effluents into drains, and the emission of smoke from boiler furnaces and flues. Controls can also extend to an office use or road access being available for a limited period only.

The cost in use of the building is the main purpose of the report on a commercial property, if the client intends to operate or use the building. Where the building is being considered for investment purposes, this may be of lesser importance, but it will influence the current and future open market value and therefore should never be far from the mind of the decision makers.

The cost in use will involve the remedial repair and maintenance costs based on the information contained in the first three sections of the report. These sections will have picked up the cost of carrying out the repairs required or proposed to the building, the cost of alterations, repairs and extensions to the service installation, and the cost of complying with current legislative and other controls.

That is the static analysis of the costs related to building. The dynamic analysis requires the projected cost of looking after the building and maintaining it in use for the next 15 or 20 years.

In order to obtain this it will be necessary to determine the life span of the materials which have been used within the construction of the building. For example, felt roofs should have a life span of 15 to 25 years, timber windows one of 30 to 40 years, electrical components and services one of 40 years. The cost of the replacement of the various components of the property should be included in the analysis.

The services will have limited life spans. For example, any major works of repair and maintenance on lifts should be referred to.

Consideration will also be given to the likely change in techniques and technology over a limited period of time. No surveyor is provided with a crystal ball, irrespective of what most clients believe. Criticism of reports based on hindsight is the prerogative of the poor executive.

The reasoned appraisal of future energy consumption should form part of this report, and anticipated trends of the future cost of the supply of gas, water and electricity should be made. With the decline in the world stocks of various commodities it is important that cost fluctuations are referred to when one is considering the long term life of the property. There is little point in referring to the cost of replacing a roof with a 25 year life span unless the cost of gas as a fuel is also referred to when the North Sea stocks are due to expire within that period.

The cleaning of the exterior of the building is now becoming, for environmental and prestige purposes, a regular event. The cost of cleaning a building on a ten year cycle should also be incorporated in the report. The daily cleaning or maintenance of the interior of a building is a major cost these days, and is part of the cost in use for the building. The provision of this service has now become big business; one notices that the cigarette and alcohol advertisements seen on sporting occasions are now being replaced by those advertising the activities of office cleaning services.

Insurance premiums will also be referred to. The cost of insuring of the property will need to be assessed. The reinstatement cost of the building in the event of fire or other major calamity should be recorded within the report.

Clients responsible for declining businesses may be fully aware of the possibilities of starting fires but few of them have mastered the same technique with floods.

As various equipment declines variations will take place in the premium for the reinstatement of the property. The inclusion of an adequate sprinkler system or other means of fire fighting will also affect the premium payment and will vary the cost in use of the building.

Vandalism is becoming a substantial problem. The cost of replacing broken windows or removing graffiti has to be assessed. Certain locations are more vulnerable than others. The surveyor will be able to report on the problems that he saw during his inspection as well as advise on future risks. The vulnerability of the property may be a function of the design. A building located next to a public car park in an urban environment will be more vulnerable than a property situated in a rural locale miles away from major roads.

The proximity of the property to human activity (or inhuman activity in this case) will be a major factor in assessing the vulnerability of the property to vandalism. Comment must be made on the lighting around the buildings, the presence of dark corners, and ease of access to the rear of the building. All these factors affect the risk of damage taking place.

All reports which are over five pages should include a summary. The main findings within each of the various sections will be set out with specific advice and the relevant costs. For a report of up to 15 pages this summary should be located at the back, but when the benefit of your considered opinion has extended beyond this size the summary and conclusions should be located at the front. This is probably all that a busy client will have the opportunity of reading. At that point the bulk of the report becomes an appendix to the main conclusions and is available if greater detail is required.

The report should become a working document for the maintenance and repair of the building over a number of years, and if any information has been collected which may be of service it is worthwhile attaching it to the report as an appendix.

The best report is written on one side of a sheet of paper, but whilst this may be ideal for the clients it tends to upset the professional liability insurers. The aim should be to provide clear, logical, simple and readable advice and to back it up with a document which enables the party who wishes to read the substance of the report the opportunity of whiling away the twilight hours.

Checklist of contents of a commercial survey report

The building
Condition and defects
Costings
Summary and recommendations

Services
Electricity
Gas
Hot and cold water plumbing
Heating
Air circulation/air conditioning
Lifts
Escalators
Hoists/moving platforms
Waste and drainage
Window cleaning
Fire alarm
Intruder alarm
Telephone
Communication equipment
Summary and recommendations

Legislative and other controls
Town and Country Planning Acts
Building regulations
Fire precautions
Office, Shops and Railway Premises Act
Health and Safety at Work Act
Licences/controls

Cost in use
Immediate expenditure on repairs
Annual expenditure in use
Planned maintenance
Future alterations and repairs
Expenditure required to comply with requirements beyond control of client
Insurance assessments
Vandalism

Summary
Commercial premises in the London area may have parts licensed by the District Surveyor's service of the Greater London Council as a temporary structure. These structures have to be inspected regularly on an annual or bi-annual cycle and the cost of carrying out remedial works to comply with these licences can be substantial. The licence itself is a fairly minor expense.

THE DUTIES OF SURVEYORS

The duties of Surveyors their performance and their breach

Introduction

The standard of care which is required of a surveyor or valuer carrying out an inspection of a property should be no less than that required of the average practitioner within the profession. In trying to assess whether the surveyor has achieved the requisite standard. It is suggested that there are three tests which should be applied to any point of contention within a report.

1. Omission

The surveyor is responsible for carefully inspecting the property and reporting upon the defects which are evident to him. In the event of defects which were in existence and visible at the time of his inspection not having been reported, he will be negligent provided that it was reasonably practical for him to have noted, recorded or seen the failures in existence. It is unreasonable for a surveyor to be expected to find defects which are concealed from him by the building's construction. The extent of the inspection will have been set out and cleared in the Letters of Confirmation of Instructions and Confirmation of Appointment sent by the surveyor to his client and the vendor respectively. (See pages 10, 11 and 13).

In many cases the argument may be whether the failure which is visible at the time of a dispute was evident at the time when the inspection was carried out. If the surveyor had recorded the condition of the building by taking one or two photographs of each of the elevations, many points of issue would be easily determined by reference to the circumstances which existed at the time of the inspection.

Many disputes relating to the presence of a failure can also be eliminated by the presentation of the notes made during the course of an inspection which might indicate that particular tests were carried out to the structure at various parts revealing no defect. For instance, a failure to a window sill which is noted at a later date may not be the surveyor's responsibility if it can be shown that he carried out a routine and sensible inspection of that component and found no defect at the time of his inspection.

2. Projection

The Surveyor is responsible not only for recording defects which were evident but also for advising on the likely consequences of those failures that he has noted. He should be able to report upon the natural consequences of those deficiencies which he has noted by projecting the existing circumstances into future consequences. A report which records defects in rainwater down pipes, either due to blockage or splits, must carry on to point out that water which runs from this sort of failure into a solid brick wall will penetrate to the interior and may result in deterioration occurring to timbers to the interior. It should also report upon the necessity of carrying out remedial work to eliminate this risk. In this way the client is fully advised of the consequences of the failures which have been found and the need for remedial work.

It is not adequate to report upon an absence of air bricks to the base of a wall which should ventilate a hollow timber floor, or their being deficient or blocked up. The surveyor must continue to point out the service provided by these bricks and the natural consequences of their being blocked up. The presence of replaced rainwater down pipes or new external plumbing in the house will be an indication that the original plumbing would have been defective. Defective plumbing may have been replaced after it had soaked brickwork or masonry and this may be contributing to future failures to the property. The report must indicate at every available opportunity the natural consequences of the defects which are found and the importance of remedial work being carried out. There may be occasions when the surveyor suspects that a defect is present but is unable to identify an actual failure. in such cases he must recommend that further exposure be carried out to enable the defect to be correctly analysed and diagnosed.

When a surveyor is faced with a set of circumstances over which he is not a master, he should report to his client that he does not have the necessary knowledge or skills to diagnose a particular fault and recommend the appointment of a specialist to examine and advise upon a failure. Where such a recommendation is made, the specialist who is ultimately appointed should be deserving of that name. The appointment of a specialist to advise upon the presence of dampness, defective damp proof courses or woodworm is an abuse of the surveyor's privilege. No surveyor should regard himself as being inadequately qualified to report upon these matters and the provision of a free report provided by commercial concerns involved in selling services and products is a dereliction of the surveyor's duty.

3. Construction

The surveyor must identify the building construction which has been used in the house or other property under inspection. The surveyor must be cognisant of the history of that type of construction and know of specific failures which have been found in similar properties. If the property has a flat roof which is not built to the modern standards of design and construction, he knows that that is a defect waiting to happen. The history of failures in flat roofs to buildings constructed over the last 20 or 30 years is common knowledge within the profession and must be included in the advice that the surveyor makes to his client.

The inspection of a prefabricated re-inforced concrete property such as an Airey house would be negligent unless the surveyor reported upon the known history of failure of this type of building construction. This comment should be made even though the property that he is inspecting is free from all similar defects at the time of his inspection.

Unless the surveyor has been able to identify the particular construction used in the property he is inspecting, he is unable to report upon the likely consequences of any failures which he has found within the property.

These three tests of the work of the surveyor may help to concentrate the mind for the legal interpretation of the duties of the surveyor and of the standards which those to whom such duties are owed have a right to expect.

The principle of duty

The duty of the surveyor to his client and its effect on the surveyor's preparation for making a contract with his client has been covered in the earlier chapters. The position of the surveyor once he has accepted a professional engagement must now be considered. The quality of his work will affect others. He needs to know which people will have a remedy against him if he falls into error. He must expect his client to have a remedy. But who else? Over the years the courts have striven to establish principles on which the potential plaintiff may be identified.

It is now well settled that, apart from contract, no action for negligence will lie unless the plaintiff can show that the defendant owed him a duty of care at the time of the act or omission of which he complains. The duty was described by Lord Atkin in a celebrated passage as a duty to one's neighbour:

> "The rule that you are to love your neighbour becomes in law, you must not injure your neighbour; and the lawyer's question, Who is my neighbour? receives a restricted reply. You must take reasonable care to avoid acts or omissions which you can reasonably foresee would be likely to injure your neighbour. Who, then, in law is my neighbour? the answer seems to be – persons who are so closely and directly affected by my act that I ought reasonably to have them in contemplation as being so affected when I am directing my mind to the acts or omissions which are called in question."[2]

According to that principle, it can never be a defence to an allegation of negligence that the circumstances of the particular case have never before been held to give rise to a duty of care. This point was stressed by Lord Wilberforce in a more recent case:

> "...the position has now been reached that in order to establish that a duty of care arises in a particular situation, it is not necessary to bring the facts of that situation within those of previous situations in which a duty of care has been held to exist."[3]

Still more recently, the House of Lords and the Privy Council have given guidance as to what does have to be established. Foreseeability of harm is always a necessary ingredient; but it is not the only one. In each case a close and direct relationship between the parties has to be demonstrated. All the circumstances must be taken into account in determining whether such a relationship existed at the relevant time. In a small number of cases (unlikely to affect surveyors) considerations of public policy may be taken into account.[4]

Application of the principle

In the majority of cases involving surveyors, the factor most likely to give rise to a duty of care is RELIANCE. The surveyor, like other professional people, holds himself out as possessing particular professional skills. His whole career is founded on the invitation that he extends to members of the public to rely on those skills. Once that invitation is accepted by a person within his reasonable contemplation it is clear, on the principles quoted above that that person will be his neighbour in law.

The invitation to rely and the reliance frequently will result in a contract between surveyor and a client. Such a contract contains an implied term that the surveyor will carry out the contractual work with the skill and care reasonably to be expected of a competent surveyor. The implication is founded on the reliance and the surveyor's status that gives rise to the reliance.

Reliance on a surveyor's work is to be found not only in those who enter into contracts with them. Later in this chapter we shall be looking at examples of third parties successfully founding actions against surveyors on their reliance on surveys and the resulting reports.

But we shall see also that reliance or the likelihood of it, while a most common foundation of a duty of care, is not a prerequisite of such a duty.

It should not be thought that those to whom a duty of care is owed are divided into two mutually exclusive groups, those who have entered into a contract with the party who owes them a duty and those who have not. It would now be difficult to dispute that a client's remedy against a defaulting professional man lies in tort as well as in contract. Lord Denning MR put it in this way:

> "...in the case of a professional man, the duty to use reasonable care arises not only in contract, but is also imposed by the law apart from contract, and is therefore actionable in tort. It is comparable to the duty of reasonable care which is owed by a master to his servant, or vice versa. It can be put either in contract or in tort."[5]

The surveyor's neighbours in law

It would be dangerous, in view of the very broad principles quoted above, to be dogmatic as to the classes of person to whom a surveyor embarking on a survey of a building will, and as to those to whom he will not, owe a duty of care. However, certain classes are coming to be identified as being owed a duty.

In point of time, the person, next after the client, to whom the surveyor should turn his attention is the owner of the building to be surveyed, unless, of course, he is the client. The surveyor needs the building owner's consent before he may carry out a survey. In a later section of this chapter suggestions will be made as to how he may protect himself against any suggestion that he has trespassed and against ultimate liability for any damage which may occur during the work. At this stage it should be stressed that the surveyor clearly owes to the building owner a duty to take care not to damage his property or to cause him personal injury.

For many years it appeared from the cases that a surveyor owed no duty of care to third parties when he wrote his report: first, because no action lay in respect of carelessness consisting of words as opposed to deeds; and secondly because any loss that they might suffer in consequence of the report was purely economic, and such loss standing on its own was not compensatable. Then the House of Lords decided that where a "special relationship" of trust and reliance existed between a plaintiff who had sought, and a defendant who had imparted, information or advice a duty of care was owed by the defendant to the plaintiff, even though the only loss that the plaintiff could foreseeably suffer in consequence of the information or advice being wrong was economic in character.[6] It remained to be seen how widely or narrowly the courts would interpret "special relationship". They might have confined the remedy within strict bounds by affording it only to those who would have been in a contractual relationship but for the lack of payment or other consideration. But that has not been the approach. Indeed, as we have seen, it is now clear that the relationship need not be "special": it is enough if it is "close and direct".[4]

It is, however, clear that where the only loss that could foreseeably be suffered by a person in consequence of a negligent misstatement is economic, the maker of the statement will owe that person no duty of care unless he knows or ought to know that that person will or is likely to rely on the statement.

Prospective purchasers of houses are the most obvious class of persons to whom a duty may be owed. It is easy enough to discern and justify the duty when the surveyor knows that his report, whether made for a vendor or to a building society, is to be shown to a prospective purchaser. But the fact that the plaintiff neither saw nor was expected to see the report has not deterred the court from finding a duty of care established. In *Bourne* v *McEvoy Timber Preservations Ltd*[7] the defendants were timber rot specialists instructed by a prospective vendor to report on fungal infestation and to provide an estimate for work to cure it. They knew that their inspection and report were in connection with a proposed sale. They did not know of the existence of the proposed purchaser, the plaintiff; nor did they know whether or not any proposed purchaser would see and rely on their report.

There was no evidence that the plaintiff ever saw the report; yet it was he who instructed the defendants to carry out the work recommended by their estimate and paid them for doing so, and he was named as the client on the guarantee that they issued. Bristow J found a duty of care to the plaintiff established on the basis that when the defendants made their inspection and reported they knew or ought to have known that the purchaser of the house might well be affected in the decision which he took by the contents of their report, and that they regarded the buyer as the beneficiary of the work if they got the job which they recommended should be done.

In that case the plaintiff's reliance on the report appears to have been indirect, and the defendants' knowledge was that it would be indirect. The report would affect the valuation of the house, and the plaintiff relied on the valuation of the house as being accurate. A similar case is *Yianni* v *Edwin Evans & Sons*.[8] The plaintiffs decided that they would like to buy a house for £15,000 and that they would do so if a building society agreed to provide £12,000 on mortgage. By a notice from the building society they were warned that the making of an advance would not imply any warranty as to the reasonableness of the purchase price. The mortgage application form contained a suggestion that applicants who desired a survey for their own information and protection should consult a surveyor on their own account. The plaintiffs made enquiries about the cost of an independent survey; on grounds of cost they decided not to commission such a survey. They paid the building society a fee for the survey to be conducted on the society's behalf by the defendants. The defendants carried out a survey and reported to the society that the house was worth £15,000. The society told the plaintiffs that it was willing to advance £12,000. Park J found that the society's statement to that effect served in the minds of the plaintiffs to confirm the fact that the house was worth at least £12,000 and that that undoubtedly was a factor which they took into account when their decision to buy was made. he also found that the defendants knew that their valuation, insofar as it stated that the property provided adequate security for an advance of £12,000, would be passed on to the plaintiffs, and that it was in their reasonable contemplation that the plaintiffs would rely on it when they decided to accept the building society's offer. The learned Judge was clearly much influenced by evidence that the proportion of mortgagors who have their own surveys carried out is very small – on one calculation less than 10%, on another (conducted by the RICS) between 10% and 15%. He found that these matters were common knowledge in the professional world of building societies and of surveyors and valuers employed or instructed by them, and that the defendants were fully aware of these matters.

In earlier editions of this book it was confidently stated that a duty of care will not be negatived by the mere fact the plaintiff is not expected to and does not see the surveyor's report. Gratifyingly, the House of Lords has shown that this confidence was well founded. In *Smith* v *Eric S. Bush* the plaintiff was shown a copy of the report. in *Harris* v *Wyre Forest D.C.* the plaintiffs were not. The appeals were heard together by the House of Lords. In each case the result was the same. There was evidence, similar to that in *Yiannni's* case, of an overwhelming probability, known to surveyors, that prospective purchasers of domestic property will rely on valuations prepared for mortgage purposes. Having paid fees for such valuations they are most unlikely to com-

Above and overleaf: Yianni v Edwin Evans & Sons — In this famous case the tie bar was not seen by the surveyor.

mission reports of their own. If a purchaser does not see the report he will still rely on it as showing that the house is worth at least the amount which the mortgagee is prepared to advance. The surveyor's knowledge of probable reliance and the fact that the prospective purchaser has paid the fee produce a relationship between them "akin to contract", which is a "close and direct" relationship as described above. Thus, subject to a successful disclaimer of liability, a duty of care will be owed by the surveyor to the prospective purchaser.

As to disclaimer, it has already been stated that the House of Lords held that in cases of the kind described any attempt to exclude or restrict liability for negligence will fail to satisfy the requirement of reasonableness, and that this is so even where the attempt is to negative the existence of a duty of care. Attempts of this kind may be made by means of contractual terms, or (as in *Smith* and *Harris,* but not in *Yianni*) by a notice brought to the attention of a prospective purchaser.

As mentioned above,[10] the House of Lords has reserved its position so far as the relationship between surveyors and prospective purchasers of industrial property, large blocks of flats, very expensive houses and the like is concerned. But it must not be assumed from this that the result in such cases will be different; such cases are likely to depend on their particular facts.

Prospective mortgagees are another class of persons to whom a duty may be owed. They will not, of course, be able to say, as could the Yiannis, that they are unlikely to commission a survey of their own. But where plaintiffs informed defendant valuers, who had been instructed by a property company to value land, that reliance would be placed on the valuation for the purpose of lending money with the land as security, it was admitted by the defendants that they owed the plaintiffs a duty of care.[11]

In *Smith* and in *Harris* the House of Lords made it clear that the surveyor, when making his inspection and writing his report, will not owe a duty of care to avoid economic loss to anyone not immediately connected with the prospective purchase which has given rise to his being instructed. A subsequent purchaser will not be able to ride on the coat tails of the initial purchaser.

So far we have concentrated on cases where economic loss is in question. It has been shown that in such cases reasonably foreseeable, and actual, reliance is an essential ingredient in a plaintiff's case. Quite different are cases where it is reasonably foreseeable that a person within a particular class may suffer personal injury or damage to his property in consequence of an error in a report. Suppose that a surveyor carelessly reports to his client that a beam, which in fact is dangerously defective, is sound. In consequence the client takes no steps to replace or repair the beam. Months or years later the client has a dinner party. The beam collapses. The client and his guests are all badly injured, and furniture and other property are ruined. So long as it was within the reasonable contemplation of the surveyor that the beam might collapse or otherwise injure them they will have a good cause of action. It is now well established that a careless mis-statement that causes physical injury or damage to property is actionable.[12] In such cases there is no question of reliance by the plaintiff being a prerequisite of his cause of action.

Another case may be imagined when the plaintiff, unlike the dinner party guests, is completely unconnected with the client. The surveyor carelessly fails to notice that guttering is insecurely fixed. He should have realised the risk that the guttering will fall to the highway and kill or injure whoever is unfortunate enough to be passing or damage one or more vehicles. In such a case a duty of care will be established at the time when the guttering is, or ought to be, inspected. It will be owed to the indeterminate class of persons who are, or whose property is, likely to pass some point in the trajectory of the gutter if it collapses. It is doubtful whether the surveyor's duty will go further, so as to attach to him when he walks along a street in a purely private capacity and fortuitously (or perhaps because he is of a nervous disposition and his eyes are always trained on the buildings he passes) notices that guttering is about to collapse into the street. He probably is entitled simply to "pass by on the other side".[13] In spite of his professional skill, he probably will be ably to rely on Lord Keith's dictum that there will be no liability in negligence on the part of one who sees another about to walk over a cliff with his head in the air and forbears to shout a warning.[14]

It is important that when a surveyor is carrying out a survey on behalf of a client who is not the occupier he observes a feature likely to cause injury to the occupier, he should inform the occupier as well as his client. If he fails to do so, he could well be vulnerable to the suggestion that by virtue of his skill, his presence in the occupier's home, and the danger that he has observed, he owed the occupier a duty to take reasonable care.

Performance and breach of duty: the standard required

What is the 'reasonable skill and care' that the professional man must exercise when he owes a duty? In many cases where his skill and care are called in question there is no, or no serious, dispute about the standard of the work. For example, in *Yianni's* case negligence was admitted. Very often, however, the identification of what is reasonable is crucial to the result of the case. The principles on which the skill and care of professional people are assessed will first be considered; and then certain examples of particular relevance to surveyors will be given.

We have seen that a case may be framed against a surveyor in contract and tort or in tort alone. Although the point is not altogether free from doubt, it is probable that there is no difference between the standards called for by the contractual duty to exercise reasonable skill and care and the duty of care in tort. Certainly it is a point not to be explored in a book for practical surveyors.

The point which is of importance derives from the fact that within any profession there will be found a wide range of competence and a substantial degree of specialisation. If a highly skilled and experienced specialist is engaged, is it enough for him to achieve the average standard achieved by the profession as a whole? If a newly qualified and nervous practitioner is engaged, is he to be excused for mistakes which are understandable in him but which the average practitioner would not make? Does the court make its own judgement as to what is to be expected or is it bound to accept expert evidence from within the profession?

The answers to these and similar questions have emerged from the cases. To take the last question first, the court is never bound to accept expert evidence any more than it is bound to accept evidence of fact. Suppose that the expert evidence is to the effect that the standard achieved by the defendant reached the standard ordinarily achieved by averagely competent members of the profession. The court might nevertheless find the defendant liable. It might consider the evidence unreliable and deduce that in fact the standard ordinarily achieved was higher. Or it might make the more radical, but in principle legitimate, decision that the standard ordinarily achieved was not good enough, or, conversely, that a practitioner not achieving it nevertheless achieved a reasonable standard. This is not to say that a judge may ride roughshod over the evidence: his decision must be judicial and not capricious. It should also be said that the writer knows of no case where evidence as to what surveying standards are or as to the quality of those standards has been rejected.

What of the case where the client knows that he is engaging a surveyor who either because of considerable skill and experience in the particular field in question or because of inexperience is in no sense average? Certainly the courts have recognised that it would be unrealistic to expect a uniform standard from all members of a profession. This is particularly so in the case of hospital doctors, who are identifiable as belonging to a particular rank and, at the more senior levels, to a particular speciality. A specialist will be judged by the reasonable standards of the speciality in which he holds himself out as being skilled. Yet even where a professional holds himself out as having especially high professional standards, the standard to be applied against him is that of the ordinarily competent practitioner of his profession. He may, however, attract liability which his average colleague would avoid. Where he has a higher degree of knowledge or awareness than that of the ordinarily competent practitioner, and he acts in a way which, in the light of that actual knowledge, he ought reasonably to foresee will cause damage, he will be liable in negligence; while the ordinarily competent practitioner, not having that knowledge, will not.[15] At the other extreme, inexperience is not an excuse for failing to conform to a standard that is reasonable for the profession as a whole. The requirement that that standard be reached was clearly expressed by Bristow J:

"The duty of a practitioner of any professional skill which he undertakes to perform... is to see the things that the average skilled professional in the field would see, draw from what he sees the conclusions that the average skilled professional would draw, and take the action that the average skilled professional would take."[16]

Thus the inexperienced practitioner will be at risk if he takes on work that is really beyond him. Even though he may do commendably well in view of his inexperience, if he does not achieve what is a reasonable standard for the profession as a whole he is negligent. This is because duty is tailored not to the actor but to the act which he elects to perform.[17] His negligence will lie primarily in taking on work that he was not competent to take on, and consequentially in his failure to achieve a reasonable standard in that work. Such a standard is not, however, one of perfection. It may be reached even though errors of observation and judgment are made. The question is always whether the errors are consistent or inconsistent with that standard.

Allowance must always be made in the case of a scientific profession for the necessity of its advancing its knowledge and its techniques. If judges were to find (which they do not) that negligence lies in the mere fact of experimentation, a judicial dead hand would stifle initiative to research. But where injury or damage has resulted from experimental work, a judge is likely to need to be satisfied that there was a reasonable need for the experiment, that adequate research preceded it, and that the client was given such warning of the risks involved as was appropriate in the circumstances.

Allowance is also made for the fact that on a particular matter professionals may belong to one of two or more reasonable schools of thought, and for the fact that some professionals are by temperament or training more cautious than others. As was said in a case involving settlement:

"The carrying out of a survey and the reporting to a client involve observation, dedication and the exercise of professional skill and judgment. The mere fact that one professional man might suffer from an excessive caution does not mean that another man, exercising his judgment to the best of his skill and ability and taking perhaps a somewhat more optimistic view, is guilty of a departure from the appropriate standard of professional care and skill."[18]

All the above discussion has related to the obligation to exercise reasonable skill and care. It must not, however, be forgotten that the contract between a surveyor and his client is likely to impose on the surveyor certain express duties, and that the surveyor will be in breach of an express duty if it is not fulfilled in terms. In that case it will not be sufficient for him to rely upon the general quality of his work

Performance and breach of duty: some examples

Caution should be exercised in drawing from an individual case involving breach of duty a conclusion as to what the result would be in a case of a similar type. Except insofar as they express new statements of principle, reported cases will be of use to the surveyor only insofar as they may bring into sharp focus some error into which he may be in danger of slipping in his professional work.

Nevertheless, the surveyor may be helped by a classification supplementary to that which is at p.29 above, of the errors that may cause him to be liable, together with a modest ration of examples.

(i) Failure to carry out instructions

This is the failure referred to at the end of the previous section. Unless the terms of the engagement between surveyor and client are clearly expressed the surveyor will be at risk of misconstruing the instructions that he has accepted. If he does so he is highly likely to break the contract while doing work that, as far as it goes, is of excellent quality. For example, in one case[19] a surveyor who had undertaken to survey a house and prepare plans of it misinterpreted his duty to survey as a duty to carry out only a measured survey. On appeal he was held liable for failure to carry out a structural survey and consequently for failure to detect the defects that such a survey, if competently carried out, would have disclosed.

(ii) Inadequate knowledge or experience for the work undertaken

Without expert advice, the court will be unable to form an opinion as to what is and what is not well known to competent surveyors. Such knowledge may be general or local. For example, in a case already referred to,[20] a surveyor was held to have been negligent in not recognising trees as poplars. An unpopular mistake. He should have done so because of the known risk of damage to buildings on shrinkable clay subsoil and near to poplar trees. That was general knowledge. Clearly a surveyor must also have and exercise a thorough knowledge of significant local characteristics. Particularly important for a structural survey is knowledge of the characteristics of the soil on which the structure in question stands.

Inexperience is no excuse. But it is only by dealing with cases that the inexperienced practitioner will mature into experience. No firm is likely to be negligent merely by putting a new recruit to work on a testing survey. Liability is likely to be incurred where the work of the recruit is not thoroughly supervised, so that errors that a recruit is likely to make are looked for, identified and corrected by a partner or senior employee of the firm.

(iii) Inadequate inspection

In some cases a surveyor has been negligent in omitting from his inspection a part of the building, for example the rafters of an old cottage,[21] the roof spaces of a manor house,[22] and the cellar of a house.[23] A survey of a flat must take some account of the structure of the building of which the flat forms part.[24]

A common complaint is of failure to uncover and open up. In a case already referred to[21] the surveyor was put on enquiry, having found beetle in the woodwork and wet rot on the ground floor. It was practicable for him to raise floorboards on the first floor and he should have done so. Where there are grounds for suspicion about the fabric, the surveyor must ascertain the extent of the defect even though this involves an investigation which a surveyor, on an inspection of the type being carried out, will not normally undertake. He must accept that sometimes he will have to spend much longer on some properties than on others.[25] In many cases the express terms of the surveyor's engagement may exempt him from any duty to uncover or open up; or the terms of the licence granted to him by the building owner may preclude him from doing so. In such a case, he must guard against failure to warn the client in clear terms as to the risk of defects lying in the uninspected parts. Indeed, if the risk is other than remote the surveyor may be negligent if he neither makes arrangements with the building owner for the uncovering or opening up of one or more of the concealed parts nor warns the client that unless such an arrangement can be made he should not commit himself to the transaction that he proposes.[26]

A failure to make proper use of instruments may lead to a finding of negligence. In *Fryer* v *Bunney*[27] damp was not visible to the human eye unless a wall was stripped; but it was detectable by use of an electric moisture meter placed in the right position. The surveyor was found negligent in not having investigated sufficiently thoroughly with his moisture meter. This case should not be regarded as laying down any rule as to what equipment is to be used on a survey or as to the manner in which such equipment is to be used. Issues of this kind are likely to develop in response to the availability of the wide range of equipment described elsewhere in this work. It will often be easy enough to prove that a particular feature, undetectable and not to be suspected with normal surveying skills, would have been disclosed had equipment available in the market been used. It is important that the courts should hear debate as to the extent to which, and the circumstances in which, the profession as a whole is using the new equipment. Otherwise judges may be too readily led to accept that what could have been done should have been done, and a substantial armamentarium, expensive to surveyor and client alike, will come to be used defensively, often in circumstances where the subject matter does not warrant it.

Another common basis for findings of negligence is a failure to observe properly. There are so many ways in which this failure may be manifested, and each case turns so much on its own facts, that examples taken from decided cases are of limited value. The later part of this book is designed to assist the surveyor by showing the signs that are likely to be of significance. It is, however, perhaps worth mentioning here that the field of observation is not to be confined to the particular building under survey. The surveyor must be careful not to miss important evidence to be derived from, for example, trees adjacent to the survey property.

It is one thing for a plaintiff to prove that a surveyor simply missed a piece of significant evidence. It is another for him to show that all competent surveyors appraising that evidence would have been led to a particular conclusion or course of action. If a competent surveyor could reasonably have concluded that a particular piece of evidence, though worthy of serious consideration, was not indicative of any defect, a plaintiff will effectively gain nothing by proving that his surveyor negligently missed the evidence altogether.

(iv) Inadequate report

Ideally, the surveyor needs to be a master of English prose. He must be careful of style as well as content. A vitally important point that has been expertly dealt with on inspection may be mishandled in the report by being ambiguously or misleadingly expressed. A few minutes spent in careful checking of the report may save a surveyor (and his client) from disaster. Checking is particularly important now that dictation of reports is common practice.

In the report the surveyor must be careful not to use words of commendation so broad that they cannot be justified. For example, "sound and substantial" has been judicially criticised as a description,[28] since it leads the reader to assume that at any rate during the foreseeable future there ought to be no trouble with regard to the outside walls and the structure of the house. Another area to be watched with particular care is that of warnings. It is vital that these should be well expressed as well as covering all the matters that call for warning. Two particular points may be stressed here. First, where a condition, for example dry rot, calls for a warning if active, it should be reported on even though dormant.[29] Secondly, the surveyor should evaluate the risk carefully and clearly. The risk may be such that he is under a duty to warn the client in terms not to proceed with the proposed transaction.[30]

Any general caveat as to the limits of the inspection will be looked at by a court in the context of the report of which it forms part. If a court took the view that a particular part of the building could and should have been inspected and reported on, a surveyor will not be protected by a statement that he had not inspected that part. He cannot, at the stage of writing his report, introduce into the contract terms that it did not contain at its formation. Nor can he rebut an allegation of negligence simply by reciting places that he has not examined. He should make clear to the client the extent to which he ought, by reason of the limitations of the examination, to withhold reliance on the report. Not only should he evaluate the risk associated with features that he has seen. On the basis of what he has seen he should try to evaluate for the client the risk of defects lying in the parts that he has not seen.

Protection against liability

(i) To the client

This subject has already been touched upon insofar as the relationship between the surveyor and his client is concerned. The broad conclusion reached is to the effect that the surveyor should prevent himself from being negligent, for any attempt that he may make to protect himself against the consequences of his negligence, at least in the usual range of cases, will not succeed.[31] Indeed, where the liability sought to be avoided is that for death or personal injury the attempt is bound to fail.[32] In other cases, any clause in the contract which purports to exclude or restrict liability for negligence must satisfy the requirement of reasonableness.[33] The whole point of engaging a surveyor is to obtain an expert appraisal of evidence which cannot be appraised by diligence and common sense alone and a report in reliance on which one can take an important decision. If a surveyor negligently leads his client to rely on a report that is unreliable and to make a decision on the strength of it, it is difficult indeed to see how a term that imposes on the client all, and on the surveyor none, of the loss caused by the negligence could be described as reasonable in any conceivable type of case. In theory at least, a term purporting to restrict liability for negligence should stand a better chance of satisfying the requirements of reasonableness. But we have already seen[34] that this possibility has been excluded in the case of surveys of ordinary domestic property: it remains as a possibility (and no more) only in the case of transactions at the top end of the property market. In such cases a term seeking to restrict the surveyor's liability to the limit of his insurance should stand a fair chance of success, for the client is then given the chance of insuring himself in respect of any top slice of loss. If the circumstances are such as to preclude the client, whether because of the time available or otherwise, from obtaining insurance it is likely that the respective resources of the client and the surveyor would be of great importance. A surveyor might well be able to establish that at the time when the contract was made it was foreseeable that the top slice of loss would cripple him but would be a passing embarrassment to a large corporate client, which would anyway be partially compensated, and the liability of which to corporation tax would be reduced by the uncompensated portion of the loss.

An indemnity clause that seeks to visit on the client liability for the surveyor's own negligence will not satisfy the requirement of reasonableness, but where such a clause has the limited ambition of passing on to the client liability for accidents that occur, without negligence on the surveyor's part, in consequence of the survey work it is suggested that the clause should satisfy the requirement.

(ii) To third parties generally

In England, Wales and Northern Ireland, but not in Scotland,[35] the Unfair Contract Terms Act is relevant to the relationship between the surveyor and a third party to whom he owes a duty of care as well as to that between him and his client. Just as a person cannot by reference to any contract term exclude or restrict his liability for death or personal injury resulting from negligence, so too (unless Scottish law applies) he cannot by reference to a notice given to persons generally or to particular persons exclude or restrict that liability.[36] Similarly, just as in the case of other loss or damage a person cannot by reference to any contract term exclude or restrict his liability for negligence except insofar as the term satisfies the requirement of reasonableness, so too (with the same exception) he cannot achieve that result by reference to a notice unless that satisfies the requirement of reasonableness.[37]

In earlier editions of this work it was suggested that the chances of avoiding liability to a third party for negligence, by means of a notice brought to his attention, could not be considered to be good. Now we have seen[34] that they must be discounted altogether in the case of surveys of ordinary domestic property.

Quite apart from the problem of satisfying the requirement of reasonableness, in many cases the surveyor will have difficulty in bringing an appropriate disclaiming notice to the attention of all third parties to whom he is potentially liable.

(iii) To the building owner

The surveyor must gain the owner's consent to carry out all the procedures that he must carry out to satisfy the contract with the client. Insofar as he acts without that consent, he will be a trespasser. In the last chapter the surveyor was advised to ensure that his relationship with the client is clearly set out in writing. At pp12–13 above it is suggested that it is equally important to formalise in writing the relationship with the building owner. Only if this is done can the surveyor be confident that he has a licence to do all that he needs to do if he is to fulfil his obligations to his client

THE DUTIES OF SURVEYORS

The mortgage valuation includes the need for the valuer to advise upon the cost of reinstatement in the event of destruction by fire. The valuer has been held not to owe a duty of care for the accuracy of this valuation to the purchaser of the property. It has been assumed that the link between the valuer and the purchaser are too remote. The valuer does not contemplate, so the Court of Appeal decided, the purchaser relying upon this figure.

It is important that the valuer does not act negligently in preparing the valuation. The valuer should use contemporary costs of rebuilding for the type of property under consideration. Buildings that have extensive ornamentation, carved stone, ornamental plasterwork or decorative panelling will be difficult to cost. The figures commonly available are for conventional buildings and do not relate to property of this type. If the building is outside the knowledge of the valuer, request a specialist to carry out the assessment of rebuilding cost.

Exterior of burnt out mansion and interior of one of the corridors showing the damage caused by the fire.

and to any third party to whom he may owe a duty of care.

For reasons advanced in the case of the client and that of third parties in general, it is thought that any attempt by a surveyor to exclude or limit his liability to the building owner for negligence would be unsuccessful. Where the building owner is not the client, he would have an additional point in his favour: why should not the surveyor and his client between them bear the liability? He may so argue even where the damage is not the product of negligence. No licence acceptable to a building owner could be drafted that contained a licence to cause damage; so damage that occurs without negligence will involve the surveyor in a breach of the terms of the licence under which he has entered the property. It is thought unlikely that the surveyor could avoid liability for such breach by means of a contract term. Even if the building owner granted him a licence including such a term, he would be able to argue that it is not reasonable for the surveyor to rely on it, because it is open to him to obtain a suitable indemnity from his client for whose benefit the whole exercise is being mounted.

---

1. The English terms 'plaintiff' and 'defendant' will be used throughout, with apologies to Scottish readers, for whom 'pursuer' and 'defender' would be appropriate.
2. *Donoghue* v *Stevenson* [1932] AC 562, 580.
3. *Anns* v *Merton London Borough Council* [1978] AC 728, 751-2.
4. *Yuen Kun Yen* v *Attorney-General of Hong Kong* [1988] AC 175; *Hill* v *Chief Constable of West Yorkshire* [1989] AC 53.
5. *Esso Petroleum Co Ltd* v *Mardon* [1976] QB 801, 819. The principle is not confined to the case of a professional man: see *Batty* v *Metropolitan Property Realisations Ltd* [1978] QB 554, 566.
6. *Hedley Byrne & Co Ltd* v *Heller & Partners Ltd* [1964] AC 465.
7. (1976) 237 EG 496.
8. [1982] QB 438; (1981) 259 EG 969.
9. [1989] 2 All ER 514.
10. At p.7 above.
11. *Singer and Friedlander Ltd* v *John D. Wood & Co* (1977) 243 EG 212.
12. *Clayton* v *Woodman & Son (Builders) Ltd* [1962] 2 QB 533; *Dutton* v *Bognor Regis UDC* [1972] 1 QB 373; *Anns* v *Merton London Borough Council* [1978] AC 728.
13. St. Luke 10, 31-2
14. *Yuen Kun Yen* v *Attorney-General of Hong Kong* [1988] AC 175, 192
15. *Wimpey Construction UK Ltd* v *Poole* [1984] 2 Lloyds Rep. 499, 506-7.
16. *Daisley* v *B. S. Hall & Co.* (1973) 225 EG 1553, 1555.
17. *Wilsher* v *Essex A.H.A.* [1987] QB 730, 750.
18. *Leigh* v *Unsworth* (1974) 230 EG 501, at p. 649.
19. *Buckland* v *Watts* (1968) 208 EG 969.
20. *Daisley* v *B. S. Hall & Co* (1973) 225 EG 1553.
21. *Hill* v *Debenham, Tewson & Chinnocks* (1958) 171 EG 835.
22. *Stewart* v *H. A. Brechin & Co.* 1959 SC 306
23. *Conn* v *Munday* (1955) 166 EG 465.
24. *Drinnan* v *C. W. Ingram & Sons* 1967 SLT 205.
25. *Roberts* v *J. Hampson & Co.* [1989] 2 All ER 504.
26. *Grove* v *Jackman and Masters* (1950) 155 EG 182.
27. *Fryer* v *Bunney* (1982) 263 EG 158.
28. *Hood* v *Shaw* (1960) 176 EG 1291, 1293.
29. *Hardy* v *Wamsley Lewis* (1967) 203 EG 1039, 1043.
30. *Daisley* v *B. S Hall & Co.* (1973) 225 EG 1553, 1555.
31. See the section entitled "The drafting of the contract" on page 8.
32. Unfair Contract Terms Act 1977, ss 2(1) and 16(1) (a).
33. *Ibid*, ss. 2(2) and 16(1)(b).
34. At p.7 above.
35. ss. 15-25 of the Act relate only to Scotland, and the limitation to contracts is the principal point of difference.
36. Unfair Contract Terms Act 1977, s. 2(1).
37. *Ibid*, s.2(2).

Top: Air brick blocked – could easily be missed.
Centre: Damp penetration to interior wall.
Bottom: Dry rot.

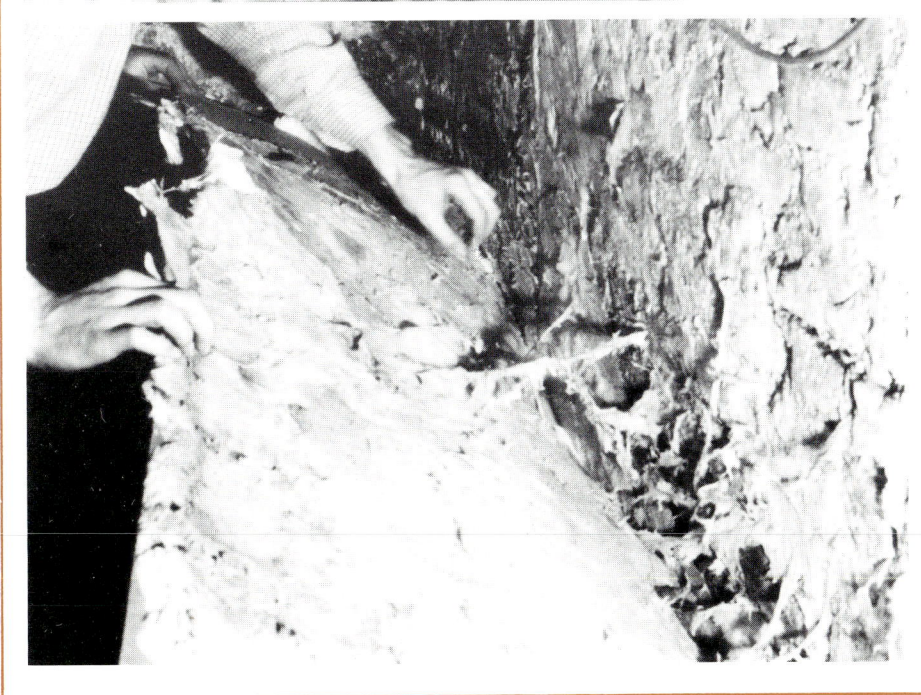

The consequences of breach of duty

37

Introduction

This chapter is based on the premise that the worst has happened, and a surveyor has failed in some way to match the standard of skill and care required of him towards some person to whom he owes a duty of care. The law relating to the consequences of a breach of duty can, if its labyrinthine passages are fully explored, provide a substantial appendage to an already complex surveyor's negligence case. In every case the attempt has to be made to identify the difference in financial terms between a state of affairs that should not have occurred but did occur and one that should have but did not. In some cases many permutations of supposed events can be postulated, each permutation carrying its own financial consequence to set against the actual position of the plaintiff; and that position itself often leaves room for substantial argument.

The full complexity of the subject will not appear in this chapter. But it is hoped that the surveyor will nevertheless be able by the end of the chapter to see, in broad terms and subject to argument that may arise on the facts of a particular case, where he will stand in the event that he is negligent.

The principle of compensation

There is one single principle that governs all cases of breach of duty where damages are, in consequence, to be awarded. It has been stated and restated on many occasions. Donaldson LJ, in a case to which we shall later return, restated it in this way:

"The general object underlying the rules for the assessment of damages is, so far as is possible by means of monetary award, to place the plaintiff in the position which he would have occupied if he had not suffered the wrong complained of, be that wrong a tort or a breach of contract."[1]

All the law that is dealt with below has been worked out on the foundation of that general object.

Contractual fees

Usually, where a surveyor has been negligent his fees are small compared with the financial consequences of his negligence. Nevertheless he ought to be able to know how his fees, whether paid or unpaid, are to be dealt with. Unfortunately, the courts have not adopted a uniform approach. In some cases the negligent surveyor has been allowed his fees, in others not. There is no clear authority as to whether part may be allowed and part disallowed. It is clear that a claim for damages may be set off against a claim for fees relating to the work that gave rise to the damage.

In one typical situation a disgruntled client does not settle the surveyor's bill. The surveyor sues for his fees. The client defends the claim, pleading that the surveyor's work was without value to him or of less value than that for which he bargained. He counterclaims damages for breach of duty and claims to be entitled to set those damages off against the surveyor's claim. The other typical situation is the other way round. The client claims damages. If he has paid the surveyor's fee he claims the return of it. If he has not paid it, the surveyor counterclaims it.

There should be a principle governing the question whether the negligent surveyor is to be paid his full fee or part of it or no fee at all. It should all depend on the bargain between him and his client. The best view of that bargain is that if the work done is effectively worthless the surveyor should have no fee for it, but that if on analysis it has some worth as well as some defects the fee should be reduced to take account of those defects but should not be altogether abated. The principle of a partial fee has never in terms been judicially approved, and it has been doubted;[2] but it is suggested that it accords with the common sense view that people would be likely to take of the bargain if they gave their minds to it. Where the client has paid the fee in the belief that the surveyor's work is up to the standard called for, he may recover it or, it is suggested, an appropriate part of it on the basis that it (or the relevant part) was paid under a mistake of fact.

What of the damages that the client recovers? The client must be compensated for the wrong done to him: he must never be better off than he would have been if the surveyor had not been negligent. So the damages that he recovers must always allow for the fact that had there been no breach of duty he would have been liable for the full fee. This may be illustrated by an example:

A surveyor (S) does work which if up to standard would command a fee of £200. His inspection and report have defects, but much about them is accurate and valuable. As a result the client (C) suffers damage calculated, without taking account of S's fee, at £2,000. C does not pay the fee and S sues him for it. C counterclaims damages. Then:
(i) Subject to set-off, S is entitled to part of his fee – say £100.
(ii) C's counterclaim must take account of the fact that had S done his work properly C would have had to pay a fee of £200 instead of one of £100. The counterclaim is thus worth £2,000 − £100 = £1,900.
(iii) C may set the £1,900 off against S's £100, so that S's claim fails altogether and C is awarded judgment for £1,900 − £100 = £1,800 on his counterclaim.

Stage (ii) was apparently not followed by the Court of Appeal in *Buckland v Watts*.[3] The surveyor failed to perform the full survey that he should have performed. Had such a survey been done the plaintiff would rightly have paid a fee of £3x and he would not have paid a deposit on the property. As it was, he paid the fee of £3x but the surveyor did a limited survey worth only £2x. The plaintiff paid a deposit, which he lost when, on discovering the defects before completion, he withdrew from the purchase. He recovered from the surveyor £x and damages equal to the whole of the deposit. Thereby, it is submitted, he was overcompensated, in that he ended up better off by £x than he would have been had the surveyor performed his engagement without fault.

Where, as in the example given above, a surveyor's breach causes damage, it makes no difference in the result whether the surveyor is entitled to his full fee, part of it or none of it. A difference would arise where, fortunately, a breach of duty causes no damage. Why should the surveyor be paid in full for a piece of work part of which was useless and was likely to, but fortunately did not, cause loss to his client? It would seem unjust that he should enjoy success in proceedings against his justifiably disgruntled client equal to that of a colleague who has done immaculate work for a client who has not bothered to pay.

Causation and remoteness

Here, the law is at its most labyrinthine. Surveyors will not want to be troubled with the intricacies of two difficult subjects. But they cannot be left out altogether. Once a plaintiff has established a breach of duty on the part of the surveyor he has to go on to demonstrate what his position after that breach was. Then he must show what it would have been but the for breach. In other words, he must identify the difference that the breach caused. Even when he has established that difference, he is not necessarily home. Some part of the loss and damage that he has suffered may remain uncompensated on the basis that it is too remote. We should look briefly at causation and remoteness in turn.

(i) Causation

When a plaintiff is trying to establish a causal link between a breach of duty and an item of loss or damage he is under the difficulty that the position in which he finds himself is never the product of a single cause. There will always have been an interplay of causes, of which the fault of which he complains will be at most one. He has to show that that fault was an effective cause in that interplay, so that it materially affected his position. In a typical case involving a surveyor, the plaintiff will have to show that he relied on the surveyor's report and that his reliance was an effective cause of the train of events that followed. There will have been other causes: for instance, the plaintiff's eagerness to buy a house will come to nothing unless the owner agrees to sell it. When defending a claim, a surveyor and his representatives will always be looking to see whether the claim can be defeated or limited on the basis that the plaintiff effectively owes his position to some cause or causes other than the surveyor's report. For example, it may emerge that part of his loss is attributable not to his reliance on the surveyor's report but to his own imprudence. In one case[4] a plaintiff in reliance on a negligent over-valuation of his property committed himself to the purchase

THE CONSEQUENCES OF BREACH OF DUTY

House, gable end showing chimney fixed to timber frame wall of house, and internal shots of infected timbers substantially eaten away.

The discovery of death watch beetle should have lead a surveyor to suspect the condition of the lining of the flues in this old building. This was the outcome of litigation in connection with this building in the New Forest.

The surveyor had failed to see the extent of the insect infestation and did not warn the client of the risk of serious damage, and major expense. Although liability was not contested over the failure to see the extent of the death watch beetle, there was a dispute over the extent of the liability in connection with damage to parts of the building that would not have been inspected by the surveyor.

Although it is not a requirement that a surveyor should test the condition of the flue lining, he was still liable for its failure. The line of argument for the inclusion of the liability for the cost of replacing the flue lining is that a prudent surveyor, seeing the extent of the insect infestation, seeing the extent of distortion of the roof timbers, should have advised that the chimney flue may have been disturbed by this movement, and could be faulty.

The plaintiff also claimed for the cost of treating oak buildings around the main building, even though the surveyor had specifically excluded outbuildings from his inspection. This exclusion was stated both in the letter of confirmation of instruction and restated in the report. The surveyor was held to be liable because a prudent surveyor, discovering the death watch beetle in the main house would have advised the client to either obtain a specialist report, or to allow for the cost of treatment of the infected timbers. In either case it would have been impossible to avoid treating the timbers in the oak barn that lay 20 metres from the house. Just because you have said that you will not inspect part of a building does not mean that you can turn your mind off to the consequences of damage which will incur expenditure for those parts that were not inspected.

For instance, if the surveyor states that he will not test the electrics, that will not mean that the surveyor will automatically avoid liability for any loss that may be suffered if the electrics are faulty.

If the defects in the electrics could have been seen or the prudent surveyor would have assumed them to be faulty on the basis of what was seen during the inspection that was contracted, the surveyor will be liable for the loss that may be suffered because of the wiring defects. The surveyor must consider the projection of the circumstances into areas of the property, even if those areas have been specifically excluded from the ambit of the inspection and report.

This decision, based upon the facts of this particular case, reinforces the requirement for the surveyor to inspect carefully, and express a warning over the extent of any damage that may occur, based on the defects that have been found. It is essential that the circumstances of the defects in the building are projected, and the client warned as to the liability that there may be for a major defect being present.

of a property costing more than he would have spent if a proper valuation had been given. He also spent money on improvements. While his claim in large part succeeded, it failed as to £8,000 on the basis that £5,000 of it were attributable to his own under-estimate of the cost of the work, and that £3,000 would in any event have been spent on fittings.

(ii) Remoteness

In one respect a plaintiff may not be placed in precisely the position which he would have occupied if he had not suffered the wrong complained of. A defendant need not pay for a consequence that he could not reasonably foresee at the time when he committed his breach of duty.[5] We have already seen reasonable foresight as a factor necessary for the existence of a duty of care. It operates also in relation to a claim for damages. In cases involving personal injury or physical damage to property it is frequently argued that, although injury or damage was foreseeable so that a duty of care was owed, the injury or damage that occurred was not reasonably foreseeable.

In practice the courts have left this door only slightly ajar in favour of defendants. In the cases involving surveyors it does not appear that remoteness of damage is at all frequently a live issue. This is because broad patterns of loss have emerged from experience. A party to whom a surveyor owes a duty of care nearly always belongs to one of three main groups – purchasers, vendors and mortgagees. Of the purchasers some complete the purchase, others withdraw without completing. The types of loss that persons within each of the groups suffer have been identified in cases over the years; and these types of loss may be taken to be reasonably foreseeable. Let us take these groups in turn, remembering that each of them is to be compensated in accordance with the general object of compensation set out above.[6] The description of the measure of damage peculiar to each group will be followed by a section on the consequential damages which may be appropriate in any case, regardless of the group to which the plaintiff belongs.

Purchasers who complete the purchase

Probably the largest group of persons who claim damages against surveyors are those who have committed themselves to the purchase of property. The question of how damages to be awarded to plaintiffs within this group are to be calculated has formed a particularly interesting issue in the law affecting surveyors.

The hypothesis is that the surveyor has negligently painted too rosy a picture of a property. He has not affected the quality of the property. The purchaser in reliance on the report enters into a contract to purchase the property at a particular price. The real worth of the property, reflecting the defects that the surveyor missed, is less than what is paid. The purchaser may decide to retain the property and make the best of a bad job by a programme of repairs. Or he may decide to cut his losses and sell. Is he to recover damages assessed by reference to the difference in value? Or can he, if he carries out repairs, recover the cost of them? Or is there no single rule?

The answers to these questions cannot be said to have been settled beyond all doubt. They appeared to be settled in 1956 by the leading case of *Philips v Ward*,[7] in which the Court of Appeal overruled previous cases in which the cost of repairs was taken as the measure of damage. In *Philips v Ward* £25,000 was the purchase price, which was consistent with the surveyor's description, £7,000 would have had to be spent to enable the house to match that description, and £21,000 was the value of the house as it stood. Now the essence of the case against the surveyor is that he is responsible, not for the condition of the property, but for a wrong description of its condition. The cost of repairs is likely to be an appropriate measure of damages as against a defendant who is responsible for the condition that calls for repair. But the Court of Appeal decided that the true measure of the economic loss suffered by a purchaser in consequence of a defective report is the difference between the value of the property in the condition described in the report and its actual value. So the plaintiff recovered £4,000, and not £7,000. Shortly we shall look at the detailed way in which the principle established by *Philips v Ward* has been applied. But first there should be recorded a decision in which the extent of the application of the principle has been called in question. In *Perry v Sidney Phillips & Son*[8] Mr Patrick Bennett QC, sitting as a deputy High Court Judge, awarded the plaintiff by way of damages the cost of the repairs that were necessary to bring the property to its described condition. He did not reject the principle of *Philips v Ward*; but he felt entitled to conclude that it did not express a universally applicable method of calculation. He founded himself in part on words of Donaldson LJ in *Dodd Properties (Kent) Ltd v Canterbury City Council* that immediately follow those quoted above:[9]

"In the case of a tort causing damage to real property, this object" [sc. the object of placing the plaintiff in the position which he would have occupied if he had not suffered the wrong complained of] "is achieved by the application of one or other of two quite different measures of damage, or, occasionally, a combination of the two. The first is to take the capital value of the property in an undamaged state and to compare it with its value in a damaged state. The second is to take the cost of repair or reinstatement. Which is appropriate will depend upon a number of factors, such as the plaintiff's future intentions as to the use of the property and the reasonableness of those intentions. If he reasonably intends to sell the property in its damaged state, clearly the diminution in capital value is the true measure of damage. If he reasonably intends to continue to occupy it and to repair the damage, clearly the cost of repairs is the true measure. And there may be in-between situations."[10]

But that whole passage is governed by its opening words, "In the case of a tort causing damage to real property". One such case will be that of the local authority building inspector who fails to enforce building regulations and who thus contributes to the building of a defective property and to all subsequent damage.[11] In the typical case concerning the surveyor he is not responsible for damage to property. The damage is there already. Had the facts remained static after the hearing at first instance in *Perry v Sidney Phillips & Son* the Court of Appeal would have been bound, on the appeal that was brought, to decide whether or not *Philips v Ward* was of universal application.

But the facts changed in that before the hearing of the appeal[12] the plaintiff sold the property, and it was therefore common ground that the damages had to be measured by reference to a difference in value and not to the cost of repairs; by the time of sale repairs had not been carried out. In their judgments Oliver and Kerr LJJ expressly left undecided the question whether the damages could ever be measured by the cost of repair; but Lord Denning MR, undeterred by the change of circumstances, declared in an *obiter dictum* that they could not: in other words, that *Philips v Ward* was of universal application in cases where a purchaser in reliance on a surveyor's report completes the purchase of property.

So the question cannot be said to have been finally determined. It did not have to be decided in *Perry v Sidney Phillips & Son*; and it has not yet been before the House of Lords. But it is submitted that the method enunciated in *Philips v Ward* is right and that it ought not to be dislodged in the future as the sole method to be adopted in cases of the kind under discussion. This is because of the fundamental difference between these cases and cases where the cost of repair is clearly appropriate. In the former cases the plaintiff does not complain of the property being defective, but of his not being informed of its defects; in the latter cases the very essence of his complaint is that the property is defective and requires repairs to bring it into the condition in which, but for the defendant's breach of duty, it would have been.

This is not to say that the cost of repairs is an irrelevance. Far from it: it is important evidence of what the difference in value is. In many cases the difference in value is found to be equal to the cost of repair; though it is likely to be found[13] that a purchaser in negotiating the price would probably have been content to carry the risk of a small proportion of the estimated cost of repairs.

Not all cases correspond in their facts with *Philips v Ward*. In that case the price paid by the purchaser was that vouched for by the surveyor's report. A problem has arisen where the plaintiff's misfortune is tempered by his getting the property at less than the value vouched for. It now appears clear that in such a case the measure of damage is the difference between the price paid and the actual value.[14] If a plaintiff could show that it was reasonably foreseeable to the surveyor (probably nothing short of actual knowledge would suffice) that he, the plaintiff, would be able to purchase at less than the value vouched for, he might then hope to recover, in addition, the 'bargain' element, namely the difference between the value vouched for and the price paid. But that will be a special case. In any event it will not apply if the plaintiff, if told the true condition of the property before purchase, would have declined to purchase at all rather than achieve a bargain at a price lower than the value.

The converse case, where the plaintiff pays more than the value described, has arisen. In each decided case[15] the plaintiff has recovered the difference between the value described and the actual value. But in a recent case[16] the Court of Appeal held that it was wrong to strike out as unarguable a claim for the difference between the price paid and the actual value, where the price paid was higher than the top of the range of values suggested by the surveyor.

Defendants cannot rely on the cost of repairs when it suits them, any more than can plaintiffs. In one case,[17] unusually, the cost of necessary repairs was less than the difference between the value described and the actual value. A contention that the plaintiff should recover only the cost of repairs was rejected by Bristow J on the basis that 'What is sauce for the goose is sauce for the gander'. This case illustrates the proposition that in cases of this kind the plaintiff suffers a loss not later than when he commits himself by contract to buying a property worth less than its described value. If a loss of an identifiable size had not been suffered at the point of exchange of contracts, the defendants could have shown that the plaintiff could mitigate (i.e. partly avoid) a *prospective* loss by spending a modest sum on repairs.

The precise moment as at which the loss is to be measured is not absolutely certain from the authorities; though for practical purposes this uncertainty is of little importance. As a general rule, damages are to be assessed as at the date when the cause of action arose.[18] This is far from being a universal rule; and it has certainly ceased to apply in cases where the appropriate measure of damages is the cost of repairs.[19] But it appears to be universally applied in the cases with which we are dealing.

When a cause of action arises depends on whether it arises in contract or in tort. In contract, it arises at the moment of a breach of contract. But in tort there is very rarely, and in the tort of negligence never, a cause of action until damage is suffered. In the cases with which we are concerned separate breaches of duty occur when the surveyor inspects and when he reports. Damage is suffered when the purchaser exchanges contracts.

In the previous chapter we saw[20] that when a client enjoys a cause of action in contract he also enjoys a cause of action in tort. Logically, therefore, he should be entitled to choose whether his damages should be assessed as at the date of breach or as at the date when damage occurred. But in the decided cases he appears not to have been given a choice. Nor is there unanimity as to whether the date of breach or the date when damage occurs is appropriate. In *Philips v Ward* Denning LJ took the date when damage occurs.[21] Yet in *Perry v Sidney Philips & Son*[22] he (as Lord Denning MR) took the date of breach as appropriate whether the plaintiff sued in contract or in tort. Oliver and Kerr LJJ, however, took the date on which the property was obtained as appropriate.

In most cases this lack of certainty will cause no practical problems, because of the proximity of the two dates mentioned above. Once the Court of Appeal had reconsidered the question, it seemed most unlikely that a plaintiff would ever be entitled to a calculation as at a date later than that on which contracts are exchanged. Plaintiffs had already unsuccessfully argued for a later date on the basis that since purchase they had been hit by inflation.[23] Now it appears inconceivable that a plaintiff will ever be so entitled. The House of Lords has decided,[24] in a case where defects to property, as opposed to failure to identify defects, were the basis of the claim, that a cause of action arises when the defects occur and not when they are discovered by the plaintiff or are discoverable with reasonable diligence on his part. (Of course, in some cases these times may be synchronous.) By analogy, the absence of physical signs reasonably detectable by a purchaser could certainly not now be relied on in a case founded on a surveyor's failure to identify defects.

A plaintiff who may feel that he suffers from the existing method is at least partly compensated by an award of interest.[25]

Purchasers who withdraw before completion

The measure of damage described above is clearly not appropriate where the purchaser, after exchanging contracts but before completion, discovers that the property is not as described by the surveyor and in consequence decides not to complete. If properly advised he would have incurred fees, including the surveyor's, up to receipt of the report. As it is, he pays and forfeits a deposit and probably he incurs further fees. The ultimate calculation should enable him to recover the amount of the lost deposit and of the further fees, but not the fees that he would have incurred in any event. Strictly, this result should be reached by his recovering the surveyor's fees, if he has paid for a useless report, but giving credit in the calculation of damages for the fact that if the surveyor had done his job properly the plaintiff would have had to pay the fees. In the section entitled "Contractual fees" we have already seen a case in which this credit was not given.[26]

It is foreseeable that after receiving an encouraging report a client may incur fees on a further report commissioned by prospective mortgagees. Such fees should be recoverable from the negligent surveyor; and as the courts have become more reluctant than they were to restrict a claim on the grounds of remoteness of damage it is unlikely that a plaintiff would now fail in this respect, as one did in 1952.[27]

Vendors

We have already seen the measure of the damage suffered by a purchaser who completes the purchase of an overvalued property. Logically, there should be no difference between his position and that of a vendor who completes the sale of an undervalued property. The damages to be awarded to such a vendor should be calculated in accordance with the principle set out in *Philips v Ward*.[7] Where he sells at a price consistent with the value described in the report, he should recover the difference between that price and the value that ought to have been described. That is the exact converse of the facts dealt with in *Philips v Ward*.

Where the vendor manages to sell at a price higher than that described, he should recover the difference between that price and the value that ought to have been described. That is the converse of a variation of the facts dealt with in *Philips v Ward*. The variation was the subject of a decided case, *Ford v White & Co*.[28] In one case[29] a vendor was awarded the full difference of £5,000 between the value described and the proper value, though he managed to sell at £1,730 more than the value described. That decision would have been justifiable only if it had been reasonably foreseeable to the surveyor that had the correct value been described the vendor would have been able to obtain a price £1,730 in excess of it.

Where a surveyor negligently overvalues a vendor's property there is no single method for the calculation of damages. The whole claim will depend on the plaintiff's reaction to the overvaluation and the extent to which that reaction was foreseeable and reasonable. A cautious vendor may well decline to commit himself to the purchase of another property other than simultaneously with exchange of contracts with his own purchaser. But a vendor may be found to have committed himself reasonably to a purchase on the strength of his surveyor's valuation of his own property. Once that much is established, the plaintiff has the problem of establishing the difference between his actual financial position and the position that he would have occupied had he not been given an over-optimistic valuation. The difference will depend on the facts of the individual case; but in essence it is likely to be the difference between degrees of debt and to be measured by a difference in borrowing costs.

THE CONSEQUENCES OF BREACH OF DUTY

Crack on side of building, and garden wall.

The majority of claims against surveyors concern the failure to identify damp penetration, or movement in buildings. The most serious failures are the infestation of dry rot, which is expensive to eradicate, or foundation failure that requires underpinning. Both of these defects would have to be dealt with before a building was occupied, if they were discovered in a pre-purchase inspection. The identification of a fault that has to be dealt with immediately will result in the cost of dealing with the repair having a close relationship to the amount by which the price paid will be less than the value of the building in reasonable condition.

Some cracks occur overnight. The side wall of this building developed a crack of about 8mm wide one night in July 1989. The cause was a prolonged period of dry weather and the presence of trees close to the building, which was on clay soil. Even though the surveyor may not have identified the risk of a crack at that point, he should have been aware of the risk of a crack occurring because of the combination of the ground, the assumed foundation depth and the presence of trees. The boundary wall of the property shows extensive cracking. The failures in this wall, caused by the trees alongside was a warning to the surveyor, and should have put him on his guard.

Mortgagees

In *Yianni v Edwin Evans & Sons*[30] the client who commissioned the surveyor's report was the building society which the Yiannis had approached for a mortgage. It was said on the defendants' behalf that although a duty of care was owed to the building society it had suffered no damage as a result of the defendants' negligence. No doubt under the terms of the mortgage deed the mortgagees were entitled, by means of recourse to the property and to the mortgagors personally, to avoid any ultimate loss. It was the Yiannis who were bound to suffer a loss, come what might. However, the mortgagees' entitlement will rarely be translated into money actually obtained from the mortgagor. Mortgagees who in reliance on a report advance money in excess of the value of a property may well establish that they have suffered a loss. They must establish the difference between their actual financial position and the position in which they would have been had they received a carefully prepared report.

That difference can be accounted for by a number of elements. The main element is the difference between the amount of the advance and the actual value to the mortgagees of the property, possession of which they obtain after the purchaser defaults. But a submission that the loss was restricted to that difference was rejected by the Court of Appeal in *Baxter v F W Gapp & Co Ltd*.[31] The mortgagee was compensated for the loss of interest as well as the loss of his capital. It is not clear precisely how that was justified. It would be wrong to compensate a mortgagee for the loss of interest that the purchaser fails to pay; for to do so is to give the mortgagee the benefit of an agreement into which, but for the surveyor's negligence, he would never have entered. But the decision is best to be interpreted in another way. If the mortgagee had not committed his capital on the property in question he would have put it to work in some other way. It is foreseeable to the negligent surveyor that if the property is not worth the purchase price the purchaser will default and the mortgagee will get no return on his money until he recovers possession, or a smaller return than if he had committed his money elsewhere.

In *Baxter's* case the mortgagee in addition recovered the expenses of the property when it was in his hands after the purchaser's default – the expenses of abortive sales, insurance premiums, the costs of upkeep of the property, his own expenses and disbursements and the agent's commission on the sale that was ultimately effected.

In another case, *Corisand Investments Ltd v Druce & Co*,[32] it was found that the mortgagees if properly advised would in any event have advanced money to the purchaser. They had lent £60,000. It was found that had they been properly advised they would have lent not more than £16,300. As they were second mortgagees they did not recover from the property any of the £43,700 that they lent by reason of the defendants' negligence. Against the sum of £43,700 they had to give credit for £9,075 that they had received by way of interest; for since their case was founded on the proposition that if properly advised they would not have lent £60,000, the interest that they were paid on the basis of that advance had to be brought into account. On the same basis their argument that the damages should include interest that they would have received if the borrowers had not defaulted on their contractual obligations was rejected. So the damages were £34,625, there being no claim for consequential losses of the kind for which compensation was awarded in *Baxter's* case.

Just as purchasers are precluded from recovering the cost of repairs, so too are mortgagees who take possession of property and then effect repairs on it. In *London & South of England Building Society v Stone*[33] mortgagees took possession of a property in respect of which they had advanced £11,880. Because of its true condition it was worthless. The mortgagees spent £29,000 on the property; but the principle of *Philips v Ward* was applied, and they could not recover more than the amount of their advance, subject to adjustment for income received and expenses incurred in consequence of the advance.

It may be argued that the value of any covenants of the mortgagor should be taken into account. For example, where £3,015 was advanced and there was a covenant that that principal was to be reduced by £1,500 that the mortgagor was shortly to receive, the value of that covenant, less a modest amount to allow for the possibility that it might not be fulfilled, was brought into account.[34] The same approach was adopted at first instance in *London & South of England Building Society v Stone*; but on appeal the Court of Appeal by a majority declined to reduce the damages by the amount recoverable from the mortgagors had they been pursued under their covenant. This decision was based on the general principle that the reasonableness of a plaintiff's actions in mitigation of loss suffered by him is not to be weighed too nicely at the instance of the party who has occasioned the loss, and particular principles that are consistent with, and are really illustrations of, that general principle.

Heads of damage common to all claims

We have seen above that the shape of the claim of a plaintiff within a particular group has been largely formed by decided cases. But the facts of a particular case may justify a more wide-ranging award than that suggested by the standard methods of calculation.

A purchaser may show that but for the surveyor's negligence he would have refrained from buying the property concerned, and that he would have been saved a number of consequences. Although he cannot recover the cost of repairs as such, it is eminently foreseeable that having bought a house that needs repairs he will have repairs carried out. In a proper case he may recover expenses associated with the repairs, that is the cost of accommodation for people and furniture. It will all depend on the reasonableness of the course of action that he adopts in the predicament in which the surveyor has placed him. Further, a plaintiff in a given case may recover losses that he incurs by virtue of repairs, for example the temporary loss of the value of the use of the property, whether this is expressed in a loss of rent or the loss is of a general kind. It may be considered strange that a plaintiff may be recompensed for losses associated with repairs, the cost of which he may not, as such, recover. But the cases in which these awards have been made are all justifiable on the basis of reasonable foreseeability. A plaintiff landed with a defective house suffers a financial loss at the outset and, if he stays in it, physical discomfort. No-one need be surprised if he seeks by carrying out repairs to mitigate the discomfort that he expects to suffer.

A purchaser who decides to sell a defective house that he would not have bought but for the surveyor's negligence is involved in the expense of two removals within a short time. He is likely to be in a position to prove that if properly advised he would have transferred his attentions to a satisfactory property similar to that which he eventually purchased. In such a case he should recover the difference between his actual expenses and the expenses that he would have incurred had he made only one move after receiving the surveyor's report. The argument that all these consequential losses should not be recovered by a purchaser, on the basis that *Philips v Ward*[7] shows the measure of all damage that any purchaser can incur, has been rejected.[35] Indeed, an *obiter dictumm* in *Philips v Ward* supports the proposition that consequential losses may be recoverable in a suitable case.

Plaintiffs in these cases are not confined to recovering financial expenses and losses. On several occasions they have recovered general damages, usually measured in modest amounts, in respect of inconvenience, discomfort, distress and frustration. This head of damage should always be contained within modest bounds; and it is not open to a plaintiff converting a house in order to let it for profit.[36] But these claims in respect of physical damage may well be the subject of future growth. Is it reasonably foreseeable that a purchaser who is landed with a disastrous house may suffer injury to health? In a Scottish case[37] an attempt, on a preliminary issue and without evidence, to defeat a claim in respect of injury to health failed. Such a claim has been established indirectly in an English case:[38] indirectly because the claim was against a solicitor who had engaged an unqualified surveyor whose performance matched his lack of qualifications.

On the face of it, the possibility of a plaintiff suffering illness as the result of living in a defective and unhealthy house is foreseeable. Such illness may be physical, but the possibility of mental illness cannot be ignored. Any claim under this head will always be scrutinised carefully to see whether the necessary causal link is established. The cost or proving such a link is not negligible and the exercise may involve laying open a substantial part of a plaintiff's medical history. In practice it may only emerge as an issue where the plaintiff's illness has been sufficiently serious to justify the attempt to attribute it to the negligence of the surveyor. Even then the damages ought not, in all but exceptional cases, to be large. A plaintiff who finds himself suffering badly by virtue of living in a particular defective house should mitigate his damage by selling. Only rarely is he likely to show that he cannot move, or that he has suffered some permanent injury to his health that cannot be cured or mitigated by a move to a congenial house.

The field over which an award of damages may range is therefore wide. In practice, however, many claims are not wide ranging. A plaintiff who seeks to recover damages must particularise the heads of damage in respect of which he claims; and if he has suffered a quantifiable loss he must quantify it. In a properly handled case the relevant heads of claim and the approximate size of the damages likely to be awarded should be apparent well before trial.

Contributory negligence

The damages recoverable by a plaintiff in negligence may be reduced to take account of the extent to which his own negligence contributed to the loss and damage that he suffered. In the cases with which we are concerned such a reduction is unlikely to be made. In the vast majority of the cases the plaintiff's right to compensation at all is based on his having reasonably relied on the advice of someone possessing the skill that he does not possess. The idea of those two unequal persons contributing to a single damaging result is not immediately attractive. In *Yianni v Edwin Evans & Sons*[39] it was unsuccessfully alleged that the plaintiffs contributed to their loss, particularly by not commissioning an independent survey and by not reading the literature provided by the building society. The reason for their not taking these steps was their reliance on the surveyor's report – the crucial factor on which their whole cause of action depended.

In cases where the plaintiff's prudence is found wanting he is likely to be found, not to have contributed to the loss for which the surveyor bore responsibility, but to have caused, without any contribution from the surveyor, some part of the loss that he has suffered. An example of such a case was given above in the section dealing with causation.[40]

Interest

The court has power[41] to award interest on sums awarded in proceedings. The rate of interest and the period for which it is to run are at the court's discretion. In cases where economic loss has been suffered in a business transaction, it must be expected that the plaintiff will be awarded interest from the date when the loss occurred to the date of judgment at the commercial borrowing rates that applied during that period. By the award of interest a plaintiff, denied by virtue of *Philips v Ward*[7] the recovery of the cost of repairs carried out long after purchase, will recover at least some of the amount by which his damages fall short of that cost.

General damages for any inconvenience and injury to health will be awarded in accordance with the money values obtaining at the date of the award; and it has been settled by the House of Lords[42] that for this reason these awards should carry interest at only 2% per annum from the date of service of the writ.

Limitation

Through his default or that of his advisers a plaintiff may lose a good cause of action. This is not the place to describe at length the not over-simple law of limitation. None of it can affect the way in which the surveyor carries out a professional engagement. But the subject must be mentioned, particularly because of important changes in the law introduced by means of the Latent Damage Act 1986. In the vast majority of cases with which we are concerned (except Scottish cases, which are governed by separate, and complex, statutory provisions) the plaintiff's cause of action is liable to be defeated six years after it arose, if by then he has not issued proceedings. As we have seen,[43] if ever a plaintiff could have contended for a date later than that at which he enters into the relevant transaction in reliance on the surveyor's report, the *Pirelli case*[44] would appear to make such a contention hopeless now.[45]

The exceptions to the six-year rule have now been significantly widened. There had previously been an exception to provide that where the action is based on the defendant's fraud, or any fact relevant to the plaintiff's right of action has been deliberately concealed from him by the defendant, or the action is for relief for the consequences of a mistake time shall not begin to run until the plaintiff has discovered the fraud, concealment or mistake or could with reasonable diligence have discovered it.[46] "It is purely settlement, nothing to worry about", spoken by a surveyor on a return visit, has been held to amount to deliberate concealment for these purposes.[47] The open-ended nature of this exception is preserved.[48]

The new exceptions, however, are made subject to an overall "long stop" provision,[49] which prevents an action from being brought later than fifteen years after any relevant act or omission which is alleged to constitute negligence occurred. In the case of long delayed damage, this may serve to abort a cause of action which is still gestating.

Subject to that "long stop", a cause of action is now preserved for three years after the date on which the person in whom it is vested has the knowledge required for bringing an action for damages in respect of the relevant damage and a right to bring such an action, if this period expires after the end of the six year period.[50] The knowledge referred to is defined in detailed provisions.[51] In brief summary, it relates to the seriousness of the damage, to its attributability, and to the identity of the defendant.

---

1. *Dodd Properties (Kent) Ltd v Canterbury City Council* [1980] 1 WLR 433, 456
2. See, for example, *Hutchinson v Harrris* (1978) 10 BLR 19.
3. (1958) 208 EG 969.
4. *Kenney v Hall Pain & Foster* (1976) 239 EG 355.
5. *Overseas Tankship (UK) v Morts Docks & Engineering Co (The Wagon Mound (No. 1))* [1961] AC 388.
6. See page 38.
7. [1956] 1 WLR 471.
8. [1982] 1 ALL ER 1005.
9. At p. 38
10. [1980] 1 WLR 433, 456.
11. e.g. *Dutton v Bognor Regis U.D.C.* [1972] 1 QB 373.
12. [1982] 1 WLR 1297; (1982) 263 EG 888.
13. As in *Bolton v Purley* (1983) 267 EG 1160.
14. *Ford v White & Co* [1964] 1 WLR 885, 891; *Simple Simon Catering Ltd v Binstock Miller & Co* (1973) 228 EG 527, 529.
15. *Hardy v Wamsley-Lewis* (1967) 203 EG 1039; *Hingorani v Blower* (1975) 238 EG 883.
16. *Lucas v Ogden* (1988) 32 EG 45.
17. *Daisley v B. S. Hall & Co* (1972) 225 EG 1553.
18. *Miliangos v George Frank (Textiles) Ltd* [1976] AC 443, 468.
19. See, for example, *Dodd Properties (Kent) Ltd v Canterbury City Council* [1980] 1 WLR 433.
20. At p. 30 above.
21. [1956] 1 WLR 471, 474.
22. [1982] 1 WLR 1297; (1982) 263 EG 888.
23. eg in *Morgan v Perry* (1973) 229 EG 1737.
24. In *Pirelli General Cable Works Ltd v Oscar Faber & Partners* [1983] 2 AC 1, overruling *Sparham-Souter v Town and Country Developments (Essex) Ltd* [1976] QB 858.
25. See p.42 below
26. *Buckland v Watts* (1958) 208 EG 969; see p. 38 above.
27. *Parsons v Way & Waller Ltd* (1952) 159 EG 524.
28. [1964] 1 WLR 885.
29. *Bell Hotels (1935) ltd v Motion* (1952) 159 EG 496.
30. [1982] QB 438; (1981) 259 EG 969.
31. [1939] 2 KB 271.
32. (1978) 248 EG 315.
33. (1982) 261 EG 463; and, on appeal, [1983] 1 WLR 1242; (1983) 267 EG 69.
34. *Eagle Star Insurance Co Ltd v Gale and Power* (1955) JPL 679.
35. *Hood v Shaw* (1960) 176 EG 1291.
36. *Hutchinson v Harris* (1978) 10 BLR 19.
37. *Drinnan v C. W. Ingram & Sons* 1967 SLT 205.
38. *Collard v Saunders* (1972) 221 EG 795.
39. [1982] QB 438; (1981) 259 EG 969.
40. *Kenney v Hall Pain & Foster* (1976) 239 EG 355; see p. 38 above.
41. Under s.35A of the Supreme Court Act 1981 and s.64 of the County Courts Act 1984.
42. In *Wright v British Railways Board* [1983] 2 AC 773, affirming *Birkett v Hayes* [1982] 1 WLR 817.
43. At p. 40 above
44. *Pirelli General Cable Works Ltd v Oscar Faber & Partners* [1983] 2 AC 1.
45. See *Secretary of State for the Environment v Essex, Goodman & Suggitt* [1986] 1 WLR 1432.
46. Limitation Act 1980, s.32.
47. *Westlake v Bracknell D.C.* (1987) 282 EG 868.
48. Limitation Act 1980, s.32(5), added by s.2(2) of the Latent Damage Act 1986.
49. Ibid, S.14B, added by s.1 of the 1986 Act.
50. Ibid, s.14A, added by s.1 of the 1986 Act.
51. Ibid, s.14A(6)-(10).

Surveying equipment

SURVEYING EQUIPMENT

The Royal Institution of Chartered Surveyors has issued guidance to surveyors who carry out survey inspections on a number of occasions. In 1982 they issued a Practice Note. In 1985 this was replaced by *Structural Surveys of Residential Property: A Guidance Note*.

These documents are frequently referred to in court where the standard of inspection is on test, and all surveyors must make sure that they are familiar with their contents.

The equipment that a surveyor should have available during a survey is referred to in section 3 of the Guidance Note. It suggests that the equipment required will depend to a large extent upon the preferences of the individual surveyor, and goes on to list the minimum items that are necessary for an adequate examination of a residential properly.

I have set out their recommendations here; my comments are in italics.

1. Torch

A powerful robust torch of a size, weight and design which allows its use in confined areas and which produces an even spot beam.

I have always found a need to have my hands free whilst inspecting a building. Carrying torch and note pad results in the absence of a spare hand to write your notes particularly in the dark recesses of a roof void.

I have recently discovered cave explorers' headlamps and now use them for inspections where the restricted beam, strapped to the forehead, is adequate.

2. Hammer and Bolster

For Lifting and replacing floor boards, manhole covers etc.

Manhole covers should only be lifted with keys or handles. The older cast iron cover which has corroded handles can be lifted in this way, but the surveyor must exercise great care. If you lift a cover, make sure that you have a small trowel to re-rake the groove and make sure that the cover will go back properly.

3. Ladder

A ladder of a minimum length of 3 metres (10ft).

When purchasing a ladder try and make sure that you have borrowed the type you are interested in before use. It should fit your car boot, it should be easy to handle on your own, and it must be safe and foolproof. You may not be a fool but the assistant who borrows it one day is, and there are never any instructions attached.

When you replace your car, or the accountants in your company revise the surveyors' mobile expectations, make sure that the dimensions of the ladders being used are made known. Motor manufacturers alter boot dimensions regularly when they revise car specifications, and minor changes may mean that the ladder will not fit.

4. Pocket probe

This is more commonly referred to as a penknife and used for superficial testing of mortar, joinery, fractures, depths etc.

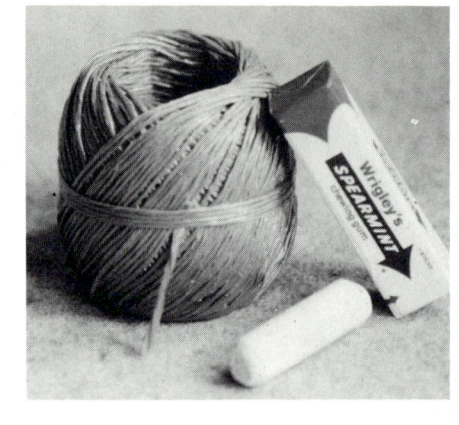

When I first started out there was little available in the field of specialist equipment for the surveyor. We developed the design of our own equipment. Experience dictated the selection of materials which were found to be the most useful.

The essentials were a piece of chalk, a length of string and some chewing gum.

The chalk was needed for touching in the marks made on the ceiling where it had been over enthusiastically prodded with a broom handle to see if it was in good order.

The string not only secured the chalk to the broom handle in order to reach the ceiling but also tied up the arm of a ball valve which had ceased to close when one had tested it in an inaccessible tank in the left.

The chewing gum? This could be used to re-seal cisterns or water tanks that had been inadvertently punctured by the over-zealous surveyor during his inspection. I understand it could also be eaten.

5. Binoculars or telescope

Not more than ×8 magnification. *If you buy one with a great magnification your handshake can be a problem. Make sure you go for a good pair of binoculars, your ability to see details of roofs, masonry, brickwork etc, depends on that quality. Binoculars are marked with two sets of numbers. For instance the pair that I use are 8 × 25. The 8 refers to the magnification of the binoculars, and the 25 to the light loss of the lenses. The higher this number the brighter the image. The quality and brightness of image is important when trying to decide whether minor variations are significant. The new zoom system on binoculars is very useful.*

6. Hand mirror

A hand mirror of minimum size 4" × 4". *Most surveyors prefer the shiny metal mirror. It saves the superstitious from concern over the next seven years after a clumsy day. Make sure you have a carrying case as they scratch. The ability to see round corners and into crevices where one cannot get to is valuable. The increasing use of the endoprobe will replace the mirror for purposes other than adjusting one's dress after a loft inspection.*

7. Moisture meter

Most meters usually measure the moisture content of timber between 16 and 28% accurately and indicate the presence of moisture in other materials inaccurately. Specialist meters are also available giving greater degrees of accuracy for timber, timber products and specific concrete mixes.

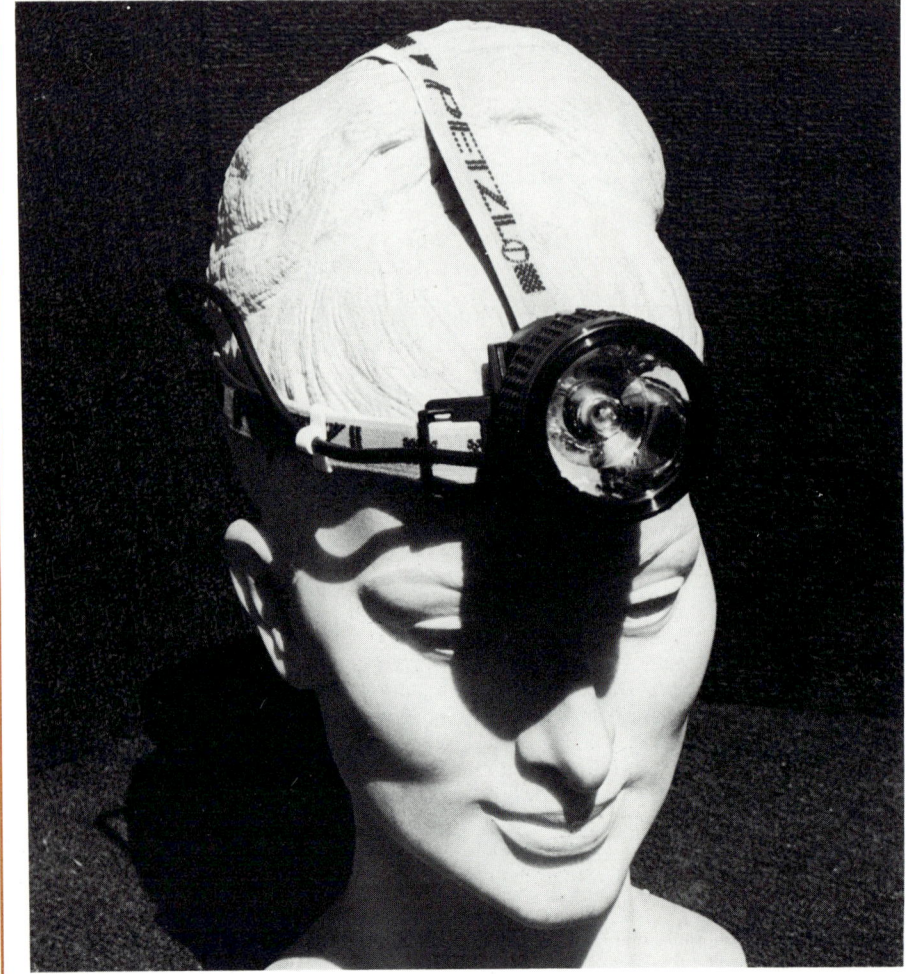

SURVEYING EQUIPMENT

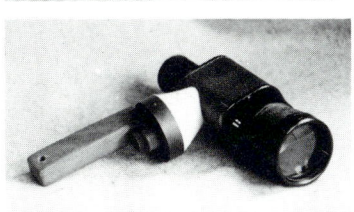

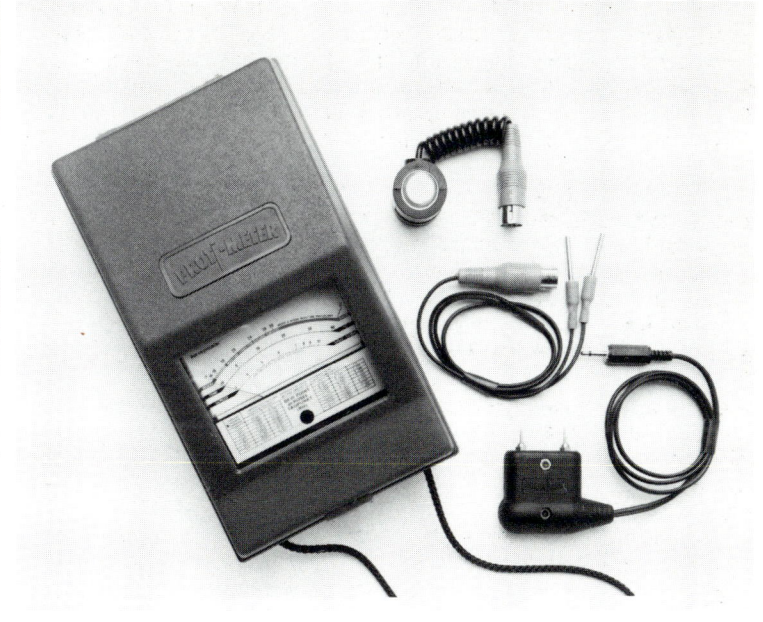

A selection of equipment: torch, binoculars, monocular and protimeter.

8. Screwdriver

The packs with interchangeable ends are useful but of little stamina. They are valuable for removing electrical covers to check wiring (make sure they are insulated or you are insured). They have no value for prizing open manhole covers, lifting floorboards, or opening the plastic wrappers of packets of biscuits! A larger more robust screwdriver is required for this work which should be undertaken with the bolster (see No. 2).

9. Measuring rod or tapes

The tape measure fits in one's pocket and is useful for dimensions of up to 1 metre (3' 4"). The rod, which is a sectional folding pack of boxwood extends to 2 metres and is useful for measured surveys but of little value in preparing a structural survey report. The new sonic measuring devices are discussed in detail later in this chapter.

10. Notebook and writing equipment

Pen and paper is the simplest, easiest and most convenient method. The pocket tape recorder is used by many people and is quick and easy to use. There are disadvantages; it is not very private and your discussion on the merits of a property do not usually go down well with an over attentive vendor. Background noise may render them unusable for notes taken in factories, schools or busy street. (Secretary's plea: please don't dictate in the car with the radio on.)

11. Plumb line

The plumb line, which is a weight fixed to a length of string, can be used to determine the alignment of walls in a vertical plane, i.e. are they bowed, badly bowed – or run for it! Their use on windy days is difficult and if working on your own they take ages to stop swinging. Placing the plumb bob in a bucket of water reduces the volatility of the swing of the line. The line should be suspended a set distance from the face of the wall to enable the bucket to fit beneath. If there is bright sunshine a photograph of the shadow of the string should record the extent of the irregularities of the wall as the shadow will duplicate the profile of the wall. You can emphasise the line by looping a boxwood staff into the line of the plumb bob to enhance the shadow. The heavier they are, the better. This is unfortunate for the surveyor who is now already overloaded with equipment.

12. Spirit level

Not for the recording the alcoholic content of the surveyor, or the level of his morale, but to check the vertical and horizontal surfaces of the building.

The level will enable you to set out the true horizontal or vertical, but will not enable you to define the extent of any fall that is found. New digital levels, however, provide an instant reading of the angle of a roof, drainage pipe or floor, and are much more useful than a simple spirit level. This is more sophisticated than the surveying equipment referred to in the Guidance note. [See page 59.]

In addition, the spirit level will not tell you whether the building was constructed straight and true and has subsided into its current state, or whether the original contractors were squint eyed and heavily inclined. Nevertheless, it can be useful in confirming what you already know. One eminent surveyor, however, pooled his resources and bought a marble; he finds it just as useful for checking the horizontal inaccuracy of buildings.

The legal profession are aware of this list published in the Guidance Note by the RICS.

In the event that a defect is missed during the inspection of a property, it will be necessary to decide whether the surveyor carried out a competent inspection. If it can be shown that an inspection with any one of the items of equipment set out on the previous page would have enabled the ordinary competent surveyor to discover the failure, and in the case of the surveyor against whom a claim is being made the surveyor did not carry the equipment with him, or the equipment was defective, the surveyor will have been negligent.

This places a considerable responsibility upon the surveyor to ensure that all equipment is maintained in good order and always has batteries that are sufficiently charged to enable test readings to be accurate, or for the torch beam to reveal concealed defects.

The list is only a suggestion that was put forward by the Royal Institution of Chartered Surveyors. It is important that the surveyor is aware of the purpose of each item and carries with him the knowledge or equipment capable of performing the analysis required.

Had that same list been produced some 20 years ago it is unlikely that the moisture meter would have been included within it. At that time it was a new toy and was not in every day use.

Surveys before the 1970s were made almost entirely without the aid of special equipment. A sharp mind, a sharp penknife, a notepad and pencil and the local builder to do all the difficult bits were all the equipment one used.

Were the surveys produced in this way less accurate than they now are? In a strange way they were less accurate but probably of more value. The surveyor knew if a wall was damp and would say so. He knew how to cure it and would so advise. The client had shorter reports which gave him an easier to understand image of the condition of the property he was involved with. If the surveyor was wrong, you didn't use him again, and made sure your friends didn't either. Those were not the litigious periods we now live in, where the mistake of the professional man is regarded as negligence.

Surveyors used to learn their trade by being articled to a professional firm. The young aspirant would follow the practitioner as he carried out his work, seeing how it was done, recognising faults, and understanding the consequences. It was an expensive education, often paid for by the aspirant's parents who had to pay the practitioner for his tuition. The apprenticeship system produced capable experienced surveyors who were

47

SURVEYING EQUIPMENT

articulate, considerate and conservative in their approach. First hand experience of the distress dry rot can bring to a household in human as well as property terms makes for a high degree of care in an inspection.

The mass production of surveyors through colleges and universities does sometimes produce a brash and more callous attitude, remote from the practical and human involvement that the work involves.

Before the surveyor becomes an adept technician, plugging, probing and computing failures with all the new equipment available he must develop the senses that he was given at birth. No one can use every piece of high technology on each part of the property he is involved with. They have to be plugged in to check or confirm what the surveyor anticipates to be a problem.

There is one marvellous moment within "Doctor in the House" written by Richard Gordon in which the eminent surgeon at the hospital reprimands the young doctor with the phrase: "The first rule of surgery, gentlemen – eyes first and most, hands next and least, tongue not at all".

The qualified surveyor should exercise the same degree of care and eminence as Sir Lancelot Spratt and should commence his inspection with his eyes alone.

The eyes have the advantage of being set some two or three inches apart and are able to determine the three dimensional configuration of the objects that are being inspected. Minor failures and blemishes on paintwork, woodwork, steel and concrete, which may form the first indications of problems, are easily seen.

The fingers are more likely to be able to pick up minor irregularities in a surface which is not lit by a strong cross light.

When it comes to inspecting damp surfaces the back of the hand is more sensitive than the front. The clammy dampness may be felt before any testing is put into effect with a meter.

Tapping surfaces is remarkably useful because the varying sounds can tell an enormous amount about the construction and its stability. Tapping wall plaster can indicate failure, such as, where the plaster has lost its key and has become loose. It reveals the area of hollows to the back of the plaster and will indicate whether the walls are of solid construction or of stud framework. The varying types of construction this will reveal can give you some insight into the cause of fractures and cracks which may be visible to the surface.

The tapping of timbers where there are indications of fungal decay produces a completely different form of resonance in the timber where it is damaged. In the defective timber the sound is slightly hollow as opposed to the unaffected wood which gives you a substantial note and sounds solid.

The sound material has a ring to the note produced by the vibration of the wood when struck. The diseased timber which has damaged fibre and cell construction is unable to produce the same note and sounds dull.

Working in an area which is known to be defective it is fairly easy to accurately identify the full extent of the infestation without having removed any of the surface material. This is a method which can only be acquired with practice but it can be accurate.

Various materials and failures have very distinctive smells. The nose's ability to identify a smell is only possible on a comparative basis. The various smells that can indicate dampness within a property are easily recognised but that smell will be less familiar to a new practitioner. In discussing the failures within this book the smell is described wherever possible. You are able to identify by smell variations in humidity within a property. This may be caused by a high moisture content or a failure in water carrying apparatus.

Settlement or movement in a property may be felt as you walk around the building. The balance of the body is very sensitive and minor slopes in floors and the direction of the slope can be located.

The creaking and movement of timber floors beneath the feet should be noted. This can reveal the stability or structural weaknesses of a property. Gently bouncing on your feet will reveal the springiness of the floor joists and the amount of deflection in many old houses, even where the floor surfaces are covered in carpets. Joists are often of an inadequate size and may be overstressed in many old properties.

The moisture meter some 20 years ago was not in every surveyor's bag of equipment. In future years, particularly with the breakthrough in micro-technology, more and more equipment will become available which will assist the surveyor in diagnosing the failures in property. At the present time the borescope enables an inspection of the condition of the components of buildings even where they are located behind wall faces or in other areas which were previously inaccessible.

The use of infra-red and x-ray technology has revealed much vital and interesting information relating to historical buildings. The use of video equipment and cameras for photographing a property is at the present time used by estate agents to display the advantage of the property but the future may well see this sort of equipment in use by the survey. Many future clients may decide that they would prefer to sit in front of their television and have the main defects of a property explained to them visually rather than read some 15 or 20 pages of a frequently boring report.

All this tends to indicate that the surveyor's bag of equipment is going to get larger and larger as the years progress, but it will still be the equipment that we are born with that will be of the greatest value. It is only through the use and interpretation of the information one's senses provide that one can decide which piece of equipment to get out and point at the property, to either verify or discount one's theories. There is insufficient time to check a property with each piece of equipment that is available. You cannot rely entirely upon specialised equipment.

There is a difference between an inspection to determine whether there is a defect in a building and a more detailed diagnostic inspection. The latter will be required if the surveyor is preparing a Schedule of Dilapidations, when advice is required upon the breaches of the covenants of the lease. In such an inspection the surveyor must locate and identify all the failures that he believes are the responsibility of the tenant. The surveyor may also have been asked to diagnose the cause of a failure, and to advise upon the cost of correcting the problem. In such cases he should be aware of the more complex equipment that is now available and the information that it can provide.

The use of Infra Red thermography, to locate deficiencies in cavity wall insulation, or Impulse Radar to locate fissures within walls 3 metres thick are just two examples of the technical assistance now available to us.

The value of the surveyor is his ability to understand the information that the equipment will produce and be able to interpret the data and advise upon its relevance. The interpretation of data, and the decision about the remedial work that has to take place is a value judgement that at the moment cannot be made by machines.

The machines available at the moment only deal with specific areas of the property and individual components.

It is hoped that the future will see a balanced partnership between the brain and the machine.

Detailed analysis of equipment

Access equipment

The surveyor requires a ladder which is easy to handle, easy to carry and provides him with access onto low roofs, into loft spaces, etc.

If a surveyor requires access to a part of a building beyond the capacity provided by his ladder he makes alternative arrangements. The surveyor's ladder is not the only access available to him but is a reasonable standard to provide for carrying out domestic residential surveys.

For the inspection of commercial buildings or multistorey properties a preliminary inspection would be necessary. This will reveal any special problems of access, and the appropriate equipment may be hired, and the necessary assistance organised.

A residential survey of domestic properties is usually carried out by a single surveyor acting on his own.

The survey of a commercial property, with the greater complexity of building materials and services is undertaken by a team of specialists, with the final report including all their reports on the various aspects of the property.

There are two main types of ladder available to the surveyor. They both will reduce in size to fit a large sized car boot.

1. Multi section ladder

This ladder has a number of sections which slot together. Wing nuts to each side lock one length to another, made of aluminium, it has a 4-section length of 3.6m (12'0"). Each length is about 1m long but part of that length is lost when it interlocks into the next length. They are usually sold with four lengths but extra lengths may be obtained.

Advantages

The lengths are easy to assemble, and you can use any number of sections up to about five dependent on the height you require. The unassembled ladder fits into most car boots. It is light (8 kg (18 lbs) for four sections).

Disadvantages

Unless you clip the sections together with shockcord (that large scale rubber band sailors are always using on their yachts) the ladder rattles unpleasantly in a car boot. It is cumbersome to carry because the lengths are separate, and slippery.

It can only be assembled as a straight line. Access into lofts is often easier with a set of steps rather than a ladder. When assembled it sags in the middle and has a narrow base which means that it is not very stable.

Effectiveness

It costs in the region of £75 for four sections. If you are going to decide on this design buy two sets of hooked shock cord from any chandler to lock the lengths together.

2. Articulated ladder

This type of ladder is a more recent development. Each section is connected by a locking hinge. This means that the ladder's sections are all joined together. Fully extended it has a length of 4.2m (14'0") and closed it is 1.140m (3'10"). The joint between sections has been so designed that the ladder may also be used as a set of steps giving heights of 1m, and 2.04m (6'10"), or as a bridge or staging if you want to paint a ceiling over a weekend.

Advantages

It does not rattle in a car boot, nor does it drop into individual sections as you enter the property to be inspected. The base of the ladder is wider (0.6m – 2'0") than the sectional ladders (0.356m – 1'2") and is much more stable. In use it is rigid, and does not droop or bend in the middle. If you have to pull the ladder up after you, there is no risk of a section falling off.

Disadvantages

It is much heavier than a multi sectional ladder (16 kg as opposed to 8 kg). It is easier to carry over your shoulder but this may damage your suit (or dress). It is longer in its folded state and will only fit the boot of a large car, or the back of a hatchback or estate.

Effectiveness

It is slightly more expensive than the multi—sectional ladder but is more adaptable, and safer in use. Its extra bulk makes it more difficult for less robust practitioners.

Below: A selection of sectional and articulated ladders.

SURVEYING EQUIPMENT

Access Plant

As buildings have become taller and more complex a greater variety of access equipment has been developed. The inspection of larger buildings means that a close inspection of the exterior may be required than that afforded by a good pair of binoculars. The use of other equipment to check on fixings, sealants, mosaic or other decorative treatments may require a working platform for the surveyor many metres (or feet) above the ground. The surveyor will rarely be able to include the cost of a scaffolding tower in his survey but should be aware of the benefits of many of the mobile platforms:

1. Mobile hydraulic platform

These machines fall into two categories.

A. Articulated boom

These are hydraulically operated with one or two joints. These can be hired and vary from the smallest machines mounted on a light truck or trailer having a 12m range, to large units on heavy lorries with an arm extension up to 52m. Care has to be taken when arranging for the hire of this equipment. The following must be checked:

(a) Capacity/load. For two surveyors this should be no problem. Heavy loads required for repairs or removal of loose masonry might be a problem.

(b) Space. Is there sufficient space for the elbow movement of the truck?

(c) Access. Is there clear headroom and access for the truck onto the requisite part of the site?

(d) Parking. Is there a clear parking area with adequate wheel loading capacity for parking and jacking of the vehicle?

(e) Reach. Can you park near enough for the horizontal arm reach to be within tolerance?

B. Telescopic boom

This is a single arm mounted on a rotating base. This results in a wide range and freedom of movement without the elbow. They are also available on trailers that can be towed behind cars. Some of the largest vehicles combine both articulated and telescopic extensions giving them a very large range. The problems involved with the selection of this equipment is the same as that for articulated platforms, except there is no elbow problem (item (b)).

2. Mobile hydraulic scissor lift

These are less flexible than the hydraulic platform as they are only equipped for vertical movement and have no horizontal extension. They can cope with greater imposed loads (up to 1800 kgs) and can rise up to 18m (60′). The working platform is much larger being about the same size as the lorry or trailer bed on which it is mounted.

Damp diagnosis

Although the back of the surveyor's hand is able to help him decide if there is moisture present in a structure there is no way that he can determine the precise percentage of moisture contained within the area he is concerned with. The exact moisture content will be required if he is to advise whether there is an increase or decrease in moisture.

In timber as will be shown in later chapters the moisture content may help identify or eliminate the possibility of certain fungi being present. In the event of the surveyor having to give evidence of the presence of moisture evidence may be required which only the reading from a meter will satisfy. Many lawyers are over indulgent on the evidence of a machine as opposed to the experience of the practitioner. If you have to give evidence about a defect which is to be backed up by the reading from a machine make sure you always start by checking the machine. For example check battery levels, and reset the zero or infinity counters as recommended. This must always be the first step in any event. When you give evidence in court start by identifying these checks.

It is therefore important that any equipment which is to be used must provide accurate and usable information. It is also important that the surveyor and his clients are well aware of any limitations there may be in any equipment that has been used.

Moisture meters

1. Carbide moisture meter

Purpose

The accurate measurement of moisture in walling materials.

Method

A small sample of the material of the wall is extracted as set out in the Appendix 1. This is then placed into the meter where it is mixed with calcium carbide. The water in the materials will react with the calcium carbide and a gas will be released. The pressure created by the volume of gas is measured and this is translated into a moisture content, and read on a dial.

Samples should be set aside if readings are also required for the hygroscopicity of the wall material. The minimum sample size required is approximately 3 grammes. Any samples removed from the site for later testing should be at least twice this amount.

The sample required for the hygroscopicity test should be well shaken to mix the dust, then the first half of the sample is tested and the readings recorded. The second unused part of the sample is placed in a 75% humidity enclosure (see Appendix 1). After about two days it will have achieved equilibrium and may then be retested in the carbide meter.

The two results are then compared to give the level of hygroscopicity (the relation of the moisture content to the maximum level of moisture the material can take in from the air). From this it is possible to eliminate hygroscopicity as the cause of the dampness, or to confirm that rising damp is present.

Advantages

The results are accurate, for moisture content and hygroscopicity.

Disadvantages

The equipment is bulky and quite heavy. It takes about 10–15 minutes to take the sample and obtain the reading. The

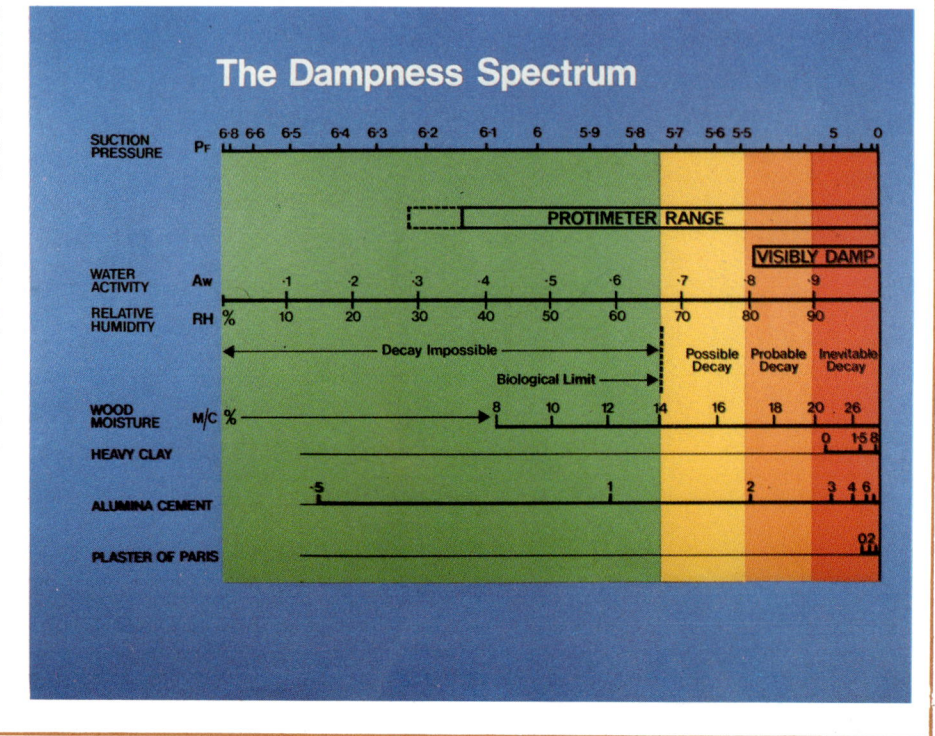

The Dampness Spectrum

samples, which are taken by drilling the wall, damage decorations. The use of electrical drills for sample taking requires a power source on site. There is a material cost of the chemicals in each test.

Notes
Testing with this equipment requires other items to be included in the inspection. A drill with a 15mm (⅝") masonry bit, tissues or similar to clean the drill bit between each sample, bottles to store the samples. The plastic film cassette container in which a 35mm film is sold, can be used, or plastic bags. A method of marking the samples for later identification and a method of recording the results on site for later cross-checking with any future results is also required.

2. Electrical resistance meter
Purpose
To accurately measure within limited ranges the moisture content of most timbers. The indication of moisture in other materials such as stone, brick, concrete, etc.

Description
The meter is a resistance meter with two prongs attached either to the body or via a cable. The two prongs are pushed into the wood sample and an electrical current passed into one prong and returned through any moisture present in the timber to the second prong. The machine then measures the resistance to the passage of the current between the prongs. As water is a high conductor of electricity the more water the higher the current passing and the lower the resistance measured. The results are given as percentage moisture contents for timber in a range from 16 to 28%. For other materials the readings are for comparative purposes only.

Below: A range of protimeters and ancillary equipment.

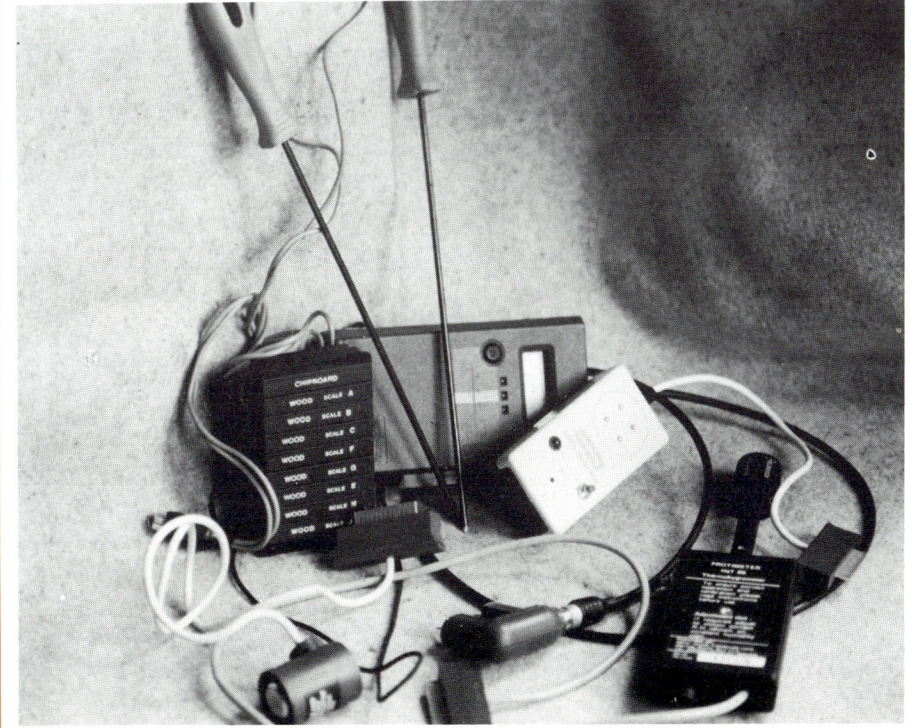

Equipment is also available which does not use the traditional two pronged head. It has the advantage of not damaging the surface which is being tested.

Specialised meters have been developed to measure moisture content below roofing surfaces. This may be useful in tracing failures in flat roofs, particularly of felt construction.

Recent development has resulted in the introduction of equipment which can give a clear indication of the presence or absence of electrically conductive salt contamination on a surface. This is available as an extension to Protimeter Surveymaster and to the new Digital Diagnostic.

Advantages
It is light, small (can be the size of a packet of cards) and easy to use. The results are easy to read and virtually no damage is done to the decorations. The machine is quick to use and there are no delays between test and result.

Disadvantages
Cannot be used to measure hygroscopicity. Because the machine is easy to use it is frequently in unskilled hands, and the results are often misrepresented.

It is only accurate within narrow limits in timber only. It does not give readings above 28% and cannot give the information required at the higher levels which would help in the analysis of various rots.

False results are frequently obtained in tests on walling materials. This may be due to foil backing on plasterboard or aluminium foil being used as a damp proof membrane.

In old walls containing salts, false readings can be obtained because it responds to the presence of salts.

Some types of breeze block contain enough carbon to give high readings even when they are dry.

Many of the problems encountered cannot be predetermined.

Comment
Provided that it is used within the limit of its ability this is a very useful tool. All surveyors must be aware of its shortcomings.

Costs range from about £50 upwards, depending on the level of sophistication and accuracy you require.

Laboratory testing
Purpose
The accurate measurement of moisture content and hygroscopicity levels in samples such as brick, plaster, concrete, tiles, etc.

Method
Samples of material are taken from the site (see Appendix 1). These are packaged and dispatched to a local laboratory who will test them and advise you of the moisture content and level of hygroscopicity in the sample. (See Appendix 2 for list of laboratories who will undertake this work.)

Advantages
The results are as accurate as possible subject to the quality control in the removal of samples and their packaging. Tolerance is fairly small over the full 100 range of percentum.

It is possible to determine moisture gradients from within the thickness of a material.

Results for court presentation are recorded by a laboratory (10 brownie points in a court).

Disadvantages
Laboratory testing is expensive. The charges are about half the cost of the electrical moisture meter. There is damage caused to decorations in removing samples. There is a time delay in obtaining the results from the laboratory.

The recording of the location where samples were taken must be accurate. There is a risk of damage in sending samples by post.

Comments
Usually only resorted to when there is some dispute about the result of earlier tests and independent evidence is required. Most inspections will not warrant this expensive method.

See Appendix 2 for methods adopted in the laboratory.

HOW TO TAKE SAMPLES
Removal of samples
Samples are obtained by using a low speed drill. This can be electric or hand powered. There are now cordless drills which are available.

A 9mm (⁵⁄₁₆") masonry drill bit is required. Samples are taken by adopting the following procedure:

1. Drill to a depth of 10mm (½").
2. Clean the drill bit and remove all dust from hole and drill.

3. Take a piece of card the size of a postcard and place horizontally below hole to catch the dust from the continued drilling.
4. Drill to a depth of 80mm (3").
5. Place sample dust in airtight container and mark to identify the location of drilling, and any other information required.

Samples should be removed every 150mm (6") in height of the suspect area. Mortar joints should be used where available as they are often damper than the brick or masonry.

Laboratory method for determining hygroscopicity

Equipment
1. Chemical balance – for very small samples, accurate scales for larger samples.
2. An oven – a domestic oven is able to satisfy the requirements.
3. Cooking salt – approximately three level teaspoons for small samples.
4. A sealed container – this can be a plastic food storage container.

Method
To determine moisture content. Minimum sample size: 1 gramme.
1. Shake up contents in sample carrier.
2. Weigh sample in container and record.
3. Open container and place half of sample on a watch glass or other oven proof dish.
4. Weigh sample and dish and record.
5. Place sample in over at gas regulo 8 or equivalent electric oven heat for one hour.
6. Remove carefully and reweigh and record.
   The moisture content is the weight Loss in the sample and container, before and after cooking (i.e. 4 – 6) expressed as a percentage of the dry weight (6).
   Example
   $$\frac{6}{4} \times 100 = \text{moisture content as \%}.$$
7. Take remainder of sample left in stage 3 above and place on a watch glass or other small oven proof dish.
8. Weigh sample and dish and record weight.
9. Prepare a container, lay the salt on the bottom and add water until all has been taken up by the salt and there is no liquid in evidence. The salt solution is then saturated. Provide a shelf for the sample and dish to rest on so that it is not touching the wet salt. Ensure that there is adequate air space.
10. Place sample and dish in container and seal. Leave for 12 hours minimum (this is in an environment of 75% moisture humidity).
11. Open box and remove sample and dish and reweigh and record weight.
12. Place sample and its dish in the oven and leave at gas regulo 8 or equivalent electric oven heat for one hour.
13. carefully remove and reweigh and record.

The hygroscopicity may be calculated by dividing the weight loss of the sample (i.e. 11 – 13) from the period after being subjected to 75% moisture humidity to the weight after removal from the oven, but the weight gain in the sample (i.e. 11 – 8). This is the increase over the sample weight in the beginning to the period after being subjected to 75% moisture humidity. Multiply by 100 to obtain the percentage.
Example
$$\frac{11-13}{11-8} \times 100 = \text{Hygroscopic moisture content at 75\% relative humidity}$$
14. Check moisture content of second sample.
Example
$$\frac{13}{8} \times 100 = \text{moisture content}.$$
compare result with first sample.
15. Weigh original storage container now empty, clean and dry and record.
16. Weigh dish or dishes used in each test.
17. Check that no excess moisture remained in container after sample removal.
$$2 = 4 + 8 - (16 + 16) + 15$$

The interpretation of dampness from the moisture content

The moisture content is the amount of water in a material divided by the weight of the material. This means that a heavy material has a much lower moisture content than a light material if they both contain the same amount of water. Dampness cannot always be measured by the moisture content in the wide range of materials that are used in building construction.

If an absorbent material is placed in a very damp atmosphere the moisture content will change until it achieves an equilibrium with the humidity of the air. If that material had been saturated prior to being placed in the damp atmosphere its moisture content could fall or, if it had been dry when placed in the damp atmosphere, its moisture content would rise. The variations in the dampness of their air around the absorbent material, or the fluctuation in the relative humidity of the air, will produce changes in the moisture content of the material that comes into contact with it. The moisture content for the varying materials used in building will not be the same when they reach a point of equilibrium with the same relative humidity in the surrounding air. The figure is different for almost every kind of building material and even between different types of the same component. For instance, the moisture content would vary from one type of brick to another and from one mortar mix to another. The general exception to this rule is wood. Although there are differences between various hard and soft woods, this is nothing like the difference that is found in other building materials. It is for this reason that moisture meters usually give a moisture content scale for wood but cannot attempt it for other materials. For example, suppose the atmosphere is at a relative humidity averaging 50%. It will be seen that the moisture content of wood is just under 11% but the bricks may cary between say 1½% and 2½%, the plaster probably around 1% and the wall board perhaps 9% or 10%. This is the moisture condition of a perfectly normal wall.

Dampness in conventional walling materials has, unhappily, been given a watershed figure. If *any* component of a wall, be it the mortar, brick or plaster, is in excess of a 5% moisture content it is to be assumed that the wall is damp. In reality the surveyor must interpret this figure by accepting that there is a grey zone around the 5% where a close examination of the cause of the dampness should be carried out. Wall or floor components with a moisture content in excess of 5% may require no action because the moisture content is within a tolerable limit whereas defects recording a moisture content below 5% may require urgent attention.

The growth of mould and decay fungi require high relative humidity. If the various components of the structure are in equilibrium with a particular relative humidity you have the connection between the moisture content of the walling component and the possibility or probability of decay occurring at that level of relative humidity.

It will be reasonable to say that dampness occurs where a material is wetter than air dry. 'Air Dry' means in equilibrium with a typical atmosphere with a relative humidity of between 40% and 65%.

The agents of decay in building materials, moulds, decay fungi, mites have a biological origin. They are able to develop in relative humidities between 75% and 85% (timber in such air conditions would have a moisture content between 17% and 20%). Above 85% relative humidity, however, these agents of disaster will develop much more quickly. It is therefore possible to say that damp occurs where building materials are in equilibrium with a relative humidity in excess of 85%. Such a statement must recognise that some materials may be damper than the supporting relative humidity of the air and be discharging water into the surrounding atmosphere. This may occur where there is water penetration taking place as a result of defects in the construction.

The relative readings on a conductance type electrical moisture meter indicate the relative dampness of different materials because they measure the free water in a materila. These measurements are a fairly close representation of the relative humidity. A high reading on such a meter would indicate a damp condition of approximately equal significance in wood, brick, plaster or wall board regardless of their very different moisture contents. It is for this reason that these meters are usually marked in zones indicating safe, intermediate risk and danger of damage or decay occurring.

SURVEYING EQUIPMENT

Humidity measurement

The accurate calculation of the level of relative humidity at a given temperature is important for the identification of condensation (see later chapter).

The *absolute* humidity is the percentage moisture content of the air at a specific temperature. The colder air becomes the less moisture it can carry. This must not be confused with the *relative* humidity which is moisture content related to the saturated content at a specific temperature. Water vapour condenses as water as the temperature falls because the cooler air can no longer support the same volume of moisture as the warmer air could. The temperature of the air and individual surfaces is taken with a thermometer. Air relative humidity may be measured with the following.

Whirling hygrometer
Purpose
This has a wet and dry bulb thermometer for the approximate determination of relative humidity
Equipment
The hygrometer looks like a football rattle.
Method
The rattle is revolved above your head in the manner of an over excited football fan. Readings are taken from the two thermometers and recorded.
Advantage
Easy to use, light in weight.
Disadvantage
Not very accurate. Fragile.
Comment
Adequate for most work on site dealing with problems of isolating condensation as a cause of damage.

Relative humidity probe
Purpose
This may be in the form of a probe linked to a moisture meter which is calibrated to give the relative humidity at a given temperature. After checking the calibration of the equipment prior to making a test, the readings may be read off with a minimum of delay.

There are a number of variations of this type of equipment all of which enable readings to be taken of both the temperature at the time and the prevailing humidity.
Equipment
This is plugged into a digital moisture meter via a programme key. After checking the equipment readings of the relative humidity are directly available.
Advantages
Very easy to use. Ideal for taking readings in confined spaces such as in suspect flat areas or below floors.
Disadvantages
Much more expensive than the whirling hygrometer.

Condensation indicator
Purpose
An electronic tell tale able to record the occurrence of condensation in a specific part of a building. The unit is reusable, powered by a small battery, with concealed rest button to fool those who may wish to influence the information being obtained.
Equipment
Small electronic unit with light indicator.
Method
A small electronic tell tale which is temporarily fixed to the wall of a property where it is thought condensation may occur at certain times during the day. In this way the surveyor can test whether condensation does or does not occur.
Advantages
Easy to use, reasonably tamper-free.
Disadvantages
The fixing which secures it to the wall face leaves marks. Not vandal or theft-proof.

Leak detection

Leakage occurs when a liquid or gas moves from a container in a manner that is not intended.

The problems may include valve and pipe failures where liquid flows in carrying amounts beyond the pipe. They may also be internal leaks where water passes through a closed gate valve.

Leakage may also occur where air escapes from a building taking with it the expensive warmth and replacing it with cold lances of air which often upsets the occupants.

The location of gas or fluid in transit may be made with the following equipment.

Ultrasonic leak detector
Purpose
To pick up the ultra sounds which are beyond the range of normal human hearing. These sounds are recorded as an audible frequence and shown on an optical indicator.

The ultrasonic range is above 20 KHz. The majority of small leaks with mid sounds within the 30 to 40 kg range.
Use
It can be used for the detection of any leaks in gas or compressed air pipelines.

It will identify a movement of water, or water penetration around a window. The accuracy of fit for doors can be checked.

The satisfactory performance of mechanical elements particularly within heating system can be checked.

The correct seating of valves can be investigated. Minor failures in a seating can result in expensive loss of gas or steam.
Advantages
It can operate in areas of high noise levels up to 80 decibels.

A whirling hygrometer.

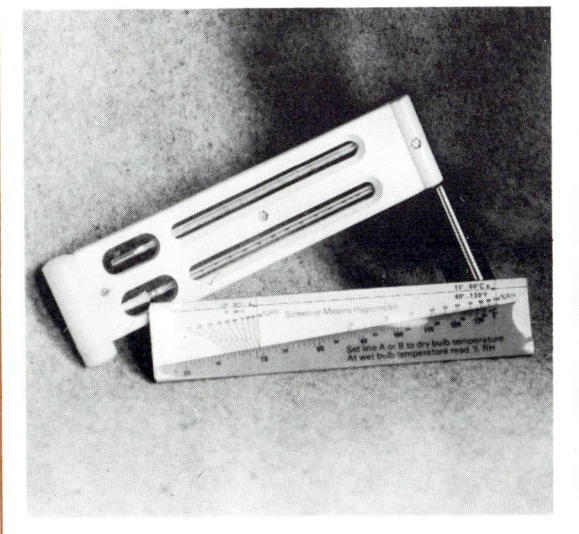

Protimeter Hammer Probe

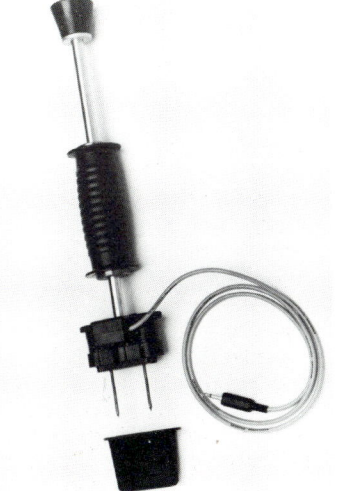

Flat roof leak detector.

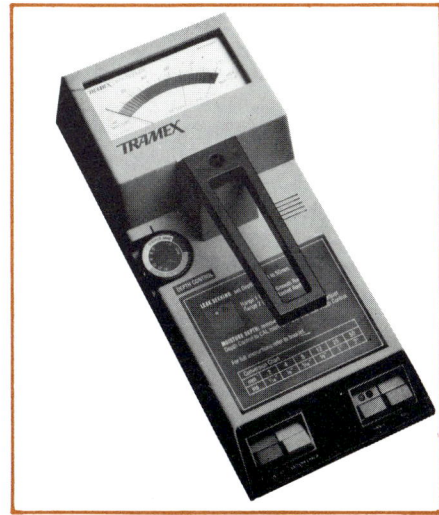

53

SURVEYING EQUIPMENT

As a non-destructive method of testing the equipment is available in a lightweight package that is easy to handle.

Disadvantages
The data recorded is not always easy to translate into a defect.

Comments
The equipment is of more value within mechanical engineering than the building industry.

Leak detection using sprays

Method
A spray of inert water soluble solution is placed over a suspect joint in a pipe. If a leak is present, foam of a very good stability will appear immediately to indicate its source.

Advantages
Easy to use and carry. Approximately 1,000 applications can be obtained from each can.

Disadvantages
Can only reveal external leakage of gas from pipes. Will not help to locate water leakage or internal faults such as bad seating of valves.
Ideal for bicycle punctures.

Comment
Of limited value. The spray is non toxic, non inflamable, non corrosive and safe.

A reinforcement cover meter.

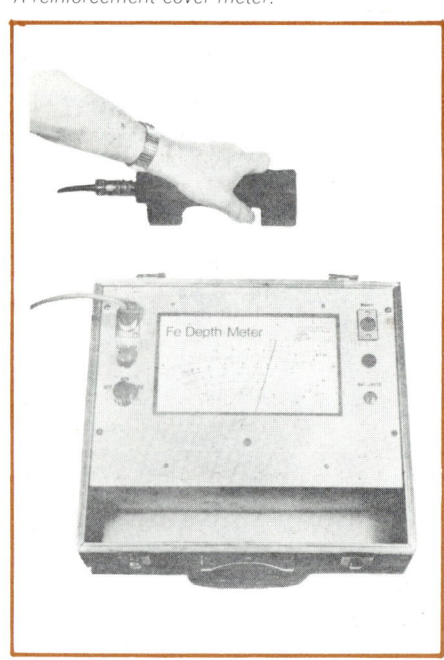

Concrete reinforcement location

Depth meters

Purpose
For the location and sizing of reinforcement bars in concrete.

Use
The equipment can be used to measure the bar size and also to measure the corrosion of embedded steel bars. It should be remembered however that in determining the diameter of a bar the accuracy is plus or minus 3mm (⅛").

It can locate any ferrous based material such as conduit pipe, wire, ducts, steel sheets or objects embedded in wood or concrete. Identification is possible up to 175mm (7").

It can be used to locate zones of concrete which are free from reinforcement to enable test core drilling to take place or for the clean cutting of a reinforcing bar when examination of the material is required.

Method
To use the meter a hand help probe is passed across the surface of the concrete structure. At the point where there is maximum deflection on the meter the probe is situated over the centre line of a reinforcing bar. If the bar size in known the cover can be read off in millimetres. If the bar size is no known it is possible to make an estimate of both bar size and depth subject to the tolerance which has already been referred to.

The equipment will operate from just above freezing to 43°C (about 110°F).

The FE depth meter is probably one of the most accurate of the steel location meters currently on the market. Although it is simple to use, to obtain accurate readings it requires some practice in its application. To gain experience it should be used on materials where the depth of cover and the size of reinforcing bars is known.

The location and measurement of individual rods embedded in concrete at reasonable distances is straightforward. Where two or more parallel bars lie within the field of the probe interpretation of the readings can become very difficult and in columns and beams of relatively small cross section where there is a multitude or reinforcement it may only be possible to locate the centre line of the bar.

Before using the equipment the meter pointer should be set at zero on the scale. If it is not, there is a considerable risk in error in the reading that will be obtained.

In order to locate reinforcing bars the concrete surface is swept systematically with the probe. The meter pointer will deflect if the probe nears a reinforcing bar.

The probe is then moved parallel to its previous direction until the meter gives a maximum reading on the linear scale. Once one locates this point the probe is then rotated to determine whether the movement of the pointer can be increased. Once the maximum reading has been obtained, after having rotated the probe and moved it horizontally, the direction of the probe is then parallel to that of the reinforcing bars or group of bars located within the material. The meter is eqipped with five parallel scales each of which relates to the concrete cover over a specific diameter of reinforcing rod. The depth of that rod can then be read off upon the appropriate scale on the assumption that you have the correct diameter of the rod.

If the exact diameter of the bar is unknown a value may be assumed and the error should be reasonably small.

The scale allows for bar sizes of millimetre diameters of 10, 16, 25, 32 and 40.

An example of the range of depth that could occur if you have a reinforcing bar of unknown dimension could run from 28mm to 40mm. These are the depths involved for bars of 10 and 40mm diameter. The greater the depth the closer together are the readings for the different thicknesses of bar and the smaller the margin for error. With intelligence one would usually be able to assume a rod diameter within a reasonably close range and not assume that in given circumstances it could be either 10mm or 40mm diameter.

Advantages
In determining concrete cover the accuracy is plus or minus 3mm (⅛").

The size of reinforcement bars and the thickness of the concrete cover over the bars is displayed on a direct reading meter. The instrument uses magnetic detection of the ferrous metal. The meter gives a direct reading of all reinforced concrete structures having covers ranging from 6mm to 200mm.

Disadvantages
The equipment when used to diagnose the size of embedded reinforcement bars is not accurate.

Considerable experience is needed in the use of the equipment, and the interpretation of any readings taken.

Corrosion in concrete reinforcement

A technique has been developed to measure the extent of corrosion which has

Concrete reinforcement location wheel.

taken place to reinforcing bars embedded in concrete. The equipment, in the form of a wheel, is passed over the area of suspect concrete and a continuous readout is given of the corrosion potential. A rotary shafting coder measures the distance which is travelled. This is coupled to a modern chart recorder driven by a stepper motor and fitted with remote drive input so that the linear movement of the wheel is directly related to the travel of the chart paper. The wheel, which is 160mm (6½") in diameter, has a foam rubber tip. A water absorbant tip is in constant contact with the rim and is kept moist by a water chamber. A silver chloride half-cell is used in the wheel instrument and its silver chloride tip is kept in contact with the water inside the chamber. The top speed at which the wheel can be drawn across the concrete is currently 200mm (8") per second.

Radiography:

X-ray and gamma-ray sources emit radiation which is able to pass through opaque materials allowing X-ray film to record hidden internal details. We are familiar with the technique which has been in use in hospitals for many generations but perhaps less aware that the same system may be used within buildings. The X-rays penetrate most building materials with varying degrees of efficiency. A clearer image is obtained from materials such as wood and plaster which the rays will pass through more easily, denser materials, such as metallic objects or concrete, will only appear in outline unless a more powerful radiating source such as cobalt 60 is used to produce gamma-rays.

The materials have to be exposed for between five and fifteen seconds and the development of the film, using the polaroid process, can take an equivalent time.

It is possible to use this method to detect failures buried deep within walls.

The system has been particularly useful in carrying out the examination of supports behind ornate or decorative surfaces such as murals or friezes. In these cases, such as "The Last Supper," it is not possible to remove the covering material to carry out sample examinations of the quality of the construction behind because it is the surface which is of greatest value.

The location of reinforcement in concrete and the determination as to whether the installation has been adequate has also been identified using this form of equipment.

There are now a number of firms who provide an on-site service at a cost comparable with those for infra-red thermography.

Ultrasonic testing

Purpose

Ultrasonic testing has become a well established method for the inspection of many different materials. The size and position of a failure may be located by measuring the time taken for a pulse of ultrasonic vibration to travel from a transducer and back to it after having been reflected as an echo from the flaw.

The measurement of pulse velocity may be used to determine the quality of the concrete, the presence of voids, cracks, or other imperfections and any changes in the concrete which may occur with time.

Estimation of the strength of concrete after fire damage may also be assessed by this type of equipment.

This method has been particularly successful in locating failures in metals. It can be used comparatively on wood, and with limitation, on concrete.

Concrete

With concrete there are problems with the use of this type of equipment. The aggregate within the concrete tends to generate multitudinous echoes and the velocity of the pulse varies with the quality and composition of the concrete.

One area where ultrasonic testing of concrete has been successful however is in the use of low frequency pulses and the measurement is of the velocity of the transmission of the pulse. (The method mentioned above as being unsuitable for concrete, related to the measurement of the time taken for the pulse to be reflected.)

The velocity of an ultrasonic pulse travelling in concrete is very closely related to the elastic modulous of the concrete. This modulous increases as its mechanical strength increases so that pulse velocity can be related to the strength of the material. The equipment used measures the time taken for the pulse to travel from the transmitting tube transducer to the receiving transducer. The distance between the transducers must be measured to a high degree of accuracy and certainly within 1%.

Method

The transducers may be placed on either side of a concrete section thus giving a direct transmission or upon the face of a section of concrete giving indirect or surface transmission.

Semi direct transmission is where the transducers are placed at right angles to each other such as on two sides of a column which share the same edge.

All the diagnosis which is possible with this equipment relies upon accuracy both in the measurement of distance and in the measurement of pulse velocity. This can only be achieved by making a good acoustic connection between the face of the transducer and the concrete. Where steel shuttering has been used in the original construction and the surfaces of the concrete are clean this is easy. In certain cases it may be necessary to apply face filling materials to give a smooth surface for the transducer to be applied to.

The homogenity of the concrete is investigated by carrying out regular measurements of pulse velocity over the surface of the material. Variations in the readings would indicate variations in the quality of the concrete.

Any defects present will be noted because the pulse arriving at the transducer will have been defracted around the edge of the failure. The transit time will be longer than in concrete with no defect. Small defects have little or no effect on transmission times. Comparative testing of the material is needed to locate the areas of the failure.

The depth of a crack can be estimated by taking readings along a line at right angles to the failure. When the transit times are plotted against the distance between the transducers those readings over the fracture will vary and the depth can be estimated.

Measurements can also be taken through formwork to try and determine the time when the concrete has reached an adequate degree of hardness to enable the formwork to be struck.

Timber

The equipment may also be used on timber.

Purpose

The pulse velocity method has been shown to provide a reasonable means of estimating the strength of timber and has been used to test a variety of timber products. It is a useful detection of rot in telegraph poles and provides an economical method of inspecting the poles whilst in service.

Method

The defects in timber materials are revealed by comparative readings and would enable the location of the extent of fungal decay in timber beams and joists by the comparative examination of the readings that have been obtained.

The examination of oak beams in old buildings where an infestation of death watch beetle has occurred will reveal the beam areas which are substantially damaged and must be replaced and those where the damage is only superficial and can remain.

As the tests are comparative, readings must be taken of sound timber and on damaged timber which is of an unacceptable strength. The analysis of the readings obtained from the site timber can be compared with those obtain from the samples used which were from an identical wood.

Advantages

Concrete. It is a non destructive method

SURVEYING EQUIPMENT

of testing, and revealing flaws. The equipment is easy to handle.

Timber. It is a non destructive method of comparative analysis.

Disadvantages

Concrete and timber. The quality of the acoustic coupling with a rough concrete face must be very high. The accuracy in measurement of the gap between electrodes is also vital.

The analysis of the results requires expertise and care. Experience in its use is important.

An electrical or battery connection is required to operate the machinery.

Comment

It is a system that has proved valuable on the testing or large steel structures where metal fatigue can be disastrous. These include bridges, planes, etc. Used carefully if is of value in determining failures in large concrete frame construction, or for quality testing of the work.

Laser measurement

Purpose

Accurate measurement over distances from 5m to 5 km.

Method

A larger pulse is emitted from a recording head in a set direction and reflected back from natural or passive reflectors, such as walls, river banks, milestones etc.

The equipment will vary dependent upon the range, but will incorporate telescopic sight and a digital display panel with continuous readout updated ten times a second.

Advantages

Accuracy is ±0.05% of measured range. They are quick and easy to use over the smaller ranges. Some are not too heavy (3.9 kg – 9 lbs).

Disadvantages

The equipment is expensive to buy and to maintain. They are cumbersome, and require good support for accuracy over long ranges.

Comment

The equipment is more suited for land surveying than property measurement, because of its accuracy over the greater distances.

It has a value where it is necessary to accurately pinpoint bulges or blemishes in wall surfaces or in panel fixings to buildings.

Sonic measurement

Purpose

The accurate measurement of distances up to 83 feet (25 metres) by a single person.

The early examples of this equipment were very unsatisfactory, giving confused and often inaccurate information. The units tested for the third edition of this book have been much more reliable for approximate measurement of rooms where accuracy within 1% of the actual is adequate.

The combination of calculator and ultrasonic tape enable rapid estimation based on room areas and volumes.

Method

The device uses a sound signal emitted on a level inaudible to the human ear. This narrow beam of sound is reflected back from a suitable surface to the unit which displays the distance on a digital display.

It can be used to measure water levels, as water is a good medium to reflect a signal.

Some units enable you to read off the dimensions in imperial or metric.

Advantages

Quick and easy to use. The units have a range of 1 foot (0.3 metres) up to 83 feet (25 metres). Tolerance is plus or minus half an inch (1 centimetre). Very good for ceiling heights or in confined spaces.

Disadvantages

Can give inaccurate readings if not used with care and intelligence.

External dimensions can be fouled in poor or wet weather. The measurements are between parallel surfaces and can be difficult to obtain externally if there are no projections to measure between. Units with a range up to 10 metres are now available for the equivalent price of a 20-metre tape.

Comment

A considerable amount of development has recently taken place, and the new units are much more flexible and reliable. The retention of the reading obtained for up to 10 seconds, and the addition of a hold button, makes it much easier to take dimensions in restricted and dark locations. The accuracy is not sufficient for the preparation of accurately-dimensioned drawings where tolerances are tight.

Photo-grammetry

Purpose

The technique of photogrammetry is extensively used for the recording of physical information. It involves the setting up of overlapping photographs to produce a three-dimensional image. It is a very precise form of measurement requiring very accurate cameras. The complexities of the lens of the camera have to be taken into account to allow measurements to be taken from the photographs.

The technique is used in two distinct areas:

a) Aerial photogrammetry;
b) Terrestrial photogrammetry.

Different equipment is required in each case. For aerial photogrammetry, cameras are mounted on an aircraft and have to have maximum definition at fast speeds. The film is frequently 225mm (9") square.

Terrestrial camera work uses smaller portable cameras producing a smaller film image. Glass plates are still in use because of the stability of the image.

Because the overlapping photographs are arranged in their relative orientation, they can produce a three-dimensional image from which it is possible to take measurements of width, breadth and depth.

Sophisticated versions of photogrammetric machines are now linked to computers. This enables the plot of the image to be converted directly into computer based data. The computer then feeds this information onto an automatic plotting table to produce the drawings of the properties or maps of the area being photographed.

Use

Photogrammetry can be used in many and varied cases. It has particular reference in drawing irregular shapes or where the object is moving.

It can be the sole method of measuring objects that are too hot, too cold, too soft, too delicate, inaccessible, too toxic or too radioactive.

SURVEYING EQUIPMENT

Infra red thermography

Purpose

Infra red thermal instruments record temperature variations within a building. Some equipment enables you to see a high quality image on a minaturised television screen. The equipment has been used to detect poor insulation in heating in property and to locate where major heat loss is taking place. The detection of leakages in heating systems, district heating, warm and cold water pipework has also been located by the use of this type of equipment. The overheating of switch gear and electrical circuits have been revealed.

The equipment can be used to examine and locate: reinforcement within concrete structures; the presence of cold bridging which could lead to condensation; air leakage from buildings resulting in heat loss; the presence of condensation inside construction or on the surface of components; water leakage on flat roofs; hidden beams and services within the building construction.

Cold stores

Thermography may be used to reveal defects in the insulation to cold stores and refrigeration equipment, and structural breakdowns.

Electrical systems

Thermography may be used in the detection of potential electrical faults. The heating of areas of failure will be shown up in the thermographic display.

It is also being used to locate the sources of water leakages particularly in flat roofs but also in concrete panels to the exterior of multi-storey prefabricated concrete structures.

Wimpey Laboratories have pioneered techniques in the use of this equipment in locating timbers buried in walls up to 610mm (2') thick.

Use

Temperature variations in materials can reveal a number of failures in certain circumstances.

The equipment is pointed at an object or a building and the operator is able to view the thermal variations either on a screen or on a viewfinder.

The image is displayed 30 times per second and the field of view is between 6 and 12 degrees. The angle of view varies as you adjust the focusing from infinity down to 3 metres. The display on, for instance, the Agema Thermovision 450 is monochrome and the higher temperatures show as a whiter image. The controls on the equipment enable one to focus, control the thermal brightness and thermal contrast. The equipment will show the difference in temperatures varying from −20°C to +500°C. The picture can produce 5, 10 or 128 grey levels.

A multicolored image can be achieved with portable equipment such as the Agema Thermovision 470 with optional colour display. Software packages are available that enable the user to store and retrieve the information taken on site. This enables comparisons to be made with readings obtained over a period of time.

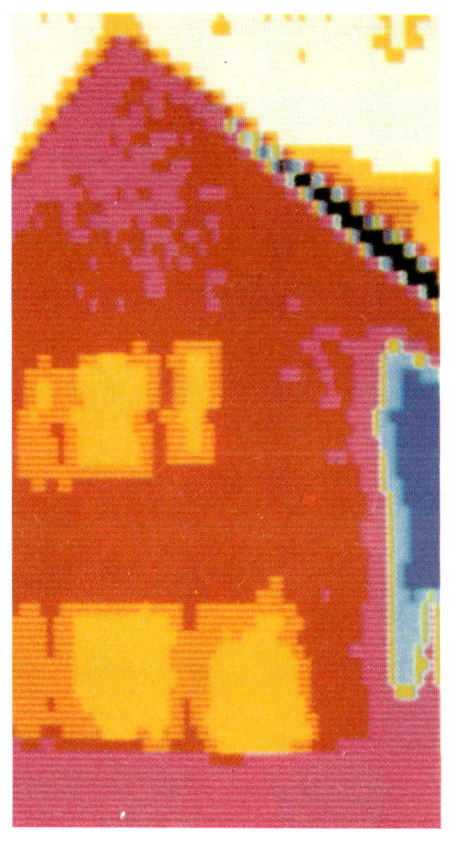

Building example

Thermography has also been used in the analysis of buildings and in the detection of half timber construction where it lies beneath a rendered plaster surface.

In West Germany, The Technical University at Damstadt were looking for a non-destructive method which would enable them to assess the construction of a number of buildings in the town ranging from ancient half timbered houses to more massive property. The temperature of each different building material used in different centuries show upon the photographic records taken of the surface of the thermogram.

This analysis of the recorded temperature distribution has given them a reliable picture of the materials used and the wall construction under the existing plaster finish. The temperature outlines are then sketched on to the facade of the building so that the once hidden building structure can be revealed. They have found that the best thermal patterns occur during the summertime where thermograms may reveal a temperature range of up to 20°C between the various materials.

Energy surveys

Thermography is a quick and cost-effective way of detecting heat loss from buildings, heating plants and industrial processes. Heat loss rates and effective U-values can be calculated. Examples: insulation deficiencies in buildings; storage vessels; pipelines; furnaces; kilns; process plant; and detection of steam losses and faulty traps.

Building surveys

Thermography may be used to check various elements of building components and locate failures in construction whether new, existing or refurbished. Examples: quality control on heating installations; welds; locations of steel; reinforcement in concrete; damaged or missing insulation.

Top: Coloured thermographic image of house exterior shows where heat loss is occurring and is of great assistance in determining design and location of necessary insulation.
Left and above: Hand held equipment giving instant readings of surface temperature.

57

SURVEYING EQUIPMENT

Infra red photography
USING 35mm CAMERAS

Infra red film is extensively used where the determination of minor colour variations is important, and this has included the detection of camouflage. This is a case where the camera clearly shows a difference between the natural green of the vegetation and the artificial green of the camouflaged areas. It is also used in forensic investigations and is particularly relevant to such matters as forgery where minor variations in the colour of ink will be revealed.

The film is able to produce clearer photographs through haze or polluted air than a conventional film and is therefore very useful in aerial photography.

The minor variations of colour particularly in the greens is relevant to the field of botany. It has been used to reveal crop infestations and can locate wild oats in fields of domestic oats. In forestry it has been used to reveal defective and damaged timbers within a woodlands area. In medicine it has been used to reveal skin diseases and has the ability to see the surface immediately below the surface of the skin.

It must be made clear that it has no ability to show any variations in heat emission and it is a popular misconception that it has any value in this context. The speed rating of the film varies according to the heat reflecting properties of the subject but generally an ASA value of about 50 seems to work.

Infra red films are available through some suppliers of Kodak film but they often require to be ordered specially. The colour reproduction from this film is not an accurate record. The colours differ, such as red for green unless filters are used.

In use the film has certain specific requirements which separate it from conventional film.

Loading should be in darkness as the felt shutter or blind which is located on the cassette is unable to prevent all light penetration to the film. There are minor variations in the focusing and most cameras carry a curve or red dot upon the lens so that allowances can be made for the use of this type of film.

The ASA or equivalent speeds on the film are far more critical and it is worthwhile taking a series of photographs of the area involved with half stop variations to each side of the setting which the light meter suggests.

There are a number of filters which are available for use with black or white infra red film to try and highlight the various images.

The unused film should be stored in the refrigerator at about −18°C to −20°C.

The processing of the developed film produces problems because most photographic shops have no experience in the handling of the material. It requires a standard E4 process which is unfortunately being phased out so one may find it easier to send it to the manufacturer for processing.

Application in building

It is able to show minor variations in colour which can highlight the influence of water penetration where discolouration to the surfaces has occurred. There is a probability that it may reveal the presence of fungal decay and worm attack in certain sections of timber.

The film is easier to handle that may be expected and the results are good. The adjustment in focussing and the ASA rating at 50 tend to work well. The pictures show clearly any bare patches on what you may have considered to be your lawn. When used during September mists a marginal improvement in depth of vision was noted but it is not an all seeing film. The specific advantage of the film is in trying to record mould growth. The conventional film is unable to highlight the minor colour variations. The change in colour reproduction is confusing at first, greens come out red, and reds come out yellow.

Comment

Photographs taken of a 14th century half timbered property produced interesting results. The minor colour variations in the woodwork that could be seen enabled the detection of infected timbers. It also located replacement timber that had been added over the last 500 years. The colour variations in the timber introduced over 400-500 years was evident. Ordinary film did not highlight these minor variations, making any interpretation very difficult.

Because rising dampness discolours brickwork these colour variations are easy to see.

This film is not a cure-all, or a see-all, but it does have a value in specific areas. Filters are available to correct the colours to those which one would regard as normal.

Left: An infra red photograph (taken on 35mm film) of a timber frame building. Colour variations indicate different types of timber used and help the surveyor to understand the repairs that have been – or may be – carried out.

Movement monitoring

The monitoring of structural movement is now of great importance in identifying the source of a failure and the seriousness of a problem.

The tell-tale

The early versions of the tell-tale were in glass. The disadvantage of the glass strip, which was secured by epoxy resin to each side of the trace, was that once movement had occurred and a fracture had taken place, calibration of the movement and the indication of which side moved in relation to the other was not possible.

Calibrated sliding perspex tell-tales

These are fixed in much the same way as the glass tell-tale, but the two sections slide independent of each other to indicate the direction and extent of movement which has taken place. These can attract the attention of young children and confusing results will occur if well-intentioned parents glue back the tell-tale which has been knocked off by their offspring!

Vernier markers

In order to use this system, studs are set into the wall to each side of the crack. These may either by glued to the face of the wall or be in the form of screws which are driven into pre-drilled pockets. Each stud has a centre punch mark and the gap between

SURVEYING EQUIPMENT

the studs is monitored using Vernier calipers. These will give the extent of movement to tolerances in the region of 0.1 of a millimetre.

By the use of the studs the triangulation of measurements will enable some determination of the direction of movement to be achieved. The small studs have the advantage over tell-tales of being less conspicuous and the results are more accurate.

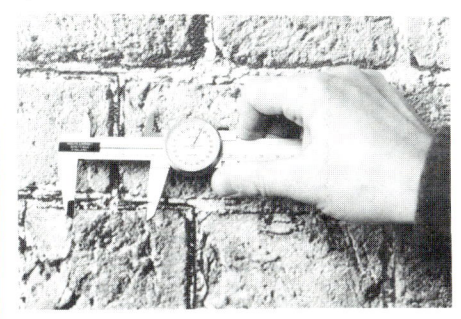

Vernier calipers.

Linear variable displacement transducers

All the previous methods rely upon the dimension recording being taken by an individual. That individual may on some occasions find it difficult to locate the pegs, tell-tales or gauges, or may mis-use the equipment at the time of his inspection. Linear Variable Differential Transducers measure voltage changes corresponding to actual physical changes in distance. They can be fixed across the cracks and provide remote electronic measurement of movement to within 0.01mm. Movement of the crack causes an electrical signal to be sent from the transducer to a remote controlled unit run from mains electricity.

The gauge itself is formed like a piston and is fixed by universal couplings over the pins around the crack. The read-out can be in many forms from graphical plotting to the recording of the dimensions of movement related to the time scale.

Equipment similar to this is currently installed in St. Paul's Cathedral to record movement in certain columns and the dome.

It would be possible for the read-out to be provided in the surveyor's office remote from the building.

Future

New techniques in laser technology will result in machinery becoming available which will be able to accurately pinpoint movement taking place to various parts of a structure.

In each case where movement is recorded the air temperature and the temperature of the construction should be recorded to help the diagnosis of the failure which is causing cracks to occur in the building.

Video

The spectacular growth of the video market over the last few years has now presented the surveyor with the challenge of considering the production of a report or an appendix to a report in a visual format recorded on a video cassette. This would enable the client to sit at home and watch the report upon his television set provided he has the necessary equipment. At the present time this may present a number of problems. There is no universally accepted cassette system. The most common in existence are the Betamax cassette and the VHS. The VHS is the more popular of the two and may eventually become the accepted format but until that happens the production of reports using cassettes which are not in universal usage may present problems for the client.

One of the other problems is that the public at large have become conditioned to the viewing of high quality television programmes. They have come to expect a high quality accurate colour television image together with good production and a competent commentary. There are probably very few surveyors who are capable of carrying out this function as well as having the technical expertise to carry out a competent survey. It is dangerous to try and compete with the media purely as a method of getting over the defects which one may find within a property.

Although there are limitations in the quality of the image which one can produce in the way of a report, video is a fast, effective and sometimes economical method of communicating the defects in a property and may give a far more complete image of the failures or advantages of a building than can be rendered within a conventional form of report.

The portable video domestic equipment can perform a valuable function for a surveyor but is vulnerable to variations in environmental circumstances. If there is inadequate or poor lighting because it is a winter's day or later in a summer's evening the limited performance of a colour video camera would be exposed. These cameras will not produce the definition, resolution and colour fidelity that people expect who have been used to watching the high standard of broadcast television such as that received in the UK.

At the present time prices range from approximately £450 to £750 for colour television cameras with portable video recording equipment ranging from £500 to £700. Floodlighting equipment would require power circuits to be connected within the property which is not always available. The main disadvantage is the problem of editing the final reportage.

It is not possible to edit by cutting the tape as one may do with cine film. One can re-record by using two compatible machines but a loss of picture quality will occur. In several large cities there are studios which can be hired but they cost about £300 per hour, with the probable minimum time of half a day. This is likely to result in the final product having an amateur appearance which is often caused by the need to get it right in one take. The high quality television programme is the result of many takes, much swearing and a considerable amount of time being expended.

The equipment is no less expensive than a secretary's complex typewriter. The material cost of a tape at £10 is not much more than the cost of photographs, paper and a presentation folder.

If an acceptable quality of film can be produced the method is cost effective over a wide range.

Training in the use of the equipment is being added to a number of college and polytechnics' curriculum. The expansion of television channels will increase demand for the techniques.

Books are now being offered on video tape so that the presentation of reports in a visual technique may not be so unusual. Most surveyors have to give a verbal report over the telephone and many are very adept communicators.

The development of duplication of tapes and simpler and cheaper editing facilities will result in an expansion in the utilisation of this medium.

Electrostatic meters

Purpose

This meter records the level of static electricity which is present in a material or its capacity to take on a static charge.

Use

The determination of static build up can be useful in checking materials prior to their inclusion in a building scheme. It can also be used to eliminate the cause of the static shock which is frequently complained of in modern office complexes.

In certain processes the presence of a static charge can be dangerous such as where explosives, or explosive gases are present. There are occasions where the light source which occurs when a static discharge takes place can be detrimental, such as in film or TV studios.

SURVEYING EQUIPMENT

Light measurement

Purpose
With the introduction of new recommendations about the level of light required within offices and shops, surveyors will have to be able to provide information about the existing levels of illumination.

A portable photometer measures light levels in various locations.

Use
Ever since Mr Bird came out on bright summers' days to protect the flower of English cricket from the ravages of defeat we have all come to have a greater respect for the alternative use and power of the light meter.

Modern meters have a range between 0 and 100,000 lux to an accuracy of ± 2%. (A lux is equivalent to 0.093 candelas.) Data hold facilities enable a reading to be retained, until you have found sufficient light to note it down.

Method
Point the light-reactive surface at the light source or the surface onto which the light is falling and quote the reading.

Advantages
The units have a high degree of accuracy and immediate response. There is long battery life, although I have found removing the batteries when not in use beneficial in prolonging it (see below).

Disadvantages
Large switch makes it easy to be switched on in its case. Remove batteries when not in use.

Manometer

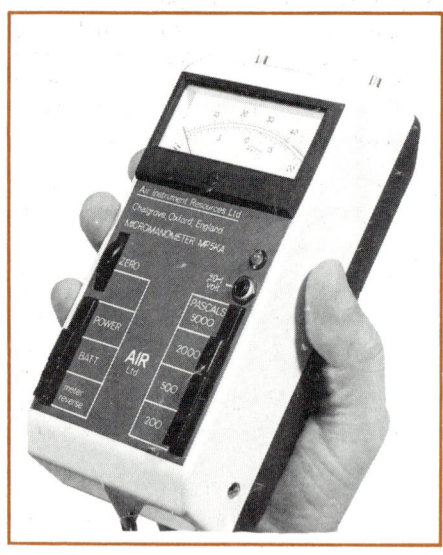

This is a meter for the measuring of pressure. They are now available as electronic meters whereas the original equipment was of the liquid-in-glass type. The liquid-in-glass type suffered from contamination, evaporation and thermal expansion which affected the accuracy of the instruments. They had to be levelled up and set prior to their use. They were also cumbersome and bulky. The electronic versions are easier to read and it is easier to determine positive and negative differentials and pilot pressures. The instruments are intended for the measurement of differential air pressures associated with fanned equipment such as ventilation and air conditioning systems. They are not intended for use with corrosive, toxic or inflammable gases and are not suitable for liquid pressure measurement. The electronic pressure measuring instruments based on the differential capacitance technique have become available. These micromanometers are specifically designed for hand use and have greatly simplified pressure measurements in situations that were previously difficult or impractical. They can be used while standing on a ladder, catwalk or cradle hoist, and can even be used in mobile applications such as where examination and marine ventilation is involved. Direct velocity readout is available from this equipment which has eliminated the need for separate calculations to be made. The equipment is small and light and has an accuracy within 1%. They automatically register negative pressures by displaying a negative sign in front of the reading. By comparison with liquid filled manometers the response of these electronic instruments is very rapid and they are capable of recording changes in pressure in less than one second. Fanned systems often exhibit slight pressure fluctuations caused by unsteady electrical supply, unstable fan characteristics, defective bearings. This can be located using this type of instrument.

Sound measurement

Purpose
The accurate recording of specific frequencies in given locations.

Sound is "mechanical vibration, propagated in an elastic medium of such character as to be capable of exciting the sensations of hearing" (BS 5643). The human ear can hear within the range 20 – 20,000 Hertz.

Measurement
There are many different ways of describing sound. A measurement is sometimes given as a maximum and sometimes as the average. It can be given for a wide range of frequencies or for a narrow band. Common units of sound measurement are Decibels (dB), Pascals (Pa) and Phons. The most common unit of measurement is the decibel. This is a logarithmic scale, meaning that increases are not represented by the unit increase in the reading. Thus, an increase in a sound level by 3 decibels will mean that the level of sound has doubled, while the doubling of a reading will represent a ten times increase in the level of sound.

The impact of sound variations differs from one person to another. We are sensitive to different frequencies within the audible range. The human ear is less sensitive to sound in the lower frequencies than it is to equal levels of sound in the upper frequencies. To deal with this problem a scale has been developed to adjust for the human reaction to each frequency, and to give them equal loudness perception.

The measurement that gives a reading of sound adjusted to compensate for differing human perception in the varying frequencies is called dBA. To achieve its purpose it has an inbuilt weighting, and measurements are taken over varying time scales and usually quoted as an average.

Noise is often referred to under different categories. Figures can be given for levels of sound which are exceeded for 10% of the time, weighted sound adjusted for noise from different sources, the sound emission from equipment, the peak levels reached in explosions, and the equivalent sound over given times scales. Equivalent sound is a definition of sound to a certain level, referred to as decibels, average per hour, six hours, twelve hours, etc. It is an indication of the level of sound which is experienced throughout the day, throughout the night, or during other periods.

Method
The meter records sound levels in decibels on three different frequencies. When switched to a frequency the machine is turned on and a needle records the sound level. This usually varies continuously so the peak sound is usually recorded.

Recordings are made of the sound level in different locations and at different times dependent upon the circumstances.

Scopes

Development

The endoscope is not a recent invention, but its application to the problems of construction and the inspection of buildings, is a comparatively recent trend. The Endo element of the word, relates to internal and its initial development was in the medical field. If you look up Endoscope in the dictionary it will advise you that it is an instrument for viewing the internal parts of the body.

Equipment

There are two main classifications of endoscope:

1. Rigid endoscope

Rigid probes may have scopes with either a warm or a cold light source. The cold light source has become possible through the development of fibre optics. This enables the light source which is usually an electric light bulb, to be placed at some considerable distance from the end of the probe. The light is transferred to the top of the probe, via a bunch of cables which are fibre optic cables. They can transfer the light with remarkably small reduction in intensity. The warm light source is where the bulb is located at the end of the probe. The advantages of the cold light source are that the top of the probe is not subjected to major temperature variations and can therefore be used in explosive atmospheres, such as may be involved if one is examining the inside of fuel storage tanks, petrol tanks, or similar volatile areas. The warm light, where there is a direct light at the end of the probe, gives better illumination, but means that the temperature of the bulb will probably exceed 200°C. This can be somewhat disastrous if it comes into contact with an explosive mixture, or if it touches the insulation around electrical cables or service pipework. It can also be the cause of fire if it came into contact with cavity insulation or cavity fill in a wall.

The difference between the rigid and flexible scopes is fairly straightforward. The rigid scope is either an individual or series of straight rods which plug together either with a screw fixing or by friction and which contain a series of lens in an arrangement which is not dissimilar to the old fashioned telescope.

There are a number of different ends available for the probe which give different angles of vision. These angles can be 0°, which is looking straight ahead of you, as would be the case with a telescope, 45°, 90° or the very popular 110°, which enables one to look slightly back from the tip of the probe. This is important if one wants to see where one has been, as opposed to the 0°, which is useful if one wants to see where one is going! These angles are incorporated as a fixed unit at the end of the probe.

If the probe has a cold light source, it is probable that it will be one single length, because of the problems of obtaining a good connection with a fibre optic light source. Maximum probe length for a 12mm diameter probe, will be in the region of 1.585m (5 ft).

For warm light probes, because the light source is located at the end of the probe, probe length can be considerably greater and their lengths can be up to 30 metres for a 24mm diameter probe. These probes are built up from a number of interlocking sections. Over the longer length they will require some form of independent support.

2. Flexible endoscope

Fully flexible probes are about five or six times more expensive than the rigid probe. This probe is capable of allowing the inspector to place the probe through a flexible route and to operate the end of the probe in such a way that it is able to move under control. This type of probe has been used extensively within medical research, for the investigation of various tracts and naughty bits of the body.

The extra expense is due to the care required to arrange the thirty or forty thousand fibre optic cables which are used. They must all remain parallel and in exactly the same arrangement of tubes through their full length, otherwise the image which is received at the extension of the probe would be jumbled by the time it was placed at the eye piece.

New techniques of investigation within buildings will be available, as a result of the extension of these probes to the building and engineering inspector. As with the development of all new techniques, the experience in their use and the understanding of the image, will take some time to acquire. It is amazingly difficult to understand the scale of what you are looking at and even in determining the

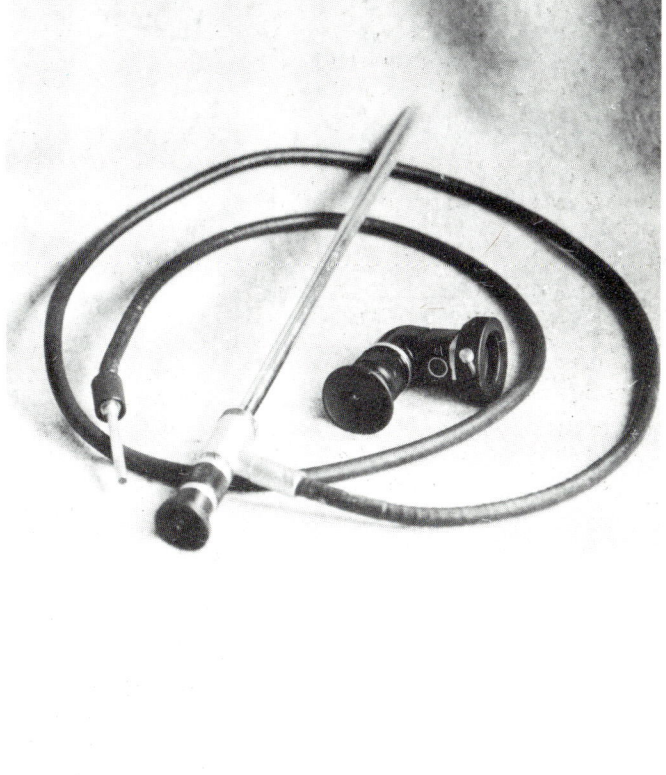

The borescope in use and repose.

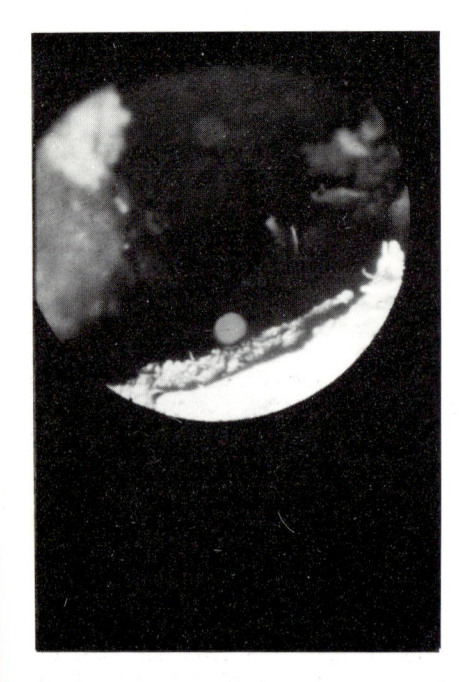

SURVEYING EQUIPMENT

Left: Using a borescope. Interior of a timber framed partition with broken edges of chipboard panels. It is not always easy to identify the size or scale of what you are looking at.
Below: A simple and inexpensive borescope.

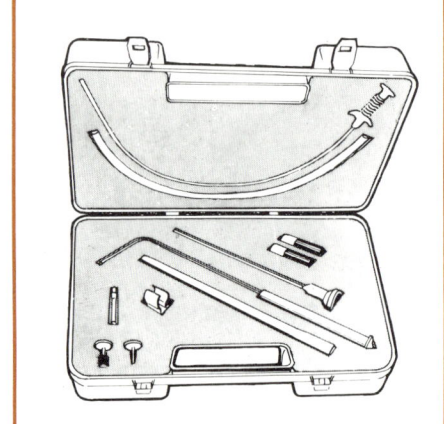

Metal detection

Purpose:

A machine to aid the location of metal covered by ground, wood, concrete, roadways, pylons etc. This metal may be wall ties, both mild steel and stainless steel, in walls, or nails in stud partioning. It may be used for the location of manhole covers and stopcock boxes hidden by tarmac or earth works.

It may also be used to locate reinforcing bars set in concrete where the intention is to drill into the concrete and one wishes to avoid striking the bars. In the timber industry it may be used to try and locate metal in timber prior to sawing.

Operation

Only an industrial detector should be used for this type of work as they are specifically designed for the purpose and are usually reasonably rugged in construction.

Ideally a detector using pulse induction principles should be used as it is deeper seeking and is unaffected by the medium between the search coil and the metal target.

The search head can be moved from wet conditions to dry, and from strong sunlight to shadow without it affecting the readings. This stability with pulse induction machines enables them to be easily used with only one control knob.

Detectors generally have an audible alarm although some have meters to indicate the presence of metal. The noise of the alarm or the reading increases as the detector passes over the target.

Different machines can have different ranges and the distance required of the detector in its search will determine the machine for the job.

Advantages

Enables the rapid organised search over a large area.

Disadvantages

The sensitivity is such that the range may be insufficient. One meter is the maximum range for 100mm diameter metal pipes or inspection covers; a 25mm diameter metal pipe, 670mm (26"); a 75mm rail, 320mm (13"). Most service pipes, particularly water pipes are below 1 meter because of frost damage risk.

(N.B. Protovale do make a live cable detector which has a range of up to 5m for load carrying cables. They also market a small unit for locating cavity wall ties, or nails in a stud partition.)

angle and direction of view that you have. It is helpful if one can practice by say drilling a hole in a floor board, which is removable, and examining the underside through the probe and then removing the board to see the full extent of what one has been looking at.

Spectron Optical Holdings produce an inspection kit which includes an endoscope with variable focusing, and a 6 volt battery pack. The equipment includes a light source, which is enclosed within the handle of the probe and a battery pack about the size of a conventional electric moisture meter. The battery gives approximately 2½ hours of use on site. The full kit comes in a standard executive type brief case and has been developed for the traditional one-handed use, that is the requirement of most surveyors. Recent development has improved the product which sells from about £700.

The Company also provides an optional extra mains adaptor with a remote light source and camera connections to the equipment, to enable the probe to be coupled to the lens of any popular single lens reflex camera.

Inspection Instruments were the pioneers in this field but the probe marketed by this company is less refined and does not suit the on site conditions found by surveyors. The probe operates with a remote light source, which may be activated by mains control or by power provided from a battery belt. The light source box is heavy and cumbersome and the battery belt weighs about 7 kg. The light source is 12 volt. They use German technology in the provision of the probe lens.

Inspections Instrument Equipment starts at £1200 and is bought in individual pieces and therefore does not have the benefit of the complete kit. The probe light source box and battery belt are cumbersome to handle, use and carry.

Flexible probes

A.O. Ophthalmic Instruments produce a flexible fibre optic probe with a battery powered source. Once again the light source is located in the handle of the probe but in this case the probe itself is completely flexible. The probe diameter of 7mm is less than the conventional probes of the products referred to. The manufacturers do market probes of smaller dimensions. The light source in this case is located to the end of the probe and it is therefore a warm light with a maximum temperature of 200°F. The tip deflection is plus or minus 120° and the probe comes with either a forward viewing head with right angle attachment or with a permanent right angled viewing head. It has an overall length of approximately 1 metre. The focus attachment is fixed.

The equipment operates from a 12 volt light source with a 50 hour life. The battery rechargeable unit has a four hour life and is about the size of the conventional electric moisture meter.

The Imp metal detector.

SURVEYING EQUIPMENT

Radon detectors

Surveyors who are operating in areas where there is a granite topography may need to be able to report upon the extent of radon emissions in their area. Radon presents a potential health hazard where excessive accumulations occur within a building.

Extensive testing has now been carried out, both within this country and in America, of houses which are built upon rock formations which are known to carry high levels of uranium and which are suspected of contributing high levels of radon to houses.

It has been found that where the houses are built directly off the bedrock or are cut into it and the rock is itself high in uranium, peak emissions have been found within the property. A covering of clay over the rock produces a substantial barrier to the emission of radioactivity and the levels of radon within properties in these circumstances have been considerably less.

Use

The qualititive interpretation of the distribution of radon. These soluble isotopes may be found in ground waters and sediments and a portable radon de-gassing unit will enable analysis to be made of the radium or radon content of water sediment or rock samples. A radon daughter detector measures the levels of radon in the atmosphere within the buildings. It easily measures ambient levels found in normal residential situations and is also capable of measuring the high concentrations often associated with uranium mining and milling. It can be used for the identification of radon gas sources as well as the evaluation of the effectiveness of any radon controlled equipment which is being used. A radon daughter detector is a portable instrument designed for the measurement of the alphaparticle activity originating from radon, its short-lived airborne daughters, and thoron. Using a simple filter collection device in series with a flow-through detector it simultaneously samples both gas and the total airborne particulate load which would include unattached radon daughters, radioactive dust and aerosols.

The sensor is a phosphate coating of silver activated zinc sulphide coupled to a specially selected stable high grain photomultiplier tube. A bright five-digit LED display retains the accumulated count in memory until it is deliberately erased.

Sampling and computational procedures are simple and straightforward. In the area to be sampled, a known volume of air is first drawn through a collection filter after which it is placed upon the scintillator tray, placed in the counting chamber and counted. Sampling takes approximately twenty minutes but may vary in different circumstances.

Following the working level measurement the gas contained in the cell is analysed over a period of approximately thirty minutes. The resulting data is then processed to yield radon concentration.

Comment

It is recommended that readings be taken over a much longer period of time. Current thinking suggests that three months monitoring of emissions is required to enable a responsible diagnosis of the presence or otherwise of excessive Radon emissions to be made.

The areas of peak emission in the UK are Devon, Cornwall and Derbyshire. The major natural isotopes are radon-222 or the biologically less important radon-220 (thoron).

Radon detection and measurement equipment.

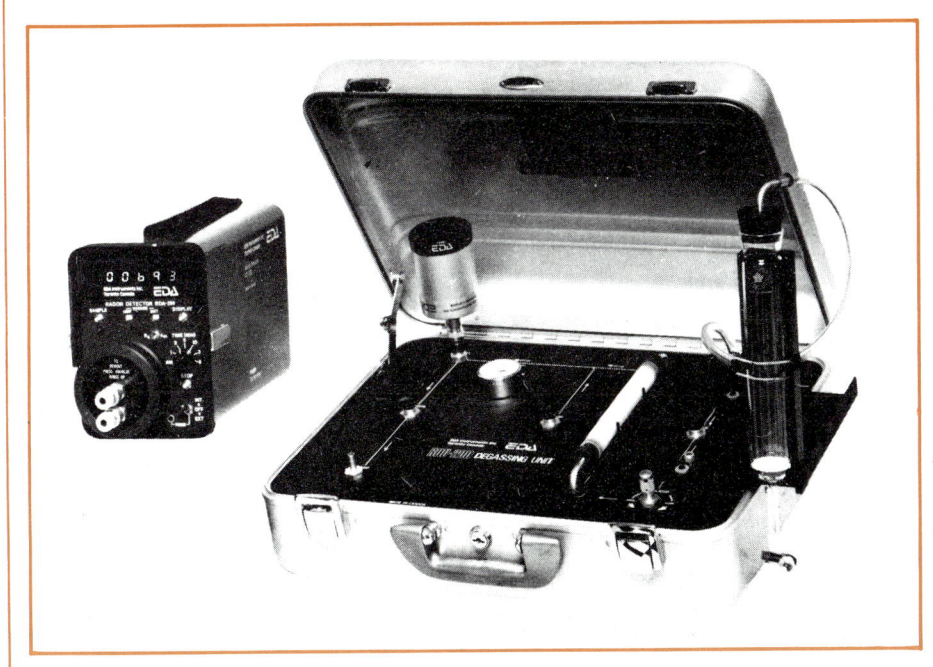

Electrical supply testing

There is a need for the inspection and testing of electrical installations to ensure that the requirements of the Institute of Electrical Engineers Regulations have been met.

No testing should be carried out by any untrained personnel.

Before equipment is put into use a visual test of an assembly and installation should be made.

The visual inspection has three main purposes:

1. To make sure that the equipment installed complies with the applicable British Standards. If the equipment is not so marked a certificate from the manufacturer or installer should be obtained.
2. To check that the installation has been carried out to comply with the IEE regulations.
3. To ensure that there is no visible damage to the installation or the equipment used.

The IEE regulations recommend that the circuit and installation be tested in a sequence.

It is important that this sequence is followed for example, the continuity or effectiveness of protective conductors must be verified before insulation resistance tests are carried out.

Continuity of ring final circuits

All conductors in ring circuits must be verified.

(a) The resistance between the open ends of each of the three rings (phase, neutral and earth) are measured individually.
(b) The two ends of each of the three rings are connect (i.e. an earth, a neutral or a live ring is made) and a reading taken from the connection to a socket outlet near the middle of the ring.
(c) The resistance of the lead is measured.
(d) Check that

$\frac{a}{4} - B - C$ (value of resistance of the loop from main position to midpoint)

Polarity

It is important that the installation is connected in the right phase. The reversal of the live and neutral will not prevent equipment working but will result in appliance remaining live when switched off.

Each socket outlet must be tested for correct connection.

SURVEYING EQUIPMENT

Equipment
Phase indicator

Earth leakage circuit breaker tests

The test consists of applying a 45 volt supply between neutral and earth. This will circulate a test current in the neutral-earth loop.

Where circuit breakers are installed the reverse current will cause them to trip.

If they do not it is an indication of a possible parallel earth path. This would have to be traced and eliminated.

Equipment
A transformer to produce the requisite voltage.

A voltameter to check the current. These are wired to a plug and connected to a power socket to carry out the test.

Earth loop impedance test

A test current, driven by the connected mains supply is introduced into a circuit via an electrical socket. The test current operates for a microsecond and measures the impedance in the circuit.

Several types of equipment are available and they usually plug into an electrical socket. The continuity of the earth loop should be verified before the test.

Equipment
Line-earth loop tester such as a Solex Seaward DL-500.

Digital level gives angle of slope as well as the ratio of slope. Ideal for checking drain falls that should be 1:40 or 5°.

Level meter

Purpose
The accurate measurement of the horizontal and vertical, and the accurate measurement of the extent of the deviation from horizontal and vertical in each case. The equipment is able to measure angle and gradient with a high degree of accuracy.

Use
The actual fall of a flat roof can be diagnosed if it is tested at the the time of laying the screed. The fall of drainage pipes can be checked at the time of laying. Variations in the level of floors within buildings can be stated as an actual fall as opposed to being "out of level".

Method
The electronic meter is placed on the surface and a reading obtained within a few seconds. The readings will be either in degrees of deviation from the horizontal, with an indication of the direction of the slope, or the grade of slope in relation to 1.

For example, a reading of the surface of my desk indicates that it has a fall of 1:017 or a fall of 1 degree to the right. An angle of 5.5 degrees is equivalent to a fall of 1:096.

Readings are shown digitally on the side of the unit and repeated on the top. Data can be retained, so readings can be taken where you cannot see the read out.

Advantages
Simple and easy to use. Accurate to within 0.5 degree, or 0.1 degree for a more expensive model.

Disadvantages
Brittle plastic case. Not suitable for use in humidity in excess of 80%.

Impulse radar

Purpose
The provision of detailed information from which can be interpreted the composition of the material, variations in thickness of layers or the presence of voids.

Operation
Specialist technicians from Harry Stanger take continuous readings from a surface, be it a roadway or a wall. These are then fed into a computer which will provide a profile of the area beneath that surface.

This technique has been used to plot the ground in a churchyard so that the feet of a tower crane could be located on solid ground, instead of over the catacombs or vaults that peppered the area.

It has been used to trace the course of dampness entering the dungeons of an ancient castle through fissures in the walls. Not only could the precise location of the narrow channels, worn over the centuries be found, but also the success of pressure filling the cracks with epoxy resin to seal them and make the rooms dry could also be checked.

The quality control of contractors can be followed up in a non-destructive set of tests. For instance, the uniformity of the hard-core and the thickness of the concrete deck beneath the tarmac surface of a road can be determined.

Advantages
The continuous profile of an area can be obtained. This is much more useful than the information gained from selective trial holes, which only tells you the condition in the area tested and not the variation that may occur in the areas not checked.

The tests do not damage the surfaces.

Disadvantages
The system is expensive, and can only be used by the specialists from the laboratory that has developed the technique.

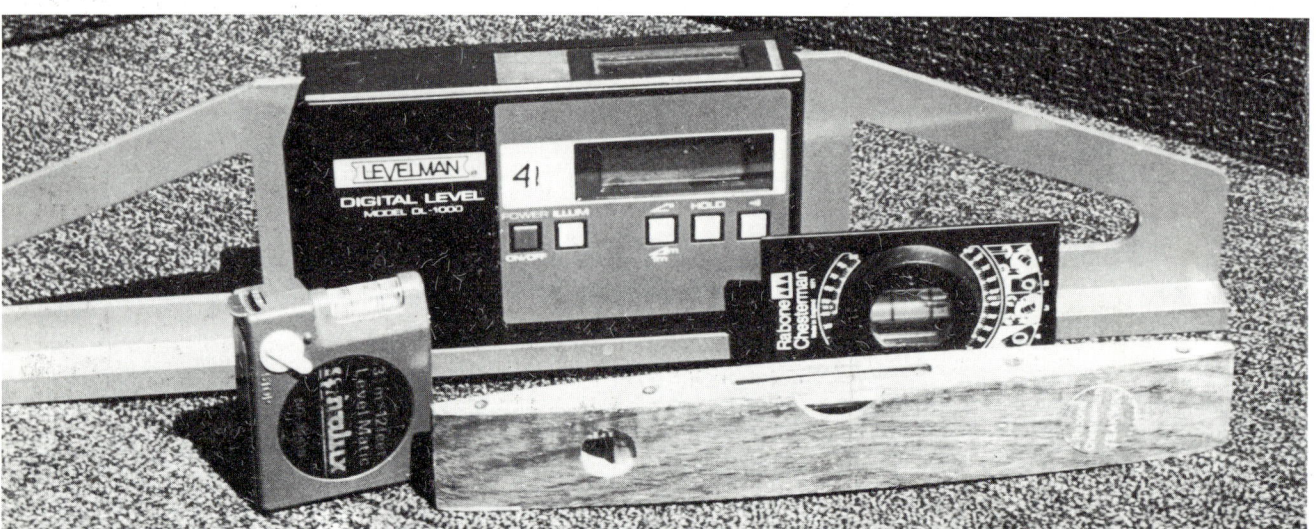

SURVEYING EQUIPMENT

On site tests

Rapid test method for cement content of mortar

Purpose
The strength of mortar for brickwork or for rendering is important in determining the long term performance of walls. Mortars which are weak can reduce the loadbearing capability of brickwork, or lead to erosion of renders. Mortars which are too strong when used with materials such as lightweight concrete can lead to cracking, or in the case of renders, pull the surface away from masonry. The Bremortest enables the actual mortar mix to be identified in new mortars, and in those which have set by up to seven days.

Use
An accurate weight of fresh mortar is stirred into a standard amount of concentrated acid. The acid reacts with the mortar and the resultant temperature rise is proportional to the amount of cement in the mix.

The method involves taking the temperature of the mortar sample, then taking the temperature of the acid, and adjust for the difference between the two temperatures. You weigh exactly 6 grams of mortar and mix into acid and note the maximum temperature reached.

The test takes about 2 to 4 minutes depending on your dexterity and temerity working with a concentrated acid.

Advantages
Produces a rapid on-site test of the mortar mix for new mortars and those up to seven days old at a cost of about £7 per test (excluding cost of purchase of the equipment).

Disadvantages
Tests on mortars containing lime are not as accurate as those on cement-sand mixes. This is less of a problem on strong mixes where the lime content is lower.

The small quantities of retarders, waterproofers and plasticisers added to mixes will not affect the temperature rise directly, but may affect the water content of the mix which slightly affects the result.

Chemical analysis of plaster, to locate nitrates as indication of rising damp.

Purpose
A test kit enables on-site tests to be carried out for soil salts in a wall and its plaster in order to pin-point the source of any dampness.

Method
The test kit involves the mixing of samples with distilled water and dissolving special tablets with the solution.

A level spoonful of sample scrapings are placed in a container and topped up with water to the 60 cc level. The mixture is then vigorously shaken and 100 cc of the liquid poured into a second container where it is mixed with a Nitrate tablet. The mixture goes yellow. A second tablet is then added. After five minutes the colour is observed. If it remains yellow there are no nitrates present, if it goes orange they are present, and if it goes red it is an indication that nitrates are present in significant amounts.

The balance of the liquid is then tested for the presence of chlorides in a similar way. The brown colour of the liquid indicates an absence of the chemical and a yellow result in the sample indicates that chlorides are present.

Advantages
Enables on-site tests to be carried out at a cost of about £4.50 each sample. The tests take about eight minutes, although part of that time is waiting for reactions, and other work can be undertaken.

Disadvantages
Colour variations are not easy to identify without experience. A sample colour chart for comparison would be an advantage.

Sundry

Ventometer
An inspection of buildings where water penetration is taking place, and the levels of exposure should be checked, can benefit from the crude information to be gained from the use of a ventometer. These simple yachting gauges give the strength and direction of the wind. Taking readings around the building and its environment can help locate any acceleration in the wind that results from the location of surrounding buildings.

Crack gauge
The Pyramus and Thisbe Club (Party Wall discussion group) has introduced a small, clear plastic card onto which lines of different width have been marked. The lines can then be compared with cracks found in a building, so that the size of crack can be referred to in the report. The pocket size and simplicity makes these cards highly prized.

Identification
Surveyors spend their time chatting their way into property, and trying to assure the occupants that they are who they claim to be and not the local thief. Identity cards are available through a subsidiary of Securicor, and carry a photograph which nearly everyone seems to find very reassuring. I now use my new European Passport, which is much smaller than the original and fits into my wallet. Regrettably, the photograph does look like me.

Face mask
I have included this, not as an escape from my appearance, but as a health protector. It fits across the mouth and reduces the quantities of dust that I inhale in roof or loft inspections, or whilst inspecting building work which is particularly dusty. Replacement filters are available. Having choked in one in the past, I now keep the mask always available in a sealed plastic bag within the survey bag.

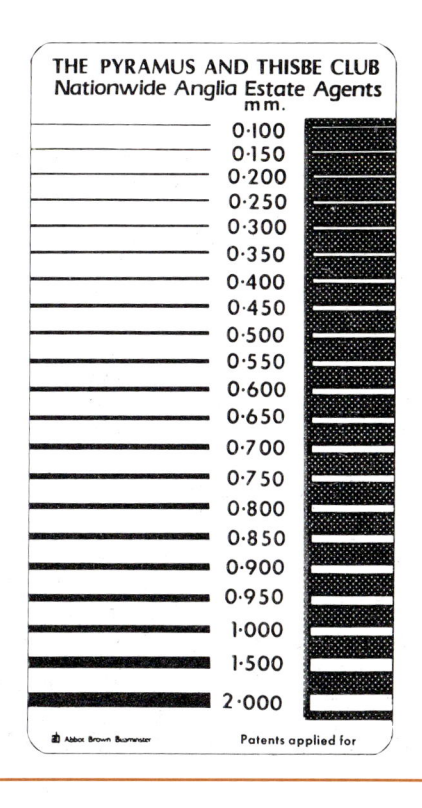

```
THE PYRAMUS AND THISBE CLUB
Nationwide Anglia Estate Agents
             mm.
             0·100
             0·150
             0·200
             0·250
             0·300
             0·350
             0·400
             0·450
             0·500
             0·550
             0·600
             0·650
             0·700
             0·750
             0·800
             0·850
             0·900
             0·950
             1·000
             1·500
             2·000
 Abbox Brown Bournmouth    Patents applied for
```

SURVEYING EQUIPMENT

Sick building syndrome

At the present time there is no suggestion that advice upon the health risks that may be present within a building is part of a surveyor's duty. However, it can only be a matter of time before he is required to comment on any risks that may be present.

Sick Building Syndrome is an emotive description of the conditions that are believed to be present when a building gains a reputation for having an adverse effect upon the people who work within it. The 'symptoms' include: the spread of infection within a building; above average absenteeism; tummy upsets or frequent headaches among the employees; or injurious circumstances that can lead to serious illness or even death.

The influence of a building upon the health of its occupants is frequently thought of as only involving office buildings. This assumption is not correct, however. Waterborne infection can also be found in the residential building.

Water-borne bacteria

A link between illness and buildings has not been made in every case. Where illness has been caused by bacteria the link has usually been made, but it cannot always be predicted or prevented.

Common design features do occur in buildings where illness has been traced to bacteria, and it will help the surveyor to have a knowledge of the circumstances that are known to have led to infection in both residential and commercial buildings.

Drinking water supply

There is a health hazard from contaminated water that may be drunk. On restaurant premises the water supply for the potable (drinking) water supply must be served from a separate tank with a bolted down and sealed lid. Ventilation is provided to the tank, but this must be provided with a screen to prevent entry by vermin. The drinking supply and the tank must be clearly identified by a notice.

In public and commercial buildings, it is a requirement that all water supplies in toilets and washrooms shall be marked with a notice if they are not suitable for drinking. Drinking water must come from a separate potable supply, which has all the necessary protection that goes with such a service unless it is served directly from the mains. An examination of public and commercial premises will include the visual inspection of the plumbing supply to ensure that it meets these requirements and has been properly maintained. If there is any doubt, water samples should be sent for analysis.

The examination of the water supply in a residential property should take note of the following:

The water tank must have a cover which is secure. The cover should not be made from chipboard or polystyrene. There should be no insulation draped over the water tank, beneath the cover or in place of a cover.

The use of unacceptable materials will lead to the contamination of the water supply, or the blockage of pipes. Most chipboard will disintegrate, and drop fibres into the water supply. (See the photograph on page 139.)

Water tanks should always be kept clean. They should be drained down every three years and thoroughly cleaned out and flushed through. The domestic water tank provides water for cooking, washing cooking utensils, cleaning teeth, and often for the provision of water for the preparation of food. The contamination of the supply, whilst not the main source of drinking water, will lead to infection.

The appearance of the water itself may provide information about the likelihood of contamination being already present.

The main danger arises when a water tank is located in an attic with a skylight. Light from this will aid the propagation of any contamination in the water supply. The interior of the tank should be free from debris on all surfaces.

The corroded surfaces of a zinc coated steel tank are an ideal breeding ground for bacteria. Discovery of these conditions should lead to a recommendation for the tank to be replaced not only on the risk of it failing, but on the ground that it may be a danger to health.

Discolouration of the water in the tank is also an indication of risk. The appearance of bubbles on the edge of water in the tank, or a slight greenish hue is sufficient for a recommendation for the immediate testing of the water.

It is common to find that the supply to a house served is by two or more tanks linked together. If the supply and service come from the same tank, there is the probability of stagnant water being present in the other tank or tanks. This then provides a breeding ground for bacteria.

The area the tanks are located in must also be kept clean. There should be adequate protection from ingress by vermin, flies and birds. Immediate action should be taken if birds are found nesting in the area of the tank, particularly if there is a defective cover on it.

Breeding Grounds for Bacteria

The following areas within the plumbing in residential and commercial property have proved to be breeding areas for bacteria:

Leather or rubber washers on taps
Shower head. The pin-holed sheet of metal which produces a fine spray of water often furs up on the reverse side due to a build up of calcium deposited by the water. This deposit collects hot or warm water which becomes stagnant between uses, and provides a breeding ground for bacteria.
Mixer tap filters on the spout. The build up of calcium on the filter is similar to the furring up of a shower head or a kettle.
Hemp. This is used as a jointing seal on metal pipes. It provides a food source for germs.
Boss white or plumber's metal. This jointing paste often has an oil base which provides a food supply for bacteria to breed. It was thinned, if required, with boiled linseed oil.
Ferrous oxide. This is the rust or sludge often found within water tanks or galvanised calorifiers.
Directional spouts on the kitchen tap. These are often flexible, and are used to direct the tap flow around the sink to help cleaning. The rubber forms a breeding ground, particularly when cracks occur. In addition, if it is connected to a mixer tap, the spout could contaminate the mains supply with the water from the hot service which could itself be contaminated.
Filters. Water is drunk from vending machines or water fountains on commercial premises. The equipment has in-line filters which remove contamination within the supply, but the filters themselves become a breeding ground for bacteria. The manufacturers stipulate that the filters must be replaced at regular intervals, but routine maintenance often slips below the required level, and a health risk may result. Examples of the equipment include:
Vending Machines. (Such as coffee/tea machines.) After the provision of a specified number of cups a new filter must be installed. The filter should have the date of installation marked upon it.
Water fountains or drinking fountains. There is a strainer which is located behind the spout: this should be replaced every two months. If one is investigating an outbreak of tummy troubles it is worth checking where the water barrel is refilled, if the fountain is not on-line to the mains. It has happened that a barrel was filled from a tap in the ladies toilet, which is not usually a potable supply.
Stagnant water. The water that collects and stands in traps will be a breeding ground for bacteria. Warm water containing skin, hair and vegetable matter will provide food and the ambient temperature for propagation.

Whirl-pool baths recirculate this water when they are used. There is a risk of contaminated water being held within the water pipes of the pumped jet system.

Where filters are fitted they must be replaced regularly.

The water traps have been known to form the breeding ground for the Legionella bacteria.

Legionella pneumophila.

Legionella pneumophila bacterium is often found in the mains water supply, but in a concentration that does not present any risk. It is important that no storage, distribution or other component part of a water system could provide conditions favourable for the multiplication of the bacteria, which if inhaled in sufficient concentration might cause a potentially fatal illness.

The bacteria thrives in the temperature ranges of 25–35°C. The ideal temperature for propagation is 36–37°C, but water in the range 20–50°C is at risk. Below 10–15°C it survives; between 20°C and 45°C it can multiply, and above 55°C it is killed.

The rapid multiplication of the bacteria occurs within the ideal temperature range, but not on every occasion. The reason for the rapid expansion of its concentration is not yet known.

The water temperature ranges suited to the growth in the numbers of bacteria can be found in many circumstances within domestic and commercial property. For example, the summer temperature of the water in a cold water storage tank can be sufficient for the propagation of the bacteria. Even river water warmed by the discarded heat of a power station has been thought to have contributed to its multiplication.

However, the main areas of risk of illness are water which has been contaminated in water-cooled air conditioning systems, in shower facilities, in whirl pool baths, or in taps. Legionella pneumophila has been known to have multiplied overnight in sealing washers, shower heads and taps.

The illness results from the inhalation of minute droplets of the infected water.

People at risk
It is currently believed that the people mainly at risk are male, and between the ages of 40 and 70. The risk of infection for other age groups of the male population is about one-fifth of that of this age group.

Between 1 and 4% of the population are at risk. Incubation takes between two and ten days. There is a 15% mortality rate. The infection is treated by drugs, such as ethromycin, anthromycin and rethromycin. Recovery should be within 48 hours if the infection is discovered in time.

Symptoms
The symptoms listed below are common to the illness. Medical checks should be undertaken quickly if an infection is suspected. The illness produces symptoms that are similar to pneumonia or double pneumonia.

It is characterised by:
- Fever
- Chills
- Headache
- Muscular aches, and pains
- Pneumonia
- Mental confusion
- Non-productive cough
- Nasal congestion
- Lethargy
- Chest pains
- Shortness of breath
- Diarrhoea
- Flu-like symptoms
- Feverish/sweating

Prevention
Avoid long pipe runs to serve shower or bath taps.

Avoid allowing stagnant water to collect in pipe branches or dead legs.

Maintain regularly any water-cooled air-conditioning systems and carry out regular sampling of water in all storage tanks.

Treatment
Water that has been stored for some time should be raised to a temperature of 55–60°C before use. This can be done within the calorifier, provided it is mixed with cooled water before use. In areas of peak risk the water should be raised to this temperature level every day.

Run the water in the shower for two minutes before using the shower. The bacteria will probably only be held in the water in the pipes. However, if the calorifier is contaminated the risk will not be eliminated.

Do not run a bath from the hot tap whilst you are close to it. There is a risk of inhaling the spray that is created, which is from a source that can be at risk.

Treat the water with chlorine; concentration 2 parts per million. This will lead to more rapid deterioration of metal plumbing pipework and storage tanks, and has to be balanced with the reduced risk of contamination that will result.

Maintain all water storage tanks and housings in good order and ensure all screens and grills are not damaged.

Method of taking water sample for analysis
Taps. The water that is required is the first flow from the tap. The minimum sample should be 1–5 litres.

Tanks. The water should be taken from the area of the tank furthest from the point of water entry. The minimum sample should be between 1 and 4 litres. If sludge is present, take a sample for analysis.

Sundry sources of contamination
Vacuum cleaners.
The fixed-bag type of vacuum cleaner must have the bag washed every month, as it is a source of infection. The through-flow of air would ensure that any infection is distributed when the machine is used. This does not apply to the disposable bag type of cleaner.

VDU's.
There is no evidence to suggest that there is any health risk from Visual Display Units.

GROUND CONDITIONS

How healthy is the house?

The energy crisis of the middle 1970's, which resulted in an increase in the cost of oil brought about a greater awareness of the cost of oil-based fuels, particularly where they provided light and heat. Much of this fuel was imported, and in an effort to reduce a balance of payments deficit, consideration was given to more efficient methods of heating buildings and of containing that heat.

The Building Regulations were amended to require a much higher level of insulation. That standard was further increased in April 1990. property.

The cold winters that followed the middle 1970's also highlighted the plight of older people who could not afford to adequately heat their homes. Hypothermia became an over-used word in the media to describe the condition frequently found in older people in cold homes, where their body temperature dropped below normal. Many deaths were ascribed to this condition. The government introduced a "Save It" campaign to encourage people to conserve energy.

They offered grants and inducements to people to upgrade the insulation in their homes. Loft insulation, cavity wall insulation and double glazing, were all promoted and advertised. Often with wild claims as to the relevant cost efficiency and benefit that would be derived from the use of various products.

Those homes which were treated with one or all of these methods of improved insulation saw a reduction in their air circulation.

There is no medical study which states the minimum number of air changes that are required in a room for healthy living. The health risk would very depending upon the age and health of the occupants. An older person is less likely to take exercise which will result in heavy or deep breathing. Any contamination or infection will be more likely to remain within the bugs than would be the case with a healthy active person.

In poorly heated houses with little ventilation the humidity of the air will tend to be quite high because the moisture created by human activity is not adequately dispersed. Condensation will occur on windows, and on wall surfaces, and mould growth will be seen on carpets, furniture and stored clothes. The use of gas and paraffin heaters, both of which emit a lot of water, will make matters worse.

As well as the health risk to the occupants these conditions may result in damage to the building. Condensation build up where timbers are present can cause deterioration. If this occurs where there is reduced or poor ventilation fungal decay, such as dry rot, can and probably will take place in woodwork. The increased moisture content of wood will increase the likelihood of certain beetles being present in the building. Higher moisture levels in the remainder of the structure will cause movement, particularly between wood and brickwork, which will leave unsightly cracks.

Mould growth is common in buildings which have condensation problems. In order to avoid mould the relative humidity should be kept below 70%. Improved insulation was advocated as a cure for this problem. It is true to say that it can contribute, but the risk of mould growth and condensation can be reduced by improving thermal insulation, increasing heating, increasing ventilation and by reducing the moisture content of the air.

The relative humidity of the air within a building is only varied to any major extent by 1½ air changes per hour. Increased air changes do not have a material effect, irrespective of the heating and insulation. In houses with no heating 1½ air changes per hour will reduce the relative humidity to below 70%.

An extractor fan run for about 50 hours per week in a two bedroom flat would control the relative humidity to a level below 70%. Internal bathrooms, where they occur have to have mechanical ventilation, capable of producing three air changes per hour, the fans are fitted with an over-run switch that is activated by turning the light on or off. This fan should run for 20 minutes after the lights are turned off to give at least one air change. How many bathrooms which have windows in them are ventilated to this level?

Radon Gas

Most of the United Kingdom has a very low level of background radioactivity. This has always been present and presents a low level of risk in most circumstances. Increased insulation can increase that risk to less acceptable levels because a reduction of air circulation by 50% would triple the condensation. (There may be a link between the spread of airborne diseases and reduced ventilation.) The concern in areas of radioactive emission is the effect of the gas radon, which occurs when uranium 238 breaks down. Uranium is present in the ground in concentrations of a few parts per million.

Uranium 238 breaks down at a constant rate and radon gas is one of the radioactive materials that result. The gas will seep out of the ground and building products and pass into the air we breathe. Polonium, which is produced as the radon gas decays, causes lung cancer. Polonium attaches itself to dust particles and will enter the lungs if they are inhaled. The polonium gives off alpha particles which penetrate the protective layers of tissue and damage living cells. Most lung cancer occurs in the top of the lungs, and it is that area where the particles are deposited.

The link between radon and lung cancer was established in the last century where 75% of the deaths of miners, in an area of high concentration of the gas, were due to lung cancer. Further research into the health and mortality of miners in uranium mines has confirmed the link. It is proposed that at an annual average exposure of 0.16 units, 600 deaths will occur each year in this country.

The average annual radiation dose received in homes in the United Kingdom is 0.16 units. (This is based on a survey carried out in 1977.)

The National Radiological Protection Board and a Royal Commission on Environmental Pollution are currently reviewing the situation, and reports will be presented over the next few years.

In homes with low air circulation the levels of radon gas will be above average, and the risk of lung cancer will be correspondingly greater. In an old timber floored house which has been insulated, draught-proofed and double glazed the concentration of radioactivity could be excessive.

Most of the gas will enter a property from the ground; smaller parts, about a quarter each, will be airborne or will percolate out of the materials used in the building construction.

If one can reduce the ground seepage of radon entering the building it is possible to reduce the risk of radon-inspired lung cancers. This can be achieved by placing a suitable horizontal membrane beneath the property to act as a vapour barrier.

In areas of higher concentrations of uranium, such as occur in granite, certain types of sandstone and shale, such a course of action may be desirable. The areas of higher emissions of radioactivity may be seen on the map.

The surveyor has a duty to advise his client of the risks and the method of reducing those risks to a tolerable level.

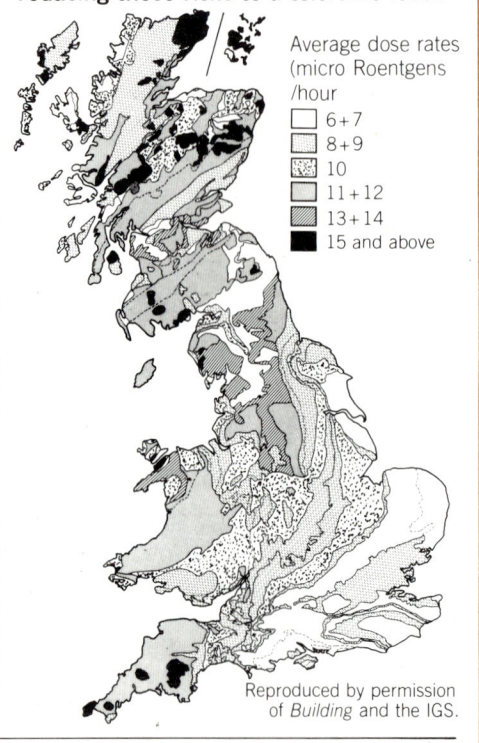

Average dose rates (micro Roentgens /hour)
- 6 + 7
- 8 + 9
- 10
- 11 + 12
- 13 + 14
- 15 and above

Reproduced by permission of *Building* and the IGS.

The section above, on radiation in houses, was prepared with the assistance of *Building*. See also "The Enemy Within" in *Building*, 29th April 1983, pp 26-27.

The inspection of prefabricated reinforced concrete dwellings

When you are faced with a building which is constructed in a non-traditional way you must not forget the rules that have been developed to inspect traditionally constructed buildings.

You must still look at the building to note any crack, mould, or failure to the fabric. You must not fail to record any defect which was available for your inspection.

You must also ensure that you have projected those defects that you have seen into the failures that are the natural consequence of the circumstances that exist.

Perhaps the most important part (one could say that of these rules, the most important is all three) in this case is construction. Not only must you correctly identify the method of construction that has been used, but you must know the history of that construction. Is this method of construction faulty? Not every method of non-traditional construction is defective. Not every method of non-traditional construction which is known to have developed faults is defective in the same place.

If you discuss the problems of a particular method of construction with surveyors from another town or part of the country, you may find that there are regional variations in the defects that are found. For this reason you must make sure that you tackle the building, not only with an open mind about the faults that you may find, but also with the research that gives you the knowledge of the failures that you expect that you will be likely to find.

Identification

The initial attempt to identify the type of construction will be on a suspicion basis. Most of the non-traditional forms of construction were developed by commercial manufacturers who displayed their wares each year in the equivalent of the Ideal Home Exhibition. I suppose that we may now think of it as an Un-ideal Home Exhibition. Many of the manufacturers updated their designs each year to encourage a new wave of enthusiasm. This means that there may be a range of variation of the appearance of each type of construction.

We need an "I-Spy Book of Building Types". Fortunately the Department of the Environment has produced such a book, except they had to call it "Housing Defects Act 1984, The Housing Defects (Prefabricated Reinforced Concrete Dwellings) (England and Wales) Designations 1984, Supplementary Information". Not a very catchy title, but a very useful pamphlet, which shows photographs of most of the more common types.

Don't forget that the local Building Inspector will have some knowledge of these buildings in his area and may be able to help you in your identification.

Research

Now that you have a reasonable idea of the type of construction, you are going to have to know about the likely problems in the building. One of the best sources of information is the series of booklets produced by the Building Research Station upon the structural condition of many of the building types.

They include Airey, Boot, Cornish, Dorran, Myton, Newland, Tarran, Orlit, Parkinson, Reema, Stent, Unity, Winget, Underdown, Wates and Woolaway.

Common problems

The Concrete Framed Prefabricated reinforced building is likely to be vulnerable in two places, the concrete and the reinforcement. These are not the only parts that may be defective. The failures that are common to all types of building must not be overlooked. Just because the construction is different does not mean that you ignore the leaking gutters, damp to ground floor, beetle in the wooden floor or the roof tiles that are broken.

Concrete is to be examined for the two relevant failures. Unsuitable materials or Carbonation. The unsuitable materials are additives that may have been introduced at the time of manufacture to accelerate the curing of the concrete. Calcium Chloride was frequently used in the manufacture and High Alumina Cement in the on-site castings of certain systems such as Orlit. The use of metal ties which have a limited life must not be ignored. Some concrete was aerated or breeze concrete. This is less resistant to the weather and a greater cover to any reinforcement is required to any of the reinforcement (greater than that provided in the Woolaway house type anyway).

Carbonation is a natural variation that occurs in concrete which comes into contact with the air. Fresh concrete is a highly alkaline material. This alkalinity protects the reinforcement that is within the concrete. The concrete reacts with parts of the air over a period of time and carbon dioxide weakens the alkaline base until it has neutralised it. The process advances into the conrete up to a depth of about 50mm. If there is a reinforcement bar within this vulnerable zone then corrosion is likely to occur. In good dense concrete carbonation takes place very slowly, but the rate of its advance can be influenced by the relative humidity in the area of the concrete. The inside of most occupied dwellings will be more humid than the outside, so inspect those concrete members that are inside the building as well as those exposed to the outside.

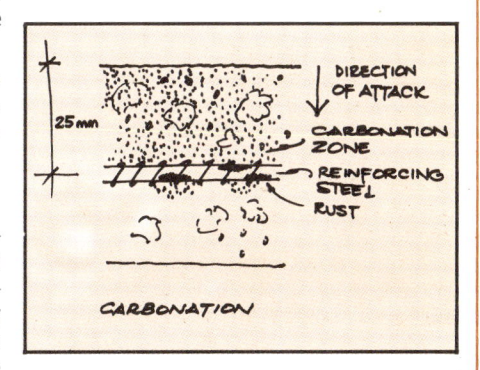

CARBONATION

The Survey

From the foregoing you will have seen that there are a number of likely failures within this type of building that cannot be investigated within the routine mortgage, house or flat buyers report or even a structural survey. Most vendors are likely to get a bit upset if the surveyor starts ripping open walls and cutting large chunks from the external concrete. There has to be a limitation to the investigation that one is carrying out. Unfortunately, in this case it will mean that you are very restricted in the extent of identification of the defects that may be present in the building. Because of this it is important that you confirm your instructions carefully and qualify your report so that your client is also aware of the limitations of your report. You may also wish to point out the advisability of a more detailed and destructive investigation of the fabric, provided the vendor will give approval.

Let us suppose that we have been given approval to do what is necessary to advise upon the condition of the building, and we have confirmed to vendor and client everything that is involved, including the cost. We have now moved out of the survey role into a diagnostic investigation. This is not a standard to be applied to the normal routine inspection of a building.

A borescope may enable you to reach a preliminary feeling about the likely condition of the columns or other concealed parts of the construction. Unless you are very familiar with the interpretation of the image revealed with such an instrument, don't rely upon it. It takes time and experience to be able to use this type of instrument to its capacity. In this case we are looking at what we will do to gain sufficient information to advise correctly. Keyhole surgery has a greater risk attached than demolition. We have to go somewhere in between. Possibly from the concern that we may have for what we think that we have seen from the borescope inspection of the base of the columns we will open up the structure in a number of places.

This must enable us to see the condition of the concrete. If it is cracked or spalled, we will want to expose the reinforcement. (Because we identified the construction we have with us the BRE guide to this type of building. These guides have details of the construction, and show the location of the reinforcement within the columns or panels). The concrete that we remove should be carefully stored so that it can be sent off for analysis to indicate the amount of carbonation as well as the presence of any problematic chemicals.

We also have to obtain enough information about the complete structure to enable us to advise upon the stability of the complete building. There is little point in our doing all this damage if we are only going to be able to say that there is some chloride attack and carbonation is well advanced.

Most clients will be waiting for the punch line because so far you have only attracted their attention. "What does this mean?"

They will rightly ask. We have to decide not only what is wrong but how wrong it is. We may have found extensive cracking to the concrete frame in the area that we have exposed. How bad is the failure in the other parts? We must expand our exposure so that we can determine the condition over a representative part of the building.

Where we have exposed fresh concrete by our activities the new surface should be painted with an indicator such as phenolphthalein. This will indicate the extent of the acidity that remains in the concrete. It is applied to the surface of the concrete and turns red when it is in contact with alkalis. The paler or colourless areas of the application are therefore carbonated and steel reinforcement is vulnerable. The reinforcement is at greater risk where moisture is present. If the concrete is completely dry then there is little risk of corrosion to the reinforcement.

The presence of water may come from any of the normal failures to the building such as failed gutters, downpipes cracked or blocked, or flower beds built up against the house. These external failures will also increase the failures that may flow from the chemicals that are in the concrete or are introduced by the water running into the building. Sulphates may enter the concrete as a result of atmospheric pollution. This is more emotionally referred to as "Acid Rain".

The information about the chemical composition and the extent of the carbonation of the concrete must be considered together with the depth of the cover of concrete over the reinforcement.

There may be a series of tests that we consider relevant to our investigation of the concrete structure:

Cover meter test to determine the depth of the reinforcement.

Electrical potential or half cell test. This will indicate the probability of reinforcement corrosion having taken place.

Vibration Testing may be carried out to locate the voids in the concrete that may be the result of expansive stress.

The use of a Schmidt Hammer. This crude test gives a limited guide to the quality of the concrete. A spring loaded plunger is fired at the concrete and the amount of rebound measured. (John McEnroe tests tennis balls in a similar way by throwing them at the ground).

Ultrasonic test. In this case a pulse is sent through the concrete. From the measurement an equivalent cube strength may be deduced.

Repairs

The practicality of repairing the structure will exercise the mind of the surveyor and engage the attention of the client. In simple terms it is possible to repair anything. We must know the cost of repairing the building. If your client is a private householder the acceptability of the repair is also an important consideration. There is no point repairing the building and finding that your client cannot sell it at a future date because no-one will accept that the repaired building has an adequate life. The economics of deciding if the building should be repaired cannot be divorced from the acceptability of the repaired building by the market.

From the information that we have collected we will be able to predict the rate of carbonation and the rate of deterioration in the building if no repairs are carried out. This, in effect, will give us an idea of the urgency of each of the repairs that we believe will be needed. It is possible to categorise them into:

Repairs needed in 6 months (i.e. URGENT).

Repairs that are anticipated in 2 years.

Repairs that will wait about 15 years.

A repair that is to be recommended to a client must deal with the problem in the long term and not be a short term expedient.

Shown here and on the following page are a representative sample of the main failures found in some of more common types of reinforced concrete buildings.

AIREY

Two storey with demented concrete ship-lapped exterior.

Failure

The concealed columns have a central hollow pole reinforcement. The cover is only 9mm in part and nil to the base. Cracks occur requiring the framework to be replaced either with a new structural frame or by providing conventional external walls.

THE INSPECTION OF PREFABRICATED REINFORCED CONCRETE DWELLINGS

BOOT
Render face with concealed column and infill panels.

Failure
Traditionally used a dense aggregate concrete. Carbonation is minimal in Scotland, but has penetrated up to 25mm in England.

ORLIT
These are either flat or pitched roof, either rendered or with a column and panel exterior. The panels are taller than the Cornish panel.

Failure
Wide range in chloride content and in carbonation.

WATES
One, two and three storey buildings, with storey height panels of varying width and ring beams to each floor. corners either butt or mitre joints.

Failure
Low chloride content. No visible signs of deterioration.

CORNISH
Visible columns with infill panels. Column spacing about 1 metre.

Failure
High chloride content found in South West England. Carbonation has reached most steel reinforcement.

UNITY
One and two storey buildings faced in blockwork type cladding with continuous vertical joints. Some render finished.

Failure
Wide range in chloride content, steel corrosion has occurred where the chloride content is in excess of 1%. In 49% of buildings inspected, corrosion of steel found.

WOOLAWAY
One, two or three storey buildings. Render finish covers the structure but bolt failure leaves a distinctive pattern.

Failure
Concrete porous and carbonation at a depth of 25mm. Depth of cover 27mm. Steel corrosion and some cracking.

THE INSPECTION OF PREFABRICATED REINFORCED CONCRETE DWELLINGS

New Scotland Yard, in Victoria Street London was built around 1963. It was faced in granite panels on a concrete frame. Problems occurred with the cladding panels because there was no provision of adequate expansion joints when the building was designed. The absence of adequate tolerance for the movement of these panels resulted in stress fractures occurring in the cladding and some sections falling away.

For many years the building was surrounded with scaffolding to catch any falling masonry. Proceedings against the designers, supervisors, builders and local authority were eventually withdrawn, however.

The cladding has now been replaced around the building with the provision of the requisite tolerance for the differing movement of the frame and panels.

A surveyor inspecting such properties must use the knowledge that is available at the time of the inspection, and not the knowledge that was current at the time the building was constructed. Accordingly, a surveyor must have up-to-date information at all times, because negligence claims can be made against him if he has failed to advise on a design deficiency at a time when failures, such as the omission of adequate movement joints, are known in buildings.

Roofs

ROOFS

Previous page: A slipped slate to the eaves course. See how thick the cast iron gutter is. They are very heavy – failures can be dangerous as well as expensive.

Below: Typical surveyor's sketch to illustrate terminology of roof components for a client. Top right: Splits in an asphalt flat roof on a timber base. There has been no allowance for movement in the surface and superstructure. The rooflight isn't very good either – look at the cut of the replacement panel of glass on the left. Bottom right: The asphalt is fixed to the parapet walls with no cover flashing.

Of all the exposed elements in the construction of a building, the roof presents the greatest difficulty to the surveyor over gaining access.

The conventional pitched roof of tiles or slates may be inspected from ground level through binoculars. This type of inspection is not possible with flat roofs, unless there are other parts of the building which look down onto the roof.

Many failures in roofs or gutters are only evident after an inspection at very close range. Most surveyors develop gymnastic abilities which are fully exploited in their attempts to gain access to various roof surfaces.

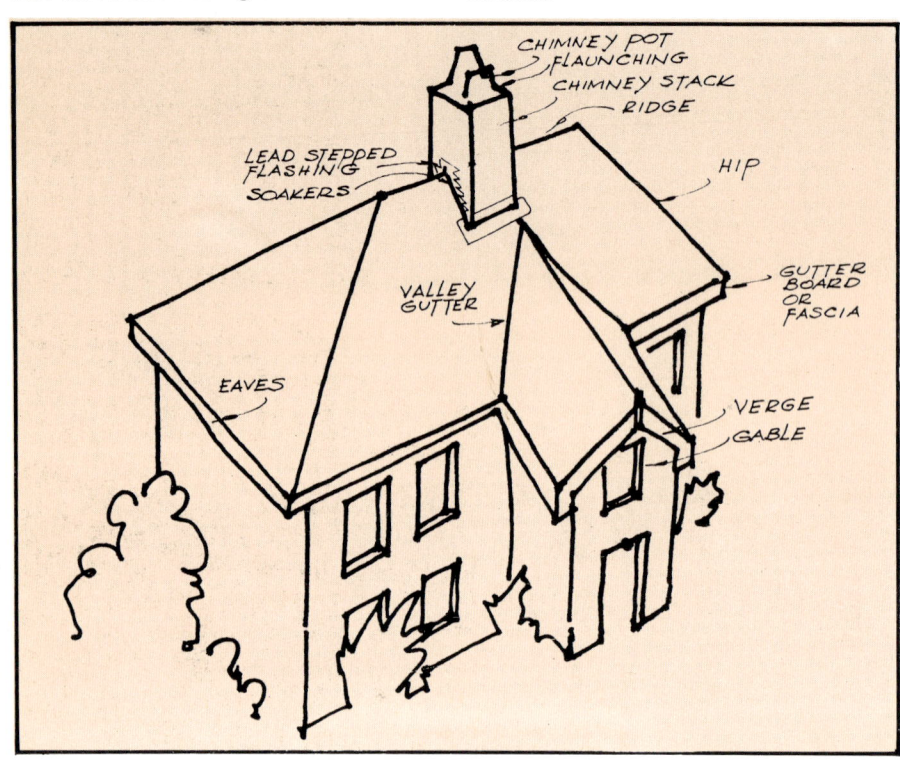

FAILURES IN FLAT ROOFS

Asphalt

Surface crazing

This is the light crazing of the surface which does not in itself necessarily allow water penetration. It may have been caused by variations in surface temperature possibly caused by water on the surface and an absence of any solar reflective treatment.

The application of a solar reflective face to the asphalt tends to prolong its life because it prevents extremes of temperature variation on the surface of the roof.

Ponding

Ponding occurs as a result of errors in the original construction or of a subsequent failure in the roof which has allowed a settlement in the finish to occur.

There is also a possibility that the face of the asphalt was not treated with sand at the time of its installation. If this does not happen there is the probability of a surface layer of bitumen lying on the face of the asphalt which will craze.

ROOFS

Top left: A ventilated asphalt flat roof on a concrete structure with a blocked outlet – now flooded. Those ventilators will now let water into the roof, and the tank room has an insufficient upstand so that water enters and then passes into the building. There should be more than one outlet from a roof.

Top right: Asphalt roof with slight crazing to the surface. The asphalt dressed up over the roof in this manner is a weak detail allowing no movement due to variations in surface temperature. How soon will it be before that slight crazing runs right through the asphalt?

Right: Asphalt upstand without metal cover flashing. Temporary adhesive flashing provided after leading began. Note damp proof course under coping stone dressed up face of chimney to prevent discharge of surface water into flue.

Bottom right: Attempts to provide seal between asphalt skirting and wall by copious use of mastic.

Fractures in the asphalt surface

These are failures which run through the material and allow water penetration. This may be the result of solar gain which causes continuous expansion and contraction of the surface. Ultra violet radiation in the atmosphere can cause oxidisation which may assist in the breakdown of the surface.

If the cracks are less than 3mm deep they may be treated by the application of a solar reflective surface but this is often a dangerous precedent to set as one cannot be convinced that every single crack is no deeper than this particular limit. Before any remedial works are carried out the actual thickness of asphalt on the roof should be found out, to ensure that there are not further layers beneath the defective surface which may have assisted in the breakdown.

In carrying out works to a balcony of a stone-built London building it was discovered that there had been previous applications of asphalt. There were three layers of covering totalling nearly 3 inches in thickness. The failures were caused by bitumen repairs in the base layer of asphalt. Each subsequent layer had reacted with the bitumen causing a weakness at the point

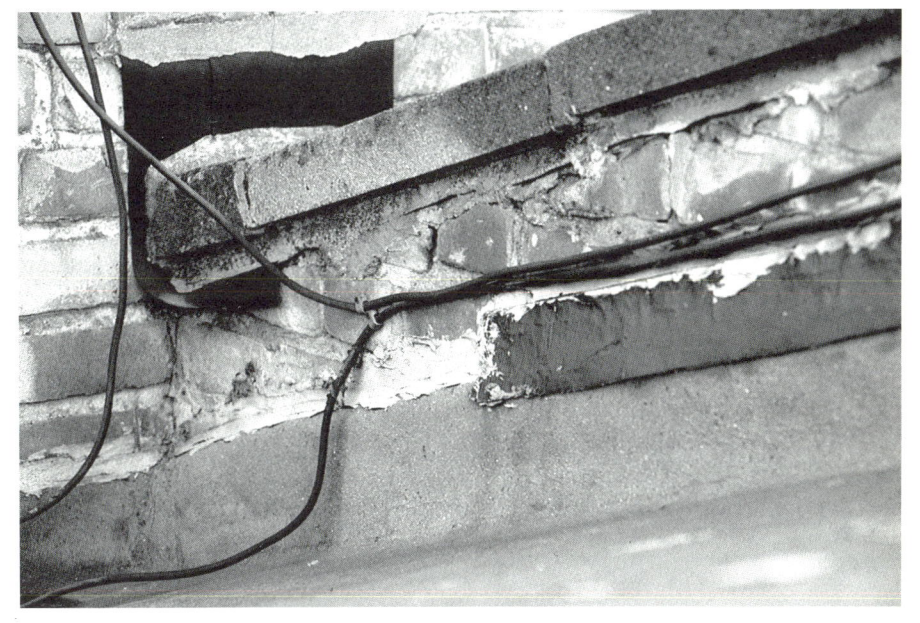

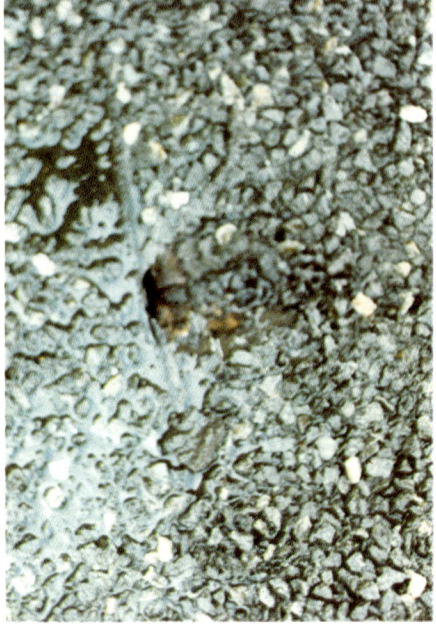

Built-up bituminous felt roofs

The design concept of roofing felt is that each of the three layers are bonded by the use of a hot bitumen so the action of the heat fuses the materials together.

Where blisters occur to the surface of the roof this is an indication that the layers of felt have not bonded together to form a single unit. This is not an acceptable finish. There can be occasions where such a roof surface may not leak for some considerable period, but this does not justify this unacceptable standard of construction.

There are a number of site circumstances which may cause blisters in the felt surfaces.

These include:
1. The bitumen was not hot enough when applied. This tends to leave a shiny finish to the bitumen which does not stick the layers of felt together. Blisters or bumps occur in the top layer.
2. Over-cooked bitumen. The bituminous surface beneath the felt is of a charcoal crusty appearance.
3. The surface of the felt may have been damp either through rain falling at the time of installation or due to the presence of dew. This tends to cause the bituminous finish to be crazed. The larger the squares in the crazed pattern the wetter the conditions in which it was laid.
4. Areas where no bitumen was applied at all.

Roofing felt

Uneven surfaces may be the result of the horizontal surface having been laid on an uneven finish such as the original boarding or more commonly where the boarding was originally designed for a metal roof covering and the ridges and rolls have been removed without adequate filling or preparation.

In many cases no remedial work is needed provided there are no failures and provided nobody walks upon the ridges causing fractures.

Where the failure is in the form of blisters which are empty and yield to pressure, this may have been caused by the poor workmanship of the contractor laying the roof or the trapping of moisture to the underside of the felt.

If no leaks are taking place an examination of the underside of the roof should be considered. If the underside is concrete construction the drying out of this surface is going to present more problems than if there is a timber roof which can be ventilated by the installation of ventilators to the void.

Distortions of this type often linked with undulations may have resulted from inadequate pressure being applied during the laying of the felt or failures in the storage of the material which may have become damaged prior to laying.

In many cases of failures there are recommendations for the applications of solar reflected treatments to reduce the problems. This is an individual decision but often can result in failures being caused by

directly above the original fracture. It was necessary to remove all the previous layers to carry out an adequate surface treatment.

Where the fractures are of a greater depth and have a definite edge they may be the result of movement within the structure of the roof.

This may have been caused by movement or the absence of expansion joints in the original construction. Sometimes the failures occur when the material used in the base meets other materials which have varying co-efficients of lineal expansion.

One section which is always vulnerable is the joint between the horizontal roof surface and the dressed up section of skirting.

Any adhesion between the skirting and the vertical enclosure may cause excessive strain on the roof covering at the point of the junction. If this occurs it will be necessary to cut out the upstand and reset it with a bond breaking barrier between the vertical surface and the upstand.

ROOFS

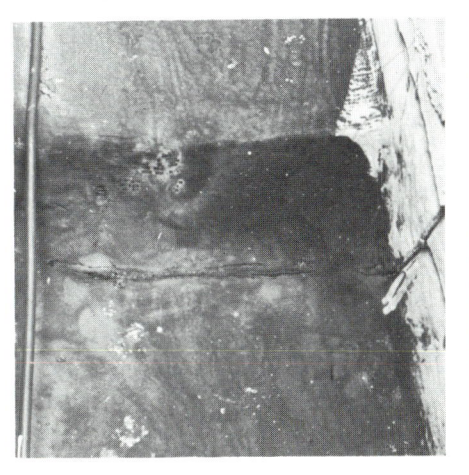

Facing page
Top and centre left: A felt roof with a bubbled surface. Note also the use of an aluminium foil faced adhesive tape as a flashing. If such a tape must be used (because of war, civil disturbance or nuclear holocaust) then dress it under the felt to provide a seal.
Centre right: A felt roof surface finished with chippings. It has a hole in it and that roof is not ventilated – so beware!
Bottom left: No roof should have that amount of water on the surface – why has it happened?
This page
Top left: A centre valley gutter. The surface must have failed as it has been 'repaired' with a treatment of bitumen and hessian dressed up to the face of the slates. Some water has run down the back of the slates and has rotted the sole plate – the timbers holding the gutter. Note also the crazing in the render to the parapet – is there a damp proof course behind that?
Top right: A mineralised roofing felt dressed up around an upstand with a discharge into the rainwater downpipe. The water is running under the felt at the point of discharge into the hopper head. Note which areas are wet and which are dry.
Above left: A parapet gutter step. This is the drop in level necessary to enable the metal finish to be joined. This is a vulnerable point because it takes a lot of movement and the metal frequently fails. This gutter has been lined with bitumen and hessian because it leaked. The fabric can be clearly seen. There is no choice but replacement.
Above right: Splits in asphalt at discharge into gutter.

the contractors movement over a delicate roof surface. The reduction in heat gain following solar paint being applied is substantial and worth the expense in many cases. The reduction in movement in the surface is considerable and should prolong the life of the material.

Parapets

Water penetration due to defects in parapets is far more common than is often assumed because most people attribute water penetration through a roof as being due to the failure in the roof surface.

In many 19th century properties the parapet is faced with a cement render to the inner, top and external faces. Any light fractures in this allow water penetration which during cold weather will expand causing an enlargement of the crack. Water expansion at 0°C is 9% of the volume of the liquid. These minor failures allow water to penetrate which eventually leads to deterioration in the members of the flat roof surfaces.

Finding a damp proof course in cavity wall parapets or solid brickwork parapets is often difficult but their presence will reduce water penetration. The inadequate location of a damp proof course may result in water penetration to the interior of the premises or into the cavity which can condense on the inner face of a cavity wall and cause damage.

In any occasion where the interior of the building has shown indications of water penetration or condensation a careful analysis of the original construction should be carried out so that one can see if there are any of the common indications of failure. These include defective installation and location of damp proof courses, failure to install damp proof courses beneath the parapet coping, the closing of the cavity by insertion of floor slab to the back of the outer face of cavity brickwork. The closing of the cavity by the construction of a floor slab right the way through the brickwork may be finished as a mosaic band or with brick slips. The comments on the risks of brick slips are covered in the chapter on Concrete Failures.

Edge treatments

Where a flat roof finishes with a verge

discharging the surface water into a gutter, failures frequently take place on the edges of the roof. These are frequently the result of differential movement in the construction and may be caused by deficiencies in the original design.

The junction at the head of a brick wall and roof is prone to failures. Sometimes timber boarding is brought up to the brick wall and the junction is made good with cement. An asphalt finish is laid with building felt, or building paper across this joint. The variations in expansion and contraction in the timber surfaces and the cement head to the brickwork are sufficient for fractures to occur horizontally or at right angles to the edge of the roof.

Failures of this type need to be closely examined in relation to the original construction. Where such information exists any repair of the fracture may only be of a temporary nature. It is possible that the construction that has been used may be so deficient that the roof surface will have to be removed and the joint reformed prior to the replacement of the waterproof membrane.

Pipes

Many failures in asphalt or felt roofs occur around the passage of pipes through the surface of the material. Although there may be a sensible dressed skirting constructed around the pipework often a small gap is left between the pipe and the asphalt. Water striking the pipe during rainfall will run down the pipe and inside the skirting. In the absence of flashings having been provided to the metal the careful selection of mastic sealants with a long term life would solve this particular problem.

Careful inspection of the detail which has been used in the dressing up of skirtings around the flat roof should be made. The seal between the head of the asphalt skirting and the vertical construction must be adequate. In all cases this ought to be covered by a metal flashing cut into the vertical surface above the termination of the skirting.

Metal covered flat roofs

All metal roofs have to take into account the high coefficient of linear expansion that the metal has. The material used in the construction of the roof is inclined to move a substantial amount because of the temperature variations that occur every day. This makes it necessary to lay the metal in small panels. The vulnerability of metal roofs is at the joints between the panels. The jointing is accomplished by seamed welts or rolls depending on the material and the method used.

Problems occur when metals are used to form parapet or valley gutters.

It is difficult to carry out a detailed examination of seams to ensure that there is a double lap and the metal is well clipped in their length.

The boarding surface is usually of timber construction beneath the metal. The timber should be laid in the direction of the fall so that if there is any curvature occurring within the boards as they weather it will not impede the discharge of water from the roof. Where lead is used it is vulnerable at all points of curvature because of the continual movement that the metal has been subjected to. Each day sunlight causes expansion and the cool of the evening causes contraction. The majority of the movement is taken up at the joint where the metal is retained. The movement results in a form of metal fatigue and surface crystalisation occurs in the lead which leads to eventual failures.

It is important to carry out a detailed examination of the junction of the flat roof and its discharge into the gutter. In many properties there is an apron flashing to prevent water running back behind the gutter in normal or windy weather. The flashings should project into the gutter. Where they are cut short water can run behind the gutter and dampen the wall.

Where internal failures have been found and it is assumed that the roof surface has failed, it is important to try and locate the area of failure. The point where water shows through the ceiling surface below often bears no relationship to the point of failure on the roof surface.

It is important to be able to carry out an examination of the boarding immediately beneath the roof surface and to do this it may be necessary to use fibre optic probes which give you access to roof voids in timber construction which cannot otherwise be achieved without the removal of the plastered surface beneath the timber joists. These are discussed in the chapters on Surveying Equipment.

Ventilation of the void between external finish and internal decorations should be provided. Felt and certain metal roofs are inclined to leak, irrespective of the precautions taken. The timbers can become very damp without a significant failure being visible to the interior or the exterior. These circumstances become a breeding ground for fungal decay, but the provision of ventilation can reduce this risk.

This page
A lead covered roof. Look at the line of the boarding beneath the lead and how the water is lying in the hollows formed by the twisting boards. These boards should follow the slope of the roof. This roof leaks at the step because of a poor detail at the junction of the cover to the rolls. Splits were also caused by the edge of the boarding stretching and splitting the lead.
Facing page
Top left: A rooflight. The metal flashings around these projections through a roof surface take a pounding. If it is an access to a roof they are frequently damaged. Make sure the glass projects beyond the frame and the water seals are sound.
Top right: A mineralised felt roof correctly dressed into a gutter to stop water running around the back of the gutter. The roof falls discharge water over the edge over the slate roof. Is the zinc vertical surface in good order, and are the flashings sound?
Centre left: The zinc roof is supported on boarding running correctly with the slope.
Centre right: A split almost along the soldered joint in a zinc roof.
Bottom left: A close-up of the zinc roof above shows the perforated condition of the edges of the cover moulding.
Bottom right: A zinc flat roof. The square shaped covers to the joints may be clearly seen. Zinc becomes perforated after only about 15 years so it must be closely inspected for early signs of failure. Note the ponding on the roof. Is this due to any failure in the timbers below? You can see the shape of the boarding below in the zinc. Note that the boarding runs across the slope and not with the slope. This can allow water to remain on the surface.

ROOFS

79

Pitched roofs

In examining any roof surface, it is important that the surveyor considers the condition of the materials and the ease of carrying out a repair. If there are a number of tiles which are fractured or damaged these would be referred to in the survey report. The surveyor should also consider how practical it will be to install sound tiles in place of the existing ones. In certain circumstances roof coverings may be so brittle and fragile as to result in further tiles being damaged by workmen trying to replace those already defective.

In other cases, such as in slate roofs, it is not possible to slip in a new slate and fix it as originally because the nailing point on the batten is inaccessible, due to the presence of those slates fixed above. Where slate roof repairs are concerned, lead or copper ties are hooked over those slates beneath the defective slate and the new slate hung on this wire. This imposes further loading on adjacent slates. The surveyor must be confident that the fixing of those sound slates is adequate to carry this extra load. He must be confident that the failure of the nails securing the slipped slates is an isolated fault. If all the nails are equally old and defective the remaining nails will not hold any replacement slates. The surveyor must consider the condition of the existing roof covering and weigh up the practicality of a repair so that he may reasonably advise when a roof requires recovering or when it can be patched for a further period. In making his judgement in his report it is sensible to set out those matters which have been considered in reaching this judgement. If the surveyor has adequately considered the alternatives and it is plain from his report that he has taken all these matters into consideration, he will not be negligent if his opinion is incorrect, provided his conclusions were reasonable.

Pitched roofs are usually of frame construction often of timbers and covered with a barrier formed of small impervious units. These may be either tiles of a clay or concrete constitution or thin layered stones which are usually slate or limestone.

Vegetable matter is also used as a roof covering and the more common examples of this are thatch and timber shingle. Most of the vegetable coverings tend to be vulnerable in the event of fire. The high insurance premium these roofs carry should be mentioned in a report.

Tile or slate surfaces are fixed onto horizontal battens which are laid across the supporting framework of the roof. The tile or slate can be either nailed onto the battens or hooked over with projections cast into the tile.

In some old buildings, clay tiles will be found which are fixed by using a peg which is located through a hole in the tile and then hooked over the bottom when they are laid. These pegs are sometimes of oak or cast iron and in repair work galvanised nails have been used. In examining a building which has this type of tile fixing, make sure that you have checked the condition of these pegs. Examples of defective cast-iron pegs have been found in modernised homes where condensation forming on the underside of the tiles has led to the rapid deterioration of the ironwork. If you are unable to inspect the pegs comment upon the possibility of their being in poor condition. If they are in poor condition the complete tiled covering to the roof will have to be replaced.

The nails fixing tiles decay over a period of time as a result of the corrosion of the metal aided by condensation forming on the back of the tile or slate surface or by water penetration. When this occurs the phrase "nail sick" is used and individual tiles or slates slip. Many of the repairs which are carried out by inserting replacement tiles or slates in a nailed roof rely on hanging the replacement tile upon the lower tile. As you will appreciate it is impossible to get to the nailing position because this is always covered by two other layers of tiles.

These repairs result in extra weight being placed upon the lower units thus speeding up further failures to the roof. A slate roof surface which is over 40 or 50 years old is liable to suffer continual failures until the surface is stripped off and the tiles or slates are renailed.

Modern sheradized nailing has a longer life than the nails which were used during Victorian times.

When replacement slates have to be

Top right: Exfoliation or lamination of slates. Note the three slates at eaves level which have already had to be replaced. An examination of the back of the slates from inside the roof space will help to determine the extent of the failure. Centre right: When you walk on a slate roof wear flexible shoes and place the ball of the foot on the top part of each slate. Avoid putting your weight on the bottom edge. In this way you reduce the risk of damage. This section is just above a valley gutter so a good view of the extent to which the metal gutter lining has been dressed up under the slates may be obtained.

ROOFS

Roofing tiles

The machine-made roofing tiles imported from Holland in the 1880s to the 1920s have been found to be a particularly brittle type. These can be identified from the shaling or lamination of their surfaces and from the failure of nibs or the tiles themselves where they have fractured or fallen apart. These tiles have a maximum life of some 60 years and where a roof of this type using this material is seen, the advice must be that recovering of the roof will be anticipated in the immediate future.

The sand-faced concrete tiles frequently produced by firms such as Marley have been found to be wearing extensively where these are of the 1930s age group. In many cases, future failures are anticipated because of the limited amount of tile which is remaining.

The 1960s sand-faced concrete tiles have now been found, in many cases, to be devoid of facing and rapid wear of the exposed surface of the concrete is anticipated and one would assume that a further 15 years may see the final deterioration of these tiles.

Top left: A workman's mark on a lead ridge section may indicate the age of the roof (1811) – or the number of visits that have been made in order to keep water out over the years!
Top right: The position of the lead soakers may be seen in this partly completed repair. These will be covered with a stepped flashing. Note how loose the bricks are below the coping and the absence of a dpc below the coping.
Centre left: A dormer projection through the roof surface. The lead flashings are well dressed over the roof slope but have split in places.
Centre right: A lead valley between slate slopes. Not always accessible they are often carrying a lot of water. Check for any crystalisation in the lead and for dampness on the timbers inside.
Bottom left: The parapet is too small to allow a finish to be constructed. The movement in the timbers (caused by rot) has split the filler and has allowed more water in. The absence of a horizontal damp proof course caused the problem in the beginning by allowing the support timbers to become wet.
Bottom right: A cement filler at the head and side of a roof slope. This may indicate the absence of metal flashings. The cracks result in water penetration.

ROOFS

Right: Underside of a tiled roof showing the torching.
Below left: Movement on a new ridge and hip tile. Has the hip tile been supported by a metal bracket at eaves level? Is the new tiling too heavy for the original roof construction?
Below right: Render failures on chimney stacks to this 'between-the-wars' semi-detached property.

secured to the lower slates, this should be carried out with stout copper wire and not with lead slips. These lead slips are of insufficient strength and tend to fail after a short period of time. (These strips are called tingles).

Where the tiles are provided with nibs which are hooked over the back of the batten they are usually nailed every three or four courses to provide extra stability.

After a period of time these nibs themselves can fail and this can be clearly seen if the underside of the roof is exposed from the loft. The replacement of these tiles is easier as it is a case of lifting upper rows of tiles and sliding in the replacement which will then hook over the battens. The examination of the roof surface from the interior to look for sections of daylight and from the exterior to locate those tiles which are missing is a moderately easy exercise.

Where the underside of the roof surface has been lined with felt under the battens close examination of the underside of the roof covering is not possible. If this roofing felt is tapped you can often hear any of the nibs which have failed which may be lying on the felt running down the inside of the roof. This is often an indication of the condition of the nibs on the roof tiling.

In certain areas of this country the underside of the tiles were pointed to prevent water penetration caused by driving rain which can be driven back up underneath the tiles and into the roof space. This pargetting as it is called was carried out in a soft mortar mix usually including lime and most places where it is still found there will be failures in the finish.

When this was originally installed it was to prevent driving water penetration and consideration should be given of any problems that may occur if the pargetting is now in poor condition.

Many modern tiles that are installed provide only a single covering of tile. They clip on to the adjoining tile located to the side and the head. This means that the roofing felt to the underside becomes a very important part of the construction as a considerable amount of water will run through the tiling and drop onto the felt. It is therefore important that the felt is well positioned, particularly at eaves level to allow this water to discharge without causing deterioration to any timbers in the roof.

On one occasion the felt had been missed from the original roof construction and when the original clay tile surfaces were replaced with a modern concrete tile felt was nailed to the underside of the rafters. This meant that the rafters were consistently in contact with water and dry rot occurred towards the lower levels as a result of deficient weathering and flashings.

In view of the fact that the rainwater which strikes a roof surface will flow down to discharge into the gutter any tile failures close to the lower section of the roof will cause more damage because of the greater amount of water flowing over them . In many survey reports great stress is made of the defective pointing to the ridge tiles to the head of the roof where very little water penetration can occur. The replacement of the pointing to the surface of the ridge will often result in greater damage being caused to the tiled or slated surfaces than the benefit of the pointing would merit.

The great vulnerability in pitched roof surfaces occur where there are openings or projections through the roof surface. Most chimney stacks are cut through the tiled areas and the position of flashings and the method of construction should be closely inspected. The rear of chimney stacks which are often finished with a small gutter section are almost impossible to examine and are regularly a point of failure to a roof surface.

In many terraced properties the pitched roof is set between raised party walls which act as a fire barrier to prevent the spread of fire from one building to the other.

The junction between the tiled or slated surfaces and the raised party walls has to be adequately flashed to prevent moisture penetration.

This flashing must allow a limited amount of movement as the timber roof construction will move as a result of variations in moisture content in the timber and as a result of the expansion and contraction caused by thermal variations. A rigid joint will fail and water penetration will result. This joint is often finished with a cement rendered fillet. If there is no under protection in the form of metal soakers failures will occur.

Where the original tiled or slated surfaces are replaced by modern concrete tiles consideration must be given to the fact that the concrete tiles are heavier than the original roof covering. In many cases the original timbers which support the roof are of inadequate strength to carry the extra load, and unless they are either replaced or upgraded to carry the extra weight deflection will occur which will put strain on any junction between the roof surface and other vertical finishes.

On one occasion the extra weight of the replaced roof covering of a pitched roof put pressure upon a pillar of brickwork between two first floor windows in a terraced property and the brickwork failed. The descent of the brickwork upon the vehicle parked in the street did little to endear the householder to his next door neighbour whose prized possession had just been provided with a ragged sun roof.

ROOFS

Left: Rot in pole plate beneath centre valley gutter.
Centre left: The replacement of roof coverings must be linked with a strengthening of the roof framework. Do you know if this was done?

Sizes of slates and their names

Empress	660 × 406 mm (26" × 16")
Princess	610 × 356 mm (24" × 14")
Duchess	610 × 305 mm (24" × 12")
Small Duchess	559 × 305 mm (22" × 12")
Marchioness	559 × 279 mm (20" × 10")
A Wide Countess	508 × 305 mm (20" × 12")
A Countess	508 × 254 mm (20" × 10")
A Viscountess	457 × 229 mm (18" × 9")
A Wide Lady	406 × 254 mm (16" × 10")
A Lady	406 × 203 mm (16" × 8")
A Wide Header	356 × 305 mm (14" × 12")
A Header	356 × 254 mm (14" × 10")
A Small Lady	356 × 203 mm (14" × 8")
A Double	330 × 178 mm (13" × 7")
A Wide Double	305 × 254 mm (12" × 10")

Colour of materials

Natural Welsh Riven Slate	Blue grey
Westmoreland Buttermere	Fine light green
Buttermere	Olive green (coarse grained)
Kentmere	Deep olive
Delabole Random	Vary in colour from rustic red to grey-greens

Lead

Sheet lead in good work should have the following weights and dimensions:

	Weights	Dimensions
Flashings	4-5 lbs	0.067-0.084" 1.7-2.5 mm
Hips, ridges & small gutters	5-6 lbs	0.084-0.1" 2.1-2.5 mm
Soakers	3-4 lbs	0.05-0.067" 1.3-1.7 mm
Lead slates	56 lbs	0.084-0.1" 2.1-2.5 mm
Damp-proof courses to cavity walls	3-4 lbs	0.05-0.067" 1.3-1.7 mm
Flat roofing (no traffic)	4-6 lbs	0.067-0.1" 1.7-2.5 mm
Flat Roof (with traffic)	6-7 lbs	0.1-0.12" 2.5-3.0 mm

CENTRE VALLEY ROOF

The chimney stacks in these two illustrations lean. Many surveyors are familiar with the tendency of some stacks to lean towards the south. This is believed to be caused by one side of the stack being drier than the other as a result of prevailing winds and the sunshine from the south. A stack which has an incline in excess of 75-100mm (3" to 4") should be considered as requiring reconstruction and the appropriate recommendation made in a report. Do not accept the obvious cause of a failure in each case, but delve beyond the surface to make sure that the incline of the stack is not the result of the removal of the chimney breast to the inside and the consequent inadequate support maintained to the stack. This was the case in the right-hand photograph.

The brickwork in the foreground of the left-hand stack will have to be rebuilt. Once it has deteriorated to this state it is beyond repointing.

83

ROOFS

Left and below: Long straw roof thatch and detail. Note wire netting.

Thatch

Until the 17th century, thatch was the most wide-spread form of roof covering used in Britain. Its use continued after that time in areas where there was not a ready supply of slate or stratified stone which could be used as roof covering.

For centuries, the most common material used for thatching was the straw left over after the harvesting of wheat, rye, oats or barley. These were called "long straw". Rye was the favourite, but wheat straw was the most generally used. Oat and barley straw, which are more brittle, were only used as a last resort.

As a result of modern harvesting methods, good long straws have become harder to find. Modern wheat crops tend to be shorter in the stem than the older long-stemmed varieties. The difficulty of obtaining this straw has placed a greater dependence on the Norfolk reed and Devon (combed wheat) reed. The Norfolk reed (*phragmite communis*) is regarded as the finest material for thatching. It grows at its best in salt water marshes and has a length of between 1.2 to 2.4 m 4-8"). It should be harvested immediately after a frost. The leaves are removed from the main stem and harvesting takes place between late December and late February or early March. The reed is harvested annually.

Combed Wheat reed or Devon reed thatching has been carried out using straw and was pioneered in the south west of England. The straw is combed to align the stems and is then laid in the same way as Norfolk reed with the exposed ends bristling outwards so that the water runs off the reed tips.

Heather and sedge have also been used in roof coverings. Sedge is always used to form the ridge in Norfolk reed roofs. This is a marsh plant with a rush like leaf. Heather has been used in more remote districts particularly in Scotland and areas of Dartmoor.

The spars or staples are usually cut from hazelwood.

The method of thatching varies slightly depending on the materials which are going to be used. The roof must have a pitch of at least 50 degrees and the rafters must overhang the walls with a tilting fillet at the end of the joists. Battens are fixed horizontally across the roof in much the same way as for tiling and at intervals of about 125mm (5") at the eaves and about 225mm (9") over the remainder of the roof.

Thatching in Norfolk reed

For thatching with Norfolk reed, the reed when delivered to site is usually graded in to long, short and coarse categories. The short reed is needed for the eaves and barges and the coarse for general work. The reed is tied into bundles of approximately 300mm (12") in diameter and these are levelled so that the cut ends are at the same point. Work begins at the eaves, the lower lines of thatch bundles are tapped on a board to give the angle of about 45 degrees to the thatch in the bundle. These are then placed at the eaves and secured to the roof.

Fresh bundles are then laid to overlap the top part of the lower bundles and then dressed by striking the butt end so that the upper sections of the bundle are pushed up the roof surface leaving a regular surface of the cut ends of the reed. Norfolk reed will not bend over the top of the ridge. The upper layers of reed are cut and horizontal reed is laid along the ridge line to support a ridge section, which is laid over the top using a more pliable material, usually sedge. This is often finished with decorative curls and patterns which frequently are the individual signature of the thatchers.

Thatching in long straw

Thatching using combed wheat reed requires the reed to be soaked from their top end prior to installation. Once again thatching starts at the eaves and works its way up to the ridge. The ridge is based on a tight packed roll of reed about 100mm (4") thick. The top sections of reed are folded back on themselves from the roof slope unlike Norfolk reed where they are cut at ridge level. A ridge capping is then formed in the reed and this capping is usually smaller than that seen on Norfolk reed roofs.

Long straw requires the best preparation before it can be laid. It is usually spread out in layers which are thoroughly wetted and then left to soak because it is important that the straw should be pliable prior to its installation. The straws are then separated and tied into bundles which are thicker at the bottom than at the top. The straw is

ROOFS

approximately 900mm (37") long and the yealms which are these triangular sections of prepared straw are approximately 450mm (18") wide and 125mm (5") deep. The work begins at the eaves and is usually finished in sections working through the full height of the roof. The main differences with this type of roof is that with the reed roof it is the ends of the reed which are visible whereas with a straw roof it is the surface of the straw which lies on the face of the roof giving it a shaggier smoother finish. Wire netting is always necessary to protect a long straw roof from birds and this type of roofing has a life expectancy of up to 15 years. It will probably require re-ridging every ten years and patching on frequent occasions. It produces a softer roof surface than the precise lines imposed by reed.

Life expectancy

Reed roofs should last 40 to 50 years and a combed wheat roof will last approximately 25. It is very difficult to distinguish between Norfolk reed and combed wheat reed and the main difference is the treatment of the ridge. With the Norfolk reed the ridge would be in sedge whereas combed wheat reed is used for both ridge and roof surface.

Differences also occur in the treatment around the windows and eaves in a combed wheat reed.

In a proper state of repair thatch will never be wet beyond an inch or two of its thickness and the original colour of fresh straw reed is revealed underneath when the roof is stripped for repair to the thatching.

Check list

The surveyor should make sure he advises on the extra insurance premiums and the greater hazard of fire when he is reporting on property which has this form of roof covering. In 1212 it became compulsory in London to treat the surface of thatch with a coat of whitewash to reduce the risk of the spread of fire.

It should be remembered that thatch is a relatively light roof covering having an approximate weight of about 7 lbs per sq. ft. (33 kg per square metre) and should be compared with slates which weight approximately 9 lbs per sq. ft. (42.50 kg per sq. metre), and concrete tiles of around 18 lb and clay tiles of around 13 lbs per sq. ft. (85 kg and 61.50 kg per sq. metre).

The timbers for thatched roof supports are similar to conventional roofing being rafters of 100 m × 50 mm (4" ⅛ 2") but with more solid battens than are used for tiled roofs. For thatched roofs they are usually 25 mm × 50 mm (1" × 2"). The life span of a thatched roof will be dictated by the roof design. A straight roof with no valleys or dormers projecting through its surface will achieve the longest life spans noted above. If the thatch is of inadequate pitch or there are dormer windows, or roofs draining onto each other then its life will be reduced. The Surveyor must note these factors in advising his client on the life expectancy of a particular thatched roof.

Top: Two views of a long straw roof and porch. The straw lies flat on the surface giving a softer line. This type is very popular with our feathered friends for nesting material so it is essential that it is always protected with a net, preferably wire.

The design of the porch shown in this photograph is not good practice. Water will fall from the main roof onto the porch below, causing damage to the porch thatch which will become worn. The porch thatched roof will have half the life span of the main roof covering. The surveyor may want to recommend that the porch should have a tiled roof as opposed to a thatched covering, unless the householder wants to replace the thatch every ten years.

Left: The sharper lines of reed thatching.

Roof pitches

Large slates 22°. Ordinary slates 26°. Small slates 33°. Plain tiles 25°. Shingles 26°. Cedar 45°. Oak plain tiles 45°. Thatch 50°.

Shingles

Shingle are thin slabs of wood used to cover roofs and walls. They are extensively used in North America and are comparatively rare in this country. Where they do occur they are usually oak and occasionally elm and teak. They were split or rent and the hardwood slabs have to have holes drilled in them to receive the nails. Those in use now are cut from western red cedar imported from Canada. This has now superceded the original use of local timbers. Western red cedar is a very durable light-weight material with straight grain and reddish brown colour which becomes silver grey when exposed to the weather. It shrinks slightly less than other soft woods and is reasonably resistant to insect attack. Cedar shingles are either sawn or split. Neither close boarding nor roofing felt should be used where timber shingles are installed to ensure that there is free circulation of air around them. It should be noted that oak shingles are often fixed to close boarding but they are an exception. Shingles are laid in random width, those wider than 10 inches should not be used.

As the shingle's tendency is to curl it is more likely to split if it does bend, therefore it is nailed twice. They are not nailed at the head, as with a conventional slate or tile but are nailed through the middle (one batten below the top of the shingle), and a second time just above the line of the end of the shingle in the layer above.

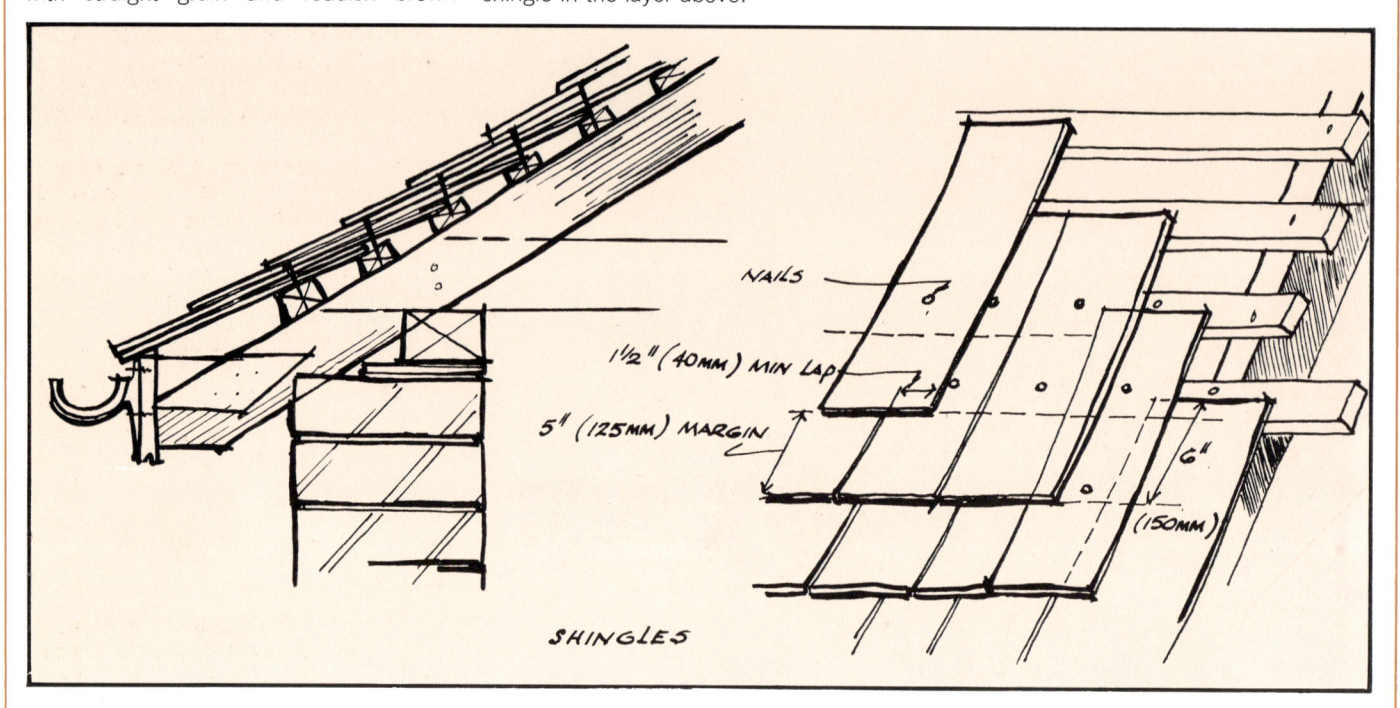

SHINGLES

A typical shingle gable.

ROOFS

Asbestos cement corrugated sheeting

This form of roof covering is extensively used in industrial buildings but has been used residentially to recover roofs which are largely concealed from view. Reference should be made to the section on asbestos which deals with the health risks in cutting and trimming asbestos based products.

Roof coverings made of this material are durable non-combustible and light in weight. The average weight of asbestos cement sheeting is approximately 3½ lbs per sq. ft. (16 kg per sq. metre). Sheets are secured usually to a metal framed roof either using hooked bolts or by screws. The bolts are secured by the use of asbestos washers and it is important that the head of the bolting be protected with a cover to prevent corrosion and deterioration in the metal.

Careful examinations of the surface of the asbestos should be made particularly to the head of the corrugation as there are a substantial number of defective sheets which have been used over recent years. Hairline fractures along the top of the corrugation are an early indication of failures and it may be necessary to carry out some research as to the manufacture and any problems that that company may have had with its product at the time that the work was carried out.

It should also be remembered that the growth of moss or lichen to the surface of the asbestos is not recommended as it tends to cause fractures within the surface of the material and lead to failure.

Chimneys

You must bear in mind that whereas old coal fires really did not do much harm, even after the parging had fallen off, such fuels as gas and oil cause rapid breakdown in the parging and then in the mortar joints. Therefore it is essential that flues for gas or oil-fired appliances should be lined.

Kopex linings were much used twenty or thirty years ago, some are still in use today. Do remember that where a flue is regularly used for, say, a heating boiler, then these Kopex linings only have a useful life of about fifteen years, after which they disintegrate. If you see patches of damp following the line of a boiler flue then this may well be the trouble.

You cannot always tell if a flue is lined but if you see a single flue stack serving, say, a boiler standing up and over, like a banana, then it is a safe guess that the flue is not properly lined.

When inspecting a balanced flue boiler make sure that the installation complies with the current Gas Regulations. Check that the outlet is properly guarded and that it does not impede and side path.

Ventilation must be provided to a boiler room irrespective of the fuel used in the boiler and whether or not there is a conventional flue or a balanced flue.

Above:
Corrugated asbestos cement roofs. The concrete gutters often leak at the joints. Lichen can do much damage to asbestos. The presence of asbestos cement should now be referred to within a report. Concern over the affect of the asbestos content upon the health of anybody who comes in contact with it means that great care has to be exercised in the removal and disposal of such sheets. Extra cost is incurred in carrying out remedial work.

Lichen on a roof surface

- Protective fungal tissue
- Green algal cells
- Tightly woven fungal hypae
- Anchorage hypae

ROOFS

Below: Bubbling on the felt enclosure to a water tank because there is no tank cover and condensation is dampening the back of the felt. The tank is against a party wall on the left. There is an inadequate fire barrier between this element and the adjoining house. The bitumen on the wall and the flashings to the base are poor. If the tank fails the water flows through the house!

Right: Slipped Welsh slates on a bay roof. Note the surface exfoliation that has just begun to occur. Extensive corrosion of the gutter can also be seen. The finish to the hip is poor with no metal flashing.

In the examination of the interior of roof space there are a number of items which must be carefully checked:

Ventilation

Make sure that there is adequate through-flow of air to remove the moist damp air which builds up within the roof space. If the roof void is completely enclosed and there is a felt lining to the underside of the roof covering, check the base of each of the rafters for deterioration. Water which settles as condensation on the underside of the roof covering will run down to the base of the slopes and can cause fungal decay or failure to the timbers.

In modern properties where ventilation is provided at eaves level make sure that any insulation provided within the roof space has not blocked the ventilation which has been provided.

Diagonal Bracing

In modern houses plated roof trusses (or trussed rafters) are frequently used. Make sure that adequate diagonal bracing has been installed to these trusses to prevent their racking. In the event that no diagonal bracing has been provided recommend that it should be installed as a matter of importance. Failure have been found in buildings which are 14 years old where no previous indication of failure had been seen.

Insulation

In checking the thickness of the insulation located within the roof space make sure that the cold water storage tank is

Above: A zinc flashing protecting the upstand on a felt roof. The zinc has corroded and dropped away from the cement filler placed over the junction with the wall. At some stage an application of hessian and bitumen has been applied to mend the hole. The whole section will have to be stripped off.

Right: An asbestos cement roof on an industrial unit. Do not walk on this type of roof as it can be incapable of supporting a person's weight. Check the bolts which secure the asbestos. Check the asbestos for splits.

adequately protected and the service pipes and connections thereto. Make sure that any wiring which is situated within the roof space lies above the insulation and not below it or within it. The heat which is built up within wiring placed beneath insulation may result in the insulation to the wiring failing with a consequent fire risk.

Strength

The examination of a roof void frequently reveals that this area is used as the warehouse to store all the unwanted items of the household. Many of these items are very heavy and are balanced upon rafters and joists within the roof space with scant regard for the capacity of those timbers to support the weight. Check the supports which have been provided to cold water storage tanks located within the roof. In a domestic house with plated rafters these must be supported over at least three trusses. The average cold water storage tank located in a domestic roof space has a weight equivalent to six or seven men.

Condition of Roof Surface

The examination of the interior of a roof void without the use of a torch will indicate the amount of daylight which is filtering

Top left: Water tank cover. External tank enclosures must be closely inspected. The fibre board on this cover has failed and small pieces of it have dropped into the tank. This can block a water system.
Top right: Blocked gutters and general debris like this are usually an indication of the 'care' someone takes of his property.
Above left: A lead flat roof panel installed with the use of too large a sheet. The lack of joints to allow expansion has resulted in fracturing. This one has been badly repaired.
Above right: Water tank enclosures must be fully inspected. If that tank fails will water enter the building? Is the water contaminated? Is the tank adequately insulated?

through faults in the roof's surface. As water seems to be able to get through holes that light cannot, some water penetration will occur where light is able to penetrate the roof covering.

In examining water storage tanks and their enclosures where they are located in an exposed position, make sure that you have referred to the condition of the insulation around the tank. The cold winter of 1984/85 resulted in the freezing up of many cold water storage tanks located on roof surfaces. This strained the seams to the underside of galvanized tanks and split some of the plastic tanks located on roof surfaces.

Inspections must be made of these seams or in the event where no access is possible, the risk to the seams should be commented upon in your report. There is no indication as to what is an acceptable level of insulation but it should be in the region of 150 mm (6") of fibre glass quilting or up to 200 mm (8") of loose-fill. Water tanks in unoccupied property in central London had a covering of ice even with 100 mm (4") of quilted insulation protecting the tank. Consider how much colder it would be in an exposed position.

Check the alignment of the overflow pipe. These should have an even fall, and no loops in which water is trapped. This water will freeze in cold weather, and in the event of a valve failure the pipe may become blocked or split.

Make sure the pipe joints are sound on an overflow pipe. Sometimes the plastic sections of pipe have not been adequately sealed.

The repair of splits in an asphalt roof surface requires the asphalt to be cut away by up to 500 mm (1'8") from the crack. Where the roof surface is extensively crazed, it will be clearly seen that this will mean the replacement of the complete roof surface. All patching to an asphalt surface must provide an adequate step in the two layers of asphalt in order to retain the integrity of the roof surface and prevent further faults occurring. If it is not possible to carry out a repair in this way the complete asphalt surface would have to be condemned.

Top: A felt surface applied in place of the original metal finish. The felt may be vulnerable to failure at the steps. Note the omission of a metal flashing over the upstand against the party wall.
Centre: A wood-wool slab used to support a flat roof which leaked. The slab expanded and now resembles a wheat field! The finish around the roof does not look as if it was adequate and this is where the water has come through.
Bottom: A split through the surface of a rendered coping. Water will go through this crack and will dampen the brickwork. How many timbers are there supported in this wall which can be damaged?

ROOFS

Above: Splits in asphalt which have been treated with a bitumen coat in an effort to stop water penetration. If another layer of asphalt is applied the original coat will bubble and weaken the new coat immediately above the weak point.

Above right: Hip tiles on this roof have no support at the bottom. On bigger roofs this will cause hip sliding. The flashing against the party wall is a sheet of felt because the slates bedded into the cement of the party wall are faulty and the owner could not afford the cost of a proper repair.

Right: A slate roof covered in hessian and bitumen. This type of treatment may last 20 years, and with further treatments of bitumen, even longer. There is no ventilation to the roof void – which can cause problems. All the slates will be damaged when the roof is eventually stripped for re-covering – so there will be no salvageable material.

WALLS

Above: Cracks in the render to the parapet wall together with small openings in the posts to the rail around the roof terrace let water into the parapet wall. The absence of a damp proof course lets water pass down into the unventilated flat roof timbers. The result is dry rot. A small failure – but an expensive repair.
Left: The asbestos tiles on the roof surface conceal the finish so that it cannot be checked; but note the damp patch over the area of rot where the roof settled and sank. Always ask why water is ponding on a roof.

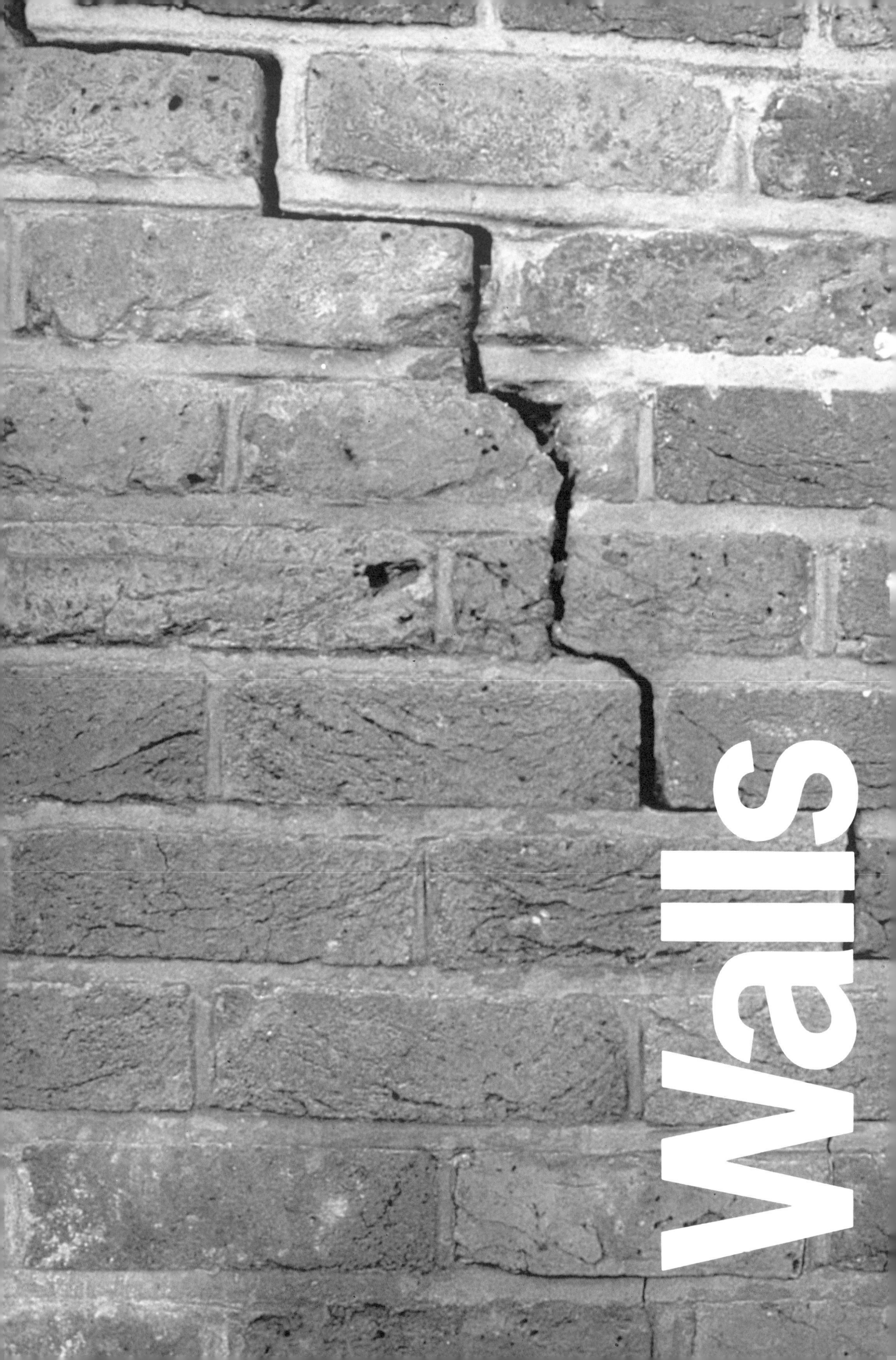

Walls

A wall can serve two purposes:
1. The structural support of the building, or its components;
2. The enclosure of the interior, a protection against the elements.

It is important to identify the functions that it is performing in a property that you are inspecting.

Brickwork

Before we go into detail about the inherent failures in brickwork it is important to carefully check on the thickness of the main external walls throughout the property that one is looking at. It is common within large cities to find many external walls of a solid brick construction being only 9″ thick. If this wall has been built using a porous brick there is a risk that it will have allowed moisture to be transmitted to the interior of the property, particularly if the wall is in an exposed position. The City Centre locations have protected this type of wall from extreme exposure and as a result they have survived for over a century with only a few problems caused to the interior of the property.

If the wall is of a solid form of construction any defect or failure in the external rainwater pipes, downpipes or other forms of pipework and flashings will result in water penetration to the interior of the property.

In certain locations external finishes have been applied to brick walls to reduce the water penetration and improve their performance. These may include rendering, tile hanging, weather boarding, harling, or similar finishes.

Pointing

A knowledge of the deficiencies of certain types of pointing to brick and stone is of assistance to the surveyor.

Tuck pointing

This is where mortar of a similar colour to the brick or stone work is applied to the joint and grooves are cut into it giving the effect of a joint. Sometimes these grooves are themselves pointed. This type of joint was used where poor quality bricks were in use and their irregular appearance would not have left the neat and precise finish required by the designer of the building. Because the mortar covers the face of the brick there is a risk of water penetration to the rear of the joint.

Recessed joints

This is not a method of pointing, which is the application of a fresh finish of cement to the joint after the brick has been laid, but part of the original finish in the laying of the brick. The recessed joint was used frequently with precise sharp-edged bricks as a design feature.

Top: Efflorescence on the face of brickwork. This is caused by salts being deposited on the brick face by the drying out of the moisture in the wall below the damp proof course.
Left: A lead flashing above a bay roof bressummer. The brickwork fracture was caused by the bending of the timber beam.

It results in water being held on the top face of the brick where the ledge is created and in some cases water penetration has taken place. Where the bricks are very dense and impervious considerable problems have occurred with substantial water penetration taking place even in buildings which are in sheltered locations.

Struck joint
This is where the finish to the joint is formed by the mason taking his trowel and striking an angled line in the mortar joint. It is the most efficient of the weathered joints in that water is thrown away from the joint by the angle of the pointing.

Bucket handle joint
This is formed by rubbing a round ended piece of wood over the joint after laying the brick giving a curved surface. Some water penetration has taken place in exposed locations with this type of joint where water has been trapped or held at the top end of the joint where the pointing reaches the face of the upper section of brickwork.

Rag joint
There are various different names given to this joint which is formed by rubbing the face of the brick and joint with a piece of sacking or similar material. The pointing ends up virtually flush with the face of the brickwork and there are instances of penetration occurring in much the same way as takes place with the tuck jointing referred to above.

The examination of brickwork and the joints should be carried out with great care. The surveyor must be looking for minor cracking between the joints and the brickwork which may indicate that in exposed locations, water can be driven into the brickwork or across a cavity.

*Above: If a damp proof course was inserted after the building was built was the internal plaster stripped off to a height of at least 1 metre? Is the new plaster the correct mix? Right: The external evidence of a recent installation will be a line of injection holes. Make sure that they provide a vertical barrier to prevent lateral penetration over the new dpc. The holes which have been drilled to insert a chemical injection damp-proof course to this wall are too far apart. The recommended spacing is 112 mm (4½") and this may mean that the damp proof course will not be effective. The holes in the vertical damp proof course at at 450 mm centres (18"). There is no way this can perform effectively.
Below: Check the lintel. The timber bressummer over these windows was rotten, but this was not evident until the removal of the fascia. Timber beams like this often support two or three floors over bay or shop windows. If you know it is there you can advise on the risk of future failure of a bay roof or a flashing.*

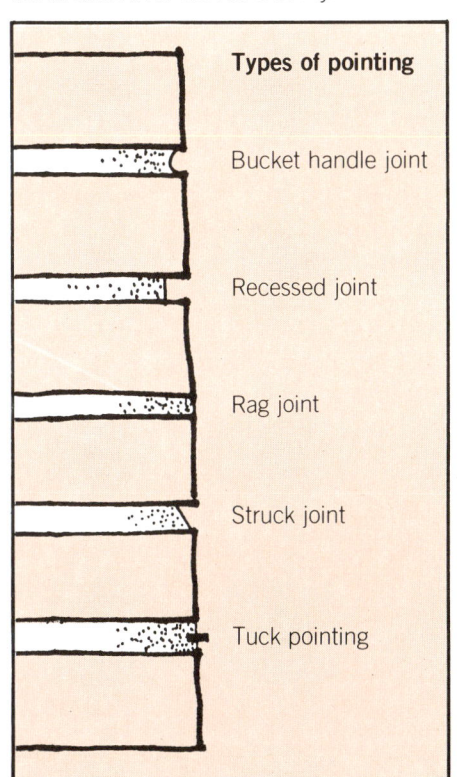

Types of pointing

- Bucket handle joint
- Recessed joint
- Rag joint
- Struck joint
- Tuck pointing

WALLS

Above left: Those bricks with holes in are supposed to ventilate a property. Blocking them increases the risk of rot if there is damp, or of condensation. The damp proof course – the wider mortar joint – is only one brick above the path. Make sure that it is not bridged.

Above: In this case the damp proof course is at path level and will be bridged by growth of vegetation.

Left: A grille should be provided for under floor ventilation. The absence of one allows free entry to vermin. Leaves and other debris could also enter and build up and rot, possibly linking the floor timbers with their dampness. That cement plinth looks as if it covers the brick course and damp proof course. This will allow water to work its way round the damp proof course.

Render

Render surfaces used to consist of a lime, sand and cement mortar of weak mix that has a flexible surface. Where renders have been applied without lime there is a risk of excessive shrinkage which produces the familiar crazing to the surface. Where this is in exposed conditions water penetration will occur which will get to the back of the render surface. A consistent wetting and drying of the render and brickwork to the rear may result in salts being brought to the surface of the brickwork which will push off the render panels.

A sound render will allow some water penetration but the water will discharge through the surface because it is able to breathe.

Pebble dash finishes are inferior to a render finish and tend to require much more maintenance.

Because the render is a monolithic surface it tends to show up any movement or structural failure in the property to which it has been applied. Where access can be obtained the surface should be tapped to try and find out if there are any areas which have lost its key or adhesion to the wall. The hollow sound which indicates the defective areas may soon be recognised after a little practice.

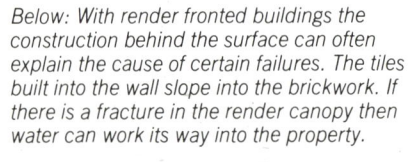

Below: With render fronted buildings the construction behind the surface can often explain the cause of certain failures. The tiles built into the wall slope into the brickwork. If there is a fracture in the render canopy then water can work its way into the property.

WALLS

Far left: The appearance of a 'roof garden' indicates dampness in the parapet resulting in cornice failures.
Left: It is probable that there is no damp proof course in the parapet.

Balconies

Make sure any repairs that may have been carried out to external render finishes or balustrading are in the correct materials. The use of gypsum plasters is not going to achieve a long-term repair – as can be seen here.

97

WALLS

Tile hanging

Tile hanging can produce a more durable and weatherproof external finish than cement render. Defects which do occur result from the trapping of water to the rear of the tiles. It is important to ensure that there is a tilting fillet placed at the bottom of the panels of tiling. Where this tilting fillet is timber there is a risk of decay unless it has been pretreated.

On no account should tiling be allowed to come down to ground level or the ground level be raised so that it covers the bottom edges of the tiling. Tile hanging must be checked for missing tiles, and for defects of deterioration in the method of fixing.

Water penetration frequency occurs if there are any failures in design or the construction of openings or edges to tiles areas.

Bottom edges must be formed in a double course of tiles. The joints with window or other edges must be formed with metal flashings. Sometimes the exposed edge of a window opening is formed with a render or cement pointing. If there is no metal flashing this must be regarded as an inferior finish, although in good condition this construction has served for over a hundred years without problems.

Lead is the most durable of the metals used in flashings. It is prone to splits caused by excessive thermal movement or failures caused by poor workmanship, often in soldering joints. (Its high scrap value can shorten its life much quicker than the elements.)

Zinc has a limited life of up to 15 years unless protected. Copper is vulnerable to reaction with cement and rainwater in industrial locations. Most of the aluminium foil adhesive materials cannot be regarded as a permanent choice.

The condition of apron flashings beneath window sills should be checked and comment made as to whether the window sill is of a single piece of wood or two pieces. As you may appreciate where there is tile hanging on the exterior of the building it is necessary to provide extra wide sills to project beyond the brick opening and the tiling on the outside. Where these sills are in two pieces the joint is vulnerable to failure if the paint surfaces are allowed to deteriorate. Water can get caught within the timber causing the sill to decay.

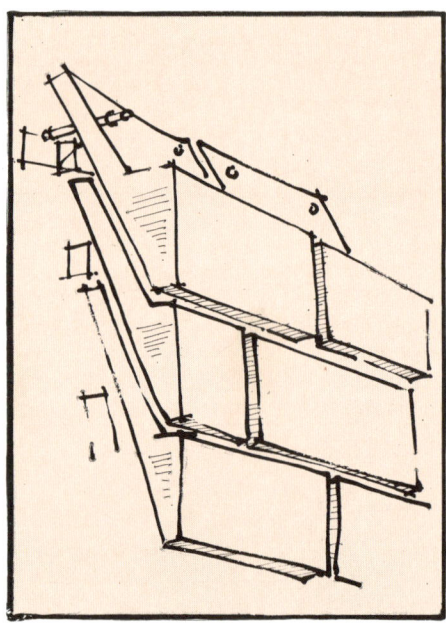

The use of mathematical tiling can produce a deceptive effect. What looks like a solid brick wall may not be!

Mathematical tiles

There are certain forms of building construction which fall into the category of 'embarrassments' looking for someone to alight upon. There are a number of external brick finishes which do not accurately reflect their construction, snapped brick headers in a 4½" external skin which most surveyors will assume to be a 9" or 13½" solid wall is just one example. Another pitfall which occurs is Mathematical Tiling.

The Tiling has no failure in the general 'brick' like appearance to alert the surveyor to the actual nature of the construction.

These tiles are supported upon a wooden framework which is battened out in the same way as conventional tile hanging, but with these tiles each end is a piece of clever design, interlocking with its neighbour to look like brickwork. The finished tiling is pointed up to confuse you even more. The wall thickness may be close to a conventional 9" so the wall thickness will not always put you on your guard.

The tile edge may be seen around openings such as doors and windows, but in many cases even these edges defy an examination by anyone other than a suspicious surveyor.

Mathematical tiling occurs on buildings, frequently in the South East of England, built in the 1700 and 1800's.

Cavity brickwork

Up to about the 1920's the majority of the walls constructed in brickwork were of solid construction. But after that date the introduction of cavity walls took place. There are some examples of cavity walls dating from 1848 but these are rare. This is the construction of a wall in two panels which are tied together by wall ties. The cavity prevented the penetration of water from the outer face to the inner skin.

A number of failures have occurred in this construction and these are usually highlighted by indications of water penetration to the inner face. Failures can have occurred for some of the following reasons:

The wall ties may have become dirty or were covered in mortar droppings during the course of construction. This produces a direct link between the exterior and interior faces of brickwork and allows water penetration. This can be investigated by the use of fibre-optic probes which enable a detailed examination of the cavity in the brickwall. Metal detectors can locate each wall tie (see section on Equipment).

The removal of sections of bricks to the outside of the internal failure is the only other method of investigation. The reduced client cost must be balanced with the equipment cost if a fibre optic probe is used.

There may be defects in the flashing around window heads or the location or condition of the damp proof course at lower level.

In some areas where there is a high degree of exposure it has been known for turbulance to be set up within the cavity which has enabled water to be blown across the gap. This is a comparatively rare failure and one which will be discounted in most calculations unless one is carrying out inspections of property in extremely exposed locations.

Moisture penetration may be noted on the inside as a result of defects and failures in the bridging ties. Where extensive corrosion has taken place the brickwork may have cracked at the point where the ties are set into the mortar.

It is usual for the end of the ties in the outer leaf to be far more corroded than that in the inner leaf and the expansion of the corroded metal shows visible defects in the external brick face.

Techniques have been developed in recent years for the reinstatement of cavity wall ties by the insertion of new ties. Where fractures are noted in the external brickwork as a reult of the corrosion in the original tie the brickwork will have to be removed and at that point the new tie can be built in. The removal of bricks in old lime or black ash mortars is not usually a problem. Resin-grouted ties can usually be installed using standard tools and those ties where there is no drip collar on the external face should be set at an angle so that the water runs down towards the outer leaf of the wall. For the detail of repairing methods reference should be made to the Building Research Establishment Digest 257 issued in January 1982. It is to be remembered that the sulphur content in black ash mortars aggravates the position relating to wall ties and speeds up the corrosion. Extreme care should be taken in any survey in a property which has this form of mortar joint.

Surveying buildings

In examining any brickwork the extent of the building's exposure to the elements must be borne in mind. A small fracture in brickwork in a property situated in the centre of town may result in less damage being caused then the equivalent fracture in a property in an exposed position such as a coastal or hill or mountain location.

Inspection of any cavity brick wall in an exposed location should be carried out with great care to try and find any indications of horizontal splitting or fractures in the mortar joints at approximately 450 mm (18") centres. This could be an indication of the failure of cavity wall ties. The alignment of the exterior and interior faces of the walls should also be checked to see if any section of the cavity has spread. This may also be an indication of the failure of the wall ties. At the present time extensive tie failure has been found in the North East of England where the mortar tends to be of a black ash type which reacts chemically with the metal of the wall tie. Many wall tie failures in these areas have occurred in the section of the wall tie between the cavities rather than within the brick joints.

This means that there is no tell-tale horizontal cracking in the joint. The inspection of properties in the area of Leeds, Sheffield and the surrounding districts should be carried out with particular reference to the risk of this type of failure which cannot be diagnosed from a surface examination of the wall.

The replacement of cavity wall ties is not difficult provided the walls are built in conventional brick or block. If the walls are of lightweight, brittle or cellular block repairs need to be undertaken with great care. There are many different types of replacement tie on the market, ranging from expanding bolts, or resin-glued ties to simple helical twist ties which are hammered into position. Each form and system has limitations and the condition and material used to each leaf must be understood before any recommendation is made. As a general guide, it costs around £5 per tie to carry out repairs and a convential three bedroom semi-detached house would require approximately at least 150 replacement ties to stablise the cavity walls.

The examination of the exterior of the brickwork should also record the quality and condition of the brick which has been used on the property. Some bricks are very soft and are prone to failure. When the face of these bricks fails, the inner section of the brick, which is very soft, will erode easily. It is not easy to cut out these soft bricks and replace them with a matching brick without perpetuating the problem. In many cases where the face failure of many bricks has occurred, the building has been rendered. Whenever you are carrying out the inspection of a brick building which has been faced in render, careful consideration should be given to the possible reasons for this treatment having been carried out. Every attempt should be made to try and determine whether the render is concealing defects such as cracks in the walls, poor brickwork, faulty bricks or non-standard forms of construction. Make sure that the render will not cause future problems by having been taken down to the ground. This will result in water rising behind the render and bypassing the damp proof course, and causing dampness to the interior. It may also prevent any moisture in a cavity wall draining out through the wall below the damp course.

Below left: Typical signs of wall-tie failure — cracks in rendering.
Below right: 'Pagoda' effect caused by wall-tie failure allowing outer skin of brick to expand.

Cavity insulation

In the design of cavity brickwork the gap between the inner and outer leaf is intended to prevent water penetration from the outside of the building to the inside. An examination of the inside of a cavity of a brick building located in an exposed position during heavy rain will show a considerable amount of water running down the inside of a cavity. In extreme locations, this will be blown across the cavity and run along the face of the inside leaf of brickwork. Wall ties which cross this cavity may become bridges to transfer moisture from one surface to another. This water runs down to the base of the wall and is then discharged at the bottom through the open joints below the damp-proof course. Traditionalists within the building industry will point out that the cavity is required as an inherent part of the construction in order to prevent failures occurring to the inside of the building.

The application of an insulation material to this cavity must bear in mind the risk of increasing the water penetration by linking the outer and inner faces of the brickwork.

Considerable problems have occurred in buildings which are in exposed locations where the cavity has been filled.

No cavity wall should be infilled where the building is situated in a coastal or hillside or hilltop position, as in each of these cases the prevailing wind will result in moisture being transmitted from one face to the other.

Many modern estates, which are constructed well away from hilltops and the coast, are equally exposed to strong winds as a result of wind tunnel effects being created by the location of buildings. It is not always easy to identify these buildings except in periods of strong winds and the surveyor carrying out his inspection may not always be able to locate the appropriate exposure indices for the building under examination.

As a result of the examination of a number of buildings which have had the cavities filled and where exposure has taken place, it has now been found that there are large pockets occurring in the infill to cavities. These may create greater problems of moisture penetration by water ponding or being trapped around these voids.

The positioning of cavity trays or damp proof courses running between one face of the cavity and the other may also increase the risk of water penetration.

The improved insulation in the wall often means that the external face of the wall is colder in winter. There is a greater risk of water freezing in the outer face of brickwork than there would be in uninsulated cavity properties. Frost action can cause failures to the brickwork.

Foam insulation cavities

The most common foam that has been in use is urea-formaldehyde. This foam is installed by drilling into the cavity through the external skin of brickwork and pumping in under low pressure a mixture of water-based resin solution and a hardener. In certain areas, the expansion of the material which results from the mixing has exerted an outward pressure on the outer leaf of the brickwork. Where cavity wall ties have been in poor condition or there have been inadequate ties installed, failures have occurred in the brickwork.

When the foam dries it shrinks, and this has led to fissures occurring in the foam within the cavity. These provide channels which water percolates through either as a result of the variations in pressure which occur to each side of the wall, or through capillary action.

The chemical reaction in the mixing of these materials releases formaldehyde gas. It is essential that there is adequate ventilation at the head of the cavity to enable this gas to disperse safely. There has been much debate over the injurious effects of this gas and in the United States this form of cavity insulation was banned for a short time.

Rigid polyurethane foam has been used with moderate success to glue together sections of brickwork where cavity wall tie failure has occurred. This is only effective where the inner faces of the brickwork are reasonably clean and dry — conditions which rarely prevail within this country!

Cement render plinths are frequently found at the base of the brick walls. Make sure that they do not overlap the damp proof course or they may allow water to by-pass the membrane. Where these render plinths become loose water can collect behind and cause damp in the wall. This, in turn, can allow frost action resulting in failure of the plinth.

Rock fibre

In this case fibres which are coated with water repellent are blown into the cavity where they form a water repellent mass. The raw material cost is higher so it is generally a more expensive process than the installation of urea-formaldehyde.

Polyurethane granules

These are irregular shaped granules which are between 5mm and 20mm and made up from the chopped waste from polyurethane foam. This polyurethane is combustible and if it catches fire it gives off toxic fumes. It is therefore important to make sure that this form of cavity fill is kept away from warm areas such as those around flues.

Expanded polystyrene loose fill

If polystyrene granules come into contact with PVC coated electrical cables plasticiser can migrate from the PVC leading to the embrittlement of the cable insulation.

WALLS

Left: Face failure in soft bricks, and movement cracks.

Below: The lack of the provision of an expansion joint in this band of brickwork caused the bond separation and lifting in this wall along the line of the damp proof course. Each end of the wall was restrained – the client's comments were not!

Provided the cable is not disturbed the electrical insulation will remain intact and consideration should be given to this risk at the time of an inspection taking place. The material is also flammable and care with its location near hot flues, or similar locations, must be taken.

In situ foam installation polyurethane

Two liquid components are mixed and injected into the cavity where they foam and rise to fill the space.

The foam adheres strongly to masonry and tends not to shrink. It has therefore been used to stabilise cavity walls where wall ties have corroded but for the density required for this type of application the material cost is very expensive.

Performance

Most of the tests indicate failures occur in urea formaldehyde foam and that there is a greater risk of failure in those properties which have their walls insulated during the course of construction than those which have been insulated some time afterwards.

Rock fibre and polystyrene beads have least water penetration; *in situ* polyurethane in intermediate density also showed good results and prevented water penetration at some wall ties which before filling had conducted water across the cavity.

One of the main factors governing failures is the exposure to driving rain. Where these failures exist consideration may have to be given to the necessity to clad external faces of walls or to remove the materials from the cavity.

In new construction care should be taken over the decoration of the inner face of the wall as the cavity fill insulation can retard the drying out process. The wall moisture can only dry inwards. As a result of this the drying process should be extended to around 18 months on all new property prior to any surface application of a skin paint or paper that will prevent the moisture passing through it.

Defects in brickwork

Clay brickwork is inclined to move from the date that it was constructed as a result of it taking up atmospheric moisture. When the clay bricks leave the kiln they are at their dryest point and they then start to take up atmospheric moisture and expand at a high rate initially and thereafter at a decreasing rate. Movement may continue at a slow and diminishing rate indefinitely. A typical brick would expand at the rate of approximately 1mm per metre in the first eight years, approximately half of this expansion takes place in the first week.

There is a small amount of movement which will take place by reversible wetting and drying but this is not likely to exceed a quarter of a millimetre per metre. It is suggested that movement joints should be provided to allow for movement of an expansion of 10mm in every 12 metres of wall length.

Brickwork is also vulnerable to sulphate attack. This occurs where ordinary Portland cement which contains trycalcium aluminate comes into contact with soluble sulphate and water. The reaction causes the expansion of the sulphate. This can cause an expansion in the brickwork as well as the possible disintegration of the mortar joints. Indications of this sort of failure include horizontal cracking on the inner face of the wall. This is due to the inner face being put in tension as a result of the expansion of the outer leaf. This defect may be seen with a fibre optic probe (see chapter on Equipment). The external walls show little signs of damage and may even have hardened slightly. This cracking is usually concentrated near the roof. Where solid construction has been used with different types of brick being used on the inner and outer faces bowing may occur in the walls.

Such failures may often show as projecting brickwork at damp proof level where the brick to the upper part has expanded and used the damp proof course as a bond breaker. The movement in the damper brickwork below the damp proof course tends to be slightly less than that above.

The condensation of flue gases within a chimney stack may result in sulphate attack on the stack itself, if the flue is unlined to the top. This risk should be commented upon when one is reporting on the condition of the property.

Sulphate expansion takes time to develop and is rarely serious in under two years. In this way it can be differentiated from ordinary expansion of brickwork which tends to occur within the first few months after the building is erected.

Where brickwork has already been damaged by sulphate attack it will be necessary to try and reduce the water penetration which is speeding up the deterioration. Repairs can be made to the brickwork when it is reasonably dry.

Soft fired bricks which have failed as a result of spalling caused by the crystallisation of sulphates behind the face may have to be cut out and replaced although a render front to the face of the bricks is often used. Where there is severe failure it may be necessary to provide some

101

form of cladding to conceal the damage which has occurred to the brickwork.

Calcium silicate bricks react in a totally different way to clay brickwork. These bricks are often known as sand limes or flint limes and consist of a uniform mixture of sand or uncrushed siliceous gravel with a lesser proportion of lime mechanically pressed and chemically bonded by the action of steam under pressure. These bricks are resistant to attack by most sulphates but it is the drying shrinkage of calcium silicate bricks which calls for special care.

The brick surface is usually smoother than that of clay bricks but textured facings are now available and this makes it very difficult to differentiate a sand face fletton from a sand lime brick.

The basic colour of the brick being white, off white or cream tends to be paler than that of the clay brick.

The use of this type of brick below damp proof course where they remain fairly wet is no problem as moisture movement is very slight.

Vertical movement joints should be provided above the damp proof course at intervals of 5 metres. Permanent movement joints should be provided by butting the bricks against a suitable separator such as polythene sheeting bituminus felt or building paper.

These joints are always visible and should be concealed with an appropriate mastic sealant.

The absence of these contraction joints will result in fractures occurring in the brickwork.

The front of this Georgian farmhouse had been altered by a Victorian builder. The brickwork suggested that this was a 13½" wall – until it collapsed to reveal two skins of 4½" not joined together, with the inner finish being a stud frame with lath and plaster. There were very few headers running through the wall to tie it together. Not everything is as it seems. (Note the broken rainwater pipe on the corner. When will builders realise how long it takes to dry out a house when this is allowed to happen?).

WALLS

Top: Reinforcement showing through where a concrete sill section has failed. Reinforcement can be seen in the pre-cast concrete sill section of this concrete framed building. There may be a number of causes which can have contributed to this failure: 1) The inadequate cover provided to the reinforcement; 2) Chemical attack, due to the presence of excess chloride ions. These were frequently added to a concrete mixture to speed up the hardening of the concrete. In structural components, it may be worth having samples taken and sent away for analysis so that the chloride content may be determined. Where the percentage of chloride exceeds 1% there is the possibility of accelerated deterioration in the reinforcement. See Appendix 2 on page 62 for a list of laboratories who will test concrete.

Centre: A broken post to a fire escape. The post bedded in the coping has rusted splitting the stone. There is no damp proof course below the stone, so water enters the brickwork.

Bottom left and right: The timber frame of this old building was not visible until the old lath and plaster surface had been removed. It is important to have a knowledge of the construction of buildings of all ages as well as knowledge of how specific house types were constructed.

103

WALLS

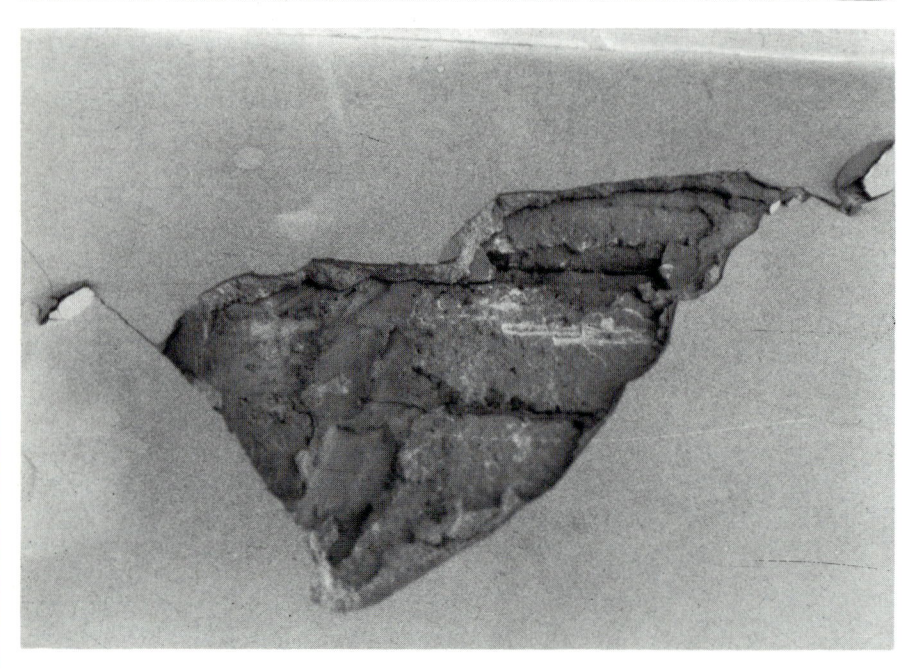

Top: The growth of vegetation will accelerate the breakdown in the render to the decoration and the copings.
Centre: Brick fractures caused by shrinkage in brickwork. The absence of a movement joint has left no point for the movement to take place without a surface failure.
Photograph centre: The brick fractures which can be seen in this photograph have resulted from the shrinkage of the bricks which are calcium silicate. The tendency is for these bricks to shrink within two or three months of them having been laid. Care should also be taken in examining this type of building to check the quality of the fixings of the slip tiles. This is the bottom row of bricks which are thin brick tiles glued to the face of the concrete beam that they are concealing. Unless these bricks are supported on non ferrous ties the slips will fail and drop off the face of the building.
Bottom: Internal plaster failures. Why did they occur? Did the concrete frame move? Has brickwork shrunk or expanded? Is there some settlement? Is the plaster properly applied? Or is there a junction between a concrete beam and brick infill where slightly different amounts of movement will occur?

WALLS

Gutters and downpipes

"We suggest that all gutters be cleaned thoroughly to enable an inspection of the fabric of the gutters and the downpipes to take place"

Examination of gutters and downpipes

The gutters and downpipes of a property carry a large volume of water away from the building each year. A failure in any component of this drainage installation will result in water running down the face of walls backing up into the roof spaces or penetrating in other ways into the interior of the property. In each case the consequent damage caused by the water penetration will be much greater than the initial penetration. It is important that any surveyor inspecting a property takes into account not only the failure which he can see but the likely consequences of that failure. It is not sufficient to report upon the number of cracked rainwater downpipes that one has found within a particular property. One must also comment upon the amount of water penetration that will already have taken place, the dampness of which may have built up within the wall fabric, the timbers which may be in connection or have come into contact with the damp wall surfaces and the possible risk of fungal decay. It is important that the surveyor is always thinking in terms of the projection of any failures that he has found into their natural consequences.

The examination of a building which has new rainwater downpipes installed should record the fact that these pipes have been replaced. This is not a piece of gratuitous information because the surveyor will also point out that this would mean in his own mind that the original pipework was faulty. He would need to elaborate on this by pointing out that the faulty pipework must have been leaking and that some sections of the walls may have become dampened and have subsequently dried out. There will be a risk of timber failures occurring where timber came in contact with these wall surfaces which had become dampened and that some deterioration may have taken place, or may be still occurring, in the inaccessible parts of the building in contact with these defective areas.

The following matters should be considered during the course of the examination of the building:

Gutters

1 Cleanliness
Are the gutters free-flowing and free from debris or build-up of material? Is water lying in the gutters indicating a poor fall?

2 What is the condition of the back of the gutter?
In metal gutters this area is rarely painted and faults occur where corrosion takes places in the metal. If you cannot examine this from a ladder or convenient window draw conclusions from the surface examination and age of the material and make such recommendations as to future inspections as you think appropriate.

Brackets
Many metal gutters are supported on brackets. Are these in good order or has corrosion taken place? Are they too widely

105

WALLS

spaced? Many plastic gutter installations are inadequately supported. If you get an opportunity, test one or two of the brackets to see what condition they are in.

Falls
Are the gutters laid to an adequate fall and will water discharge from the surface? Are the falls regular or is water expected to run uphill to the rainwater outlet?

Joints
Are there any indications of leakage from joints in the gutter sections? In metal guttering orange rust stains are frequently seen around the joints. There may be patterned staining on the face of the gutter fascia board, brickwork or other materials around the joints. In heavy rainfall one may be able to inspect the gutters under actual operating conditions. If it hasn't rained for many months clearly state this in your report as it would not be possible to pick up many of the tell-tale indications of gutter failure.

Quality of materials
Has the gutter installation been completed? Are there adequate stop-ends to gutter runs? Comment upon the deficiences of certain materials, for instance metal, the need to regularly paint the interior and exterior of cast-iron guttering, and the risk of failure. If failure of the guttering occurred what would the natural consequences be? Are children likely to be beneath heavy cast-iron guttering? Are cars parked against

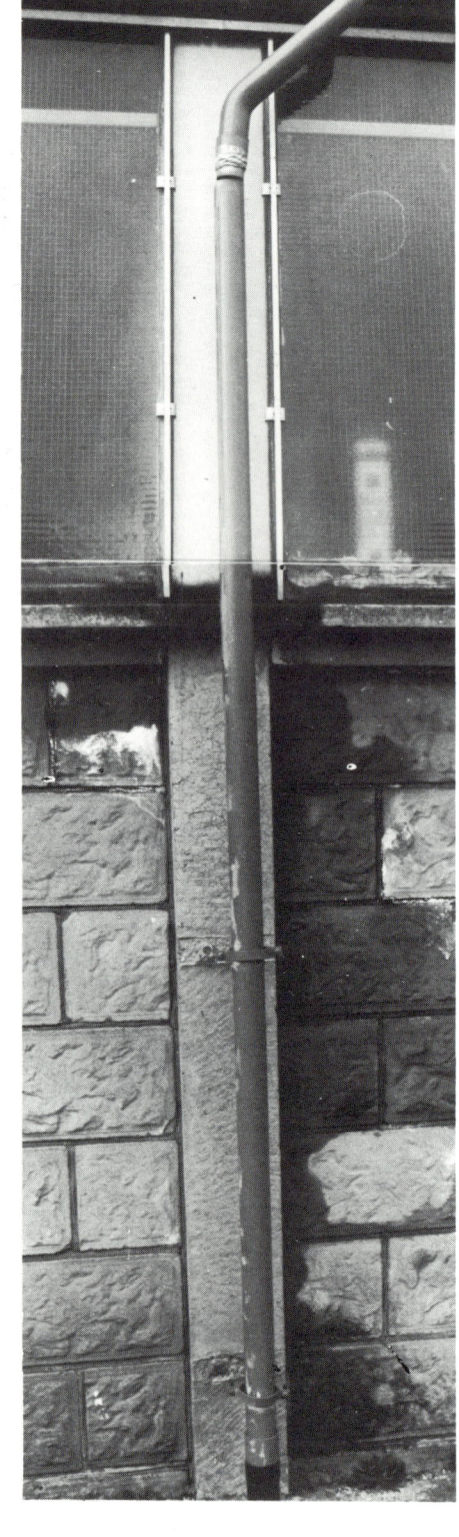

Top left: A taped repair to a lead gutter formed behind a low parapet.
Left: A slipped moulding to the side of a dormer. The metal cover to the dormer is now insecure so that the next high wind may lift it off unless the failure is repaired.
Above: A rainwater pipe on an industrial building. It is correctly bracketed but the shoe has fallen off and the pipe has dropped. Make sure that plastic pipes are securely fixed to the gutter to prevent this type of failure and subsequent soaking of the adjacent wall.

WALLS

areas where gutters drop? Are there glass roofs which are unprotected lying immediately beneath an old gutter installation? Cast-iron guttering of the O.G. type tends to have a life-span of between 40 and 50 years although this does vary. Where guttering of this age is present the risk of failure with a consequent falling of guttering from the building must be contemplated.

Asbestos cement
Asbestos cement guttering is prone to fracturing and faults particularly when maintenance takes place. The placing of ladders against the edge of the gutters may cause damage. They are also prone to deterioration where moss growth may have occurred to the face or underside of the guttering.

Consideration should be given to commenting upon the asbestos content of such guttering and the controls required in any future maintenance work.

Plastics
Many people are irritated by the noise of plastic guttering. This creaks loudly when subjected to variations in temperature. Plastic gutters are usually of inadequate strength to support ladders or other materials placed against them. Frequently they are inadequately supported and bend between the brackets.

Aluminium guttering
Many properties have been re-guttered using continuous runs of extruded aluminium guttering. Make sure that there are no materials on the surface of the roof which may reset with the aluminium and speed the deterioration of the guttering. Aluminium will degrade when coming into contact with copper.

Rainwater downpipes
During the course of your examination you should ensure that you have checked the back of the pipes. Once again this is an area of the pipe which is rarely painted and in metal pipework much of the corrosion or fracturing occurs at the back of the pipe. The integrity of the pipe can frequently be checked by tapping it with a coin. A 50p piece is of the right weight: a sound pipe leaves a clear ringing tone whereas a broken pipe gives a dull sound.

Take particular care in the inspection of square section rainwater downpipes. These are frequently fitted very close to the face of the wall and there is no way to carry out an examination of the rear metal. These are particularly vulnerable on the rear external corners and many failures are found. The replacement of such pipes is not always possible because it is difficult to get pipework of equivalent dimensions. This may mean that the complete gutter installation has to be replaced.

Capacity
Check the capacity of the downpipe and the area of roof which it is serving. It is not unusual to find that the pipes are unable to

Above: Most leaking pipes outside buildings leave tell-tale marks and failure can be readily noticed. What about all those pipes that are concealed?
Right: This may seem a sensible type of repair, but what if that hopper becomes blocked? If the property is in a conservation area or an area where properties are listed, planning or conservation bodies may be on the warpath!
Far right: External services must all be closely examined as their failure is frequently the cause of rot or decay within a building. If the wall becomes very wet, the replacement of the pipe or repair of the failure does not end the risk. The wall will have become a reservoir of water which it continues to discharge, possibly for over a year, depending on the thickness of the wall.
Lead pipes for soil or rainfall are vulnerable to the failures which are frequently caused by the shrinkage of the metal. This stretches the pipe at the joints. The connection of branches into lead often results in failures because the older lead is unable to cope with application of heat in soldering the new pipe to the old.

WALLS

cope with the amount of water running off a section of roof. These problems may have been increased by alterations to the properties where new roofs now drain onto roofs which do not have a drainage capacity capable of removing the water. A simple check is set out in the table below:

Rainwater downpipe capacity	Maximum roof area capable of being served by a pipe of this dimension
50 mm (2") diameter	(22 sq. m.) 250 sq. ft.
75 mm (3") diameter	(50 sq. m.) 550 sq. ft.
100 mm (4") diameter	(95 sq. m.) 1000 sq. ft.
125 mm (5") diameter	(150 sq. m.) 1600 sq. ft.
150 mm (6") diameter	(210 sq. m.) 2250 sq. ft.

Check that the pipes are not blocked at the time of your installation and make sure that you have examined the discharge at the base. Is there any risk of splashes dampening the face of the wall above the damp proof course?

Do the pipes run directly into the gulley, either through a back inlet gulley or by the pipe having been extended below a grating? Where this occurs there is a greater risk of spiders running through into the interior of the building.

Above: A 'window box' in a hopper head, with damaged and damp brickwork. The pipe joint below the bend isn't too clever – that bend is going to become blocked even without the garden. Make the point about the risk of future blockage in your report.
Above right: 70 feet above a London street the stone coping is being broken up by the growth of vegetation which is probably being fed by water collecting above. The damage is expensive to rectify – and in this case – potentially dangerous to passers by.
Right: Cracked pipework allowing seepage of water to the adjacent wall.
Far right: A downpipe carrying rainwater. It is blocked so water runs out of the joints.

Where rainwater pipes have been embedded in walls stress the difficulty of carrying out your examination and recommend annual maintenance checks to ensure the pipes retain their integrity. Make sure that all rainwater pipes have direct falls without bends between the gutter connection and their discharge at ground level. Any offset or bend is an area which is prone to blockage. If such offsets are present how easy would it be to gain access to the offset to clean a blockage? Is it near a window and is there an appropriate eye or point of disconnection? What are the potential causes of blockages occurring in the area of your particular property. Do trees overhang the gutters and deposit their leaves during the winter? Lime trees secrete glutinous droppings which encourage the build-up of deposits within the gutter and downpipes. Sycamore trees sending out their seeds in winter frequently sow small trees which will flourish the following spring in gutters here are not kept clean. The presence of take-away food shops can be a major contribution to gutter or rainwater downpipe blockages. The thin plastic papers used for wrapping hamburgers or convenience foods are frequently carelessly discarded and attach themselves to outlets in an irritating fashion. Where such risks are prevalent it is important that annual maintenance is recommended.

Above left: A hopper at the head of a waste pipe. Always have a close look because they are frequently blocked. They are also vulnerable to splitting at the back, or at the collar at the junction with the downpipe.
Above: Another view of a cast iron hopper. Note the assorted debris. This will eventually become blocked.
Left: A survey in dry conditions may not reveal the extent of gutter failure. In this case the short drop of rainwater pipe up to the sharp right angle bend is blocked. The render covers a timber framed wall and the seepage has resulted in dry rot. No inspection of the gutter or downpipe was carried out because it is high (7 metres/20 feet) above a flat roof.

Drainage

The purpose of the inspection of the drainage of a building is to record its present condition and location and to enable a surveyor to advise his client whether: a) it has been correctly installed and complies with modern regulations and good practice; b) is in working order and free from fractures; c) is clean and free from blockages or a build-up of solids or paper within the manholes, gullies or pipes; d) it has a reasonable life expectancy and is free from the need for major maintenance in the next five years.

The condition of the mica flag in the fresh air vent should be checked, as should the stopper on the interceptor trap. These are both found in or near the flow manhole which is between the street sewer and the house drains. Where the stopper is missing, rats or vermin from the main sewer can enter into the drainage system to the property, and have been known to enter the building through toilet traps or gullies.

Access to the drainage system is not always easy. Many surveyors have perfected techniques with old screw drivers, manhole keys, hammers or chisels for the lifting of manholes. Where the inspection cover has defeated the ministrations of the surveyor, this should be carefully noted in his report with a recommendation that the cover should be overhauled so that an inspection can take place.

The drainage system can be tested in the following ways:

Water pressure test

To carry out a water pressure test the pipes are blocked at the lower end and filled with water by discharging toilets, hand basins, baths etc until the system has filled up. The water level within the pipes should remain reasonably constant over a test period of approximately half-an-hour so that one can be assured that the pipe-work is in reasonable order. This type of test places a great strain on old drainage systems and there are occasions where failures have occurred because of the pressures exerted in this type of test. The vendor's consent should be obtained before carrying out this test.

Smoke test

Smoke tests are used to trace pipe runs and connections, but do not test for faults in a drainage system, unless linked with an air pressure test.

Air pressure test

Here the drain is sealed at both ends and air is pumped into the pipes. The pressure is checked to ensure that there is no seepage, or leak, from the system. If you inspected the system and found that the front manhole was blocked, and water had backed up to the rear manholes, you have an effective water test provided for you. Not only can you recommend the clearance of the drainage system, but also you can refer to the quality of the drainage pipes under test!

If you find such a failure, you should report it to the householder, even though he is not your client. You may be liable for damage that results from this blockage if you have not advised the householder.

Roots

The dry weather during 1989 resulted in the expansion of root activity in trees and shrubs. The examination of a drainage system must take this into account. The location of trees within 15 metres of a drain run should be referred to, as should the risk of roots getting into the drainage system.

The failure of a drainage system will result in water running into the ground. This may result in the erosion of the support beneath the foundations of a property. The damage may not be just to the drainage; it could be much more extensive.

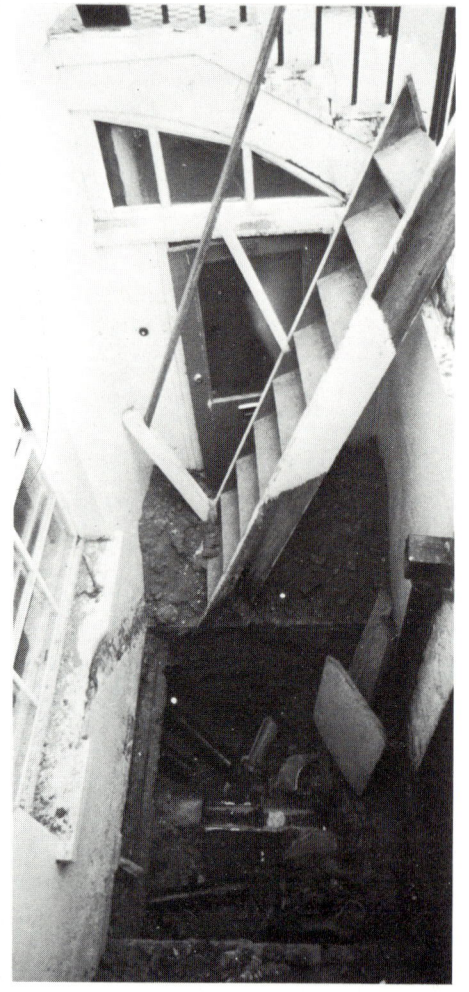

Top: The manhole can be clearly seen in this passage but the lifting ring is not standard fitting. It is unlikely that the surveyor will be able to lift this one!
Right: Is there easy access to the manhole? In this case the steps had to be taken out because they bridge the manhole.

Building materials have been obtained from many natural materials, such as timber, straw, etc. The earliest use of stone is over 5,000 years ago, and even from early days they were transported over great distances. The original choice of the architect was probably due to colour, the mason may have selected a stone for ease of working, and the client would have been influenced by the transport costs. The greater distances stone covered were by water, and buildings with easy access to dockyards or quaysides became havens for stone which was quarried 300 or 400 miles away. This applied to expensive property. The domestic architecture tended to evolve with the stone of the area. The buildings in the Cotswolds have a warmth resulting from the use of the local Oolithic limestones. The dour homes of the Welsh valleys reflect the local stones; the highly micaceous sandstones. The roofing was carried out with the split slates that could be quarried in the area.

The various stones have differing resistance to weathering. The stones from the Purbeck beds on the Isle of Purbeck are particularly resistant to weathering. They have been used in Westminster Abbey and the cathedrals of Winchester, Worcester, and Salisbury.

Portland stone, from the Isle of Portland near Weymouth, was used extensively in the rebuilding of London after the great fire. It has a high degree of resistance to weathering.

Bath stone has been quarried since Roman times. It is a soft stone which has weathered badly. It was used extensively in the Victorian development of industrial housing. Because the stone was easy to work and carve it was very popular. This very advantage was the cause of the deterioration that occurred in the following years.

To get the best from a cut stone it should be laid on its natural bed. This means that it is used in building in the same position it had occupied before quarrying. It is important that it is marked by the quarry at the time it is removed.

If the best use of the stone has been made by correct laying any failures may be due to the weakness of that particular stone.

A knowledge of the type of stone used may enable the stone and its source to be located.

In Oxford failures in the stones use on a number of colleges have been traced to the product of the Headington quarry. This stone tended to exfoliate in decay and repairs have had to be made by using other limestones such as those from Bladon or Clipsham.

Causes of decay

Water is the vehicle for the main attack on stonework, even though pure water is harmless to stone. A dry stone is unlikely to decay. Stone cannot be completely waterproofed without changing its appearance. Some of the earlier attempts to waterproof stone have caused greater damage to the material and resulted in the failure of the face.

Weathering and gradual decay is inevitable in the use of stone.

Pollution

Rainwater can become a toxic fluid when it mixes with some pollutants held in suspension in the air.

Carbon dioxide is ever present in the air and although in solution it dissolves calcium carbonate it has only a mild action on calcium carbonate stone. Calcareous sandstone whose principal binding medium is calcite may be attacked by rainwater acidified in polluted air. Sulphur dioxide combined with water to form sulphurous and sulphuric acid either of which have a particular severe action on calcium carbonate stone. The exposed face of Portland stone washed by rainwater containing these acids generally remains quite white but the face becomes extremely rough. Where stonework is not frequently washed by rain but merely remains moist a hard almost impermeable skin of calcium sulphate is formed which in time develops surface crazing and may lead to blistering or exfoliation.

All types of soluble salts damage porous building stones and the stone itself may contain excessive amounts of soluble salts before quarrying. The presence of these salts can be visually detected but some form of salt contamination is unavoidable. Sulphuric acids in the atmosphere may react with carbonates in the stone to form sulphates. Chlorides may be deposited from sea spray that has been carried considerable distances by strong winds.

Frequently salt contaminated sand is used for mortars and concrete. When the moisture evaporates near the surface of the structure, salt can easily travel from the sand to the masonry. All soils contain soluble salts and provided there is sufficient moisture these salts will move by capillary action in the unprotected porous stone. Where the stone is exposed to warm sunlight vaporisation starts beneath the surfaces where salts have been deposited internally. The crystallisation sets up forces that can damage the surface of most porous stones. This damage may result in exfoliation, in spalling or dramatic shattering. In more sheltered areas evaporation will take place closer to the surface and the salts may be deposited upon the face as efflorescence. This may corrode that face of the stone if allowed to remain in contact.

Frost action

The susceptibility to frost damage is related to the poor structure of the stone. Smaller pored stones have a greater capillary action whereas larger pored stones do not hold as much water. The Bath stones have a high capillary action as a result of the small size of the pores and are not as frost resistant. Water obtains its minimum bulk at 4°C. Between this temperature and 0°C it expands as it cools; at 0°C the increase equals 0.1% of the bulk at 4°C. Most damage that occurs is caused to buildings as a result of frost following wet weather. The volume change when liquid water becomes ice is approximately 2%.

Thermal stress

During the daytime sun on an exposed stone warms the surface more than the inside mass. At night radiation reverses the condition so that there is always a constant cycle of differential thermal stress between the surface and the mass of the stone. This constant reversal fatigues the material and minute fissures occur. Some sandstones decay by contour scaling and show a tendency to spall at regular depths to the surface. This appears to be associated with the wetting depth of rainfall and stone which has spalled, has failed due to completely blocked pores to that depth.

Wearing by wind borne solids

In coastal areas some solids, such as sand grains may be carried considerable distances by strong winds and cause some eroding of soft stone.

Algae

These are green, red or brown powder. They are sometimes slimy depending on the moisture conditions. They thrive outdoors on all types of building surface even in industrially polluted atmospheres. They do not usually destroy the stonework but they are unsightly and tend to encourage water retension within the stone fabric. Lichens are difficult to eradicate without using a substance which may be detrimental to the stone. The organisms are resistant to a certain amount of acid which may lead to the breakdown of certain vulnerable stones.

Conservation of stone: chemical treatments

The greatest dangers to stone are water borne and most chemical treatments are designed to keep the stone dry and to consolidate any friable material. No treatment produces impermeability without drastically altering the appearance of the stone.

Water repellants

Early this century it was discovered that some organic silicone compounds behaved quite differently from other synthetic compounds. Commercial manufacture began in 1953 in the United Kingdom. There are three classes of silicone water repellent.

Class A are intended for use on predominantly siliceous stones, black stone, polington etc.

Class B materials are intended for use on predominantly calcareous stones, Bath Clipsham etc. In both these classes silicone is held as an organic salt and so cannot be applied to wet or damp surfaces. As the silicone cures individual molecules come closer together because the solvent evaporates. The surface tension of the moisture on the treated surface has changed so that the moisture forms globules and runs off the surface before it can be absorbed. A silicone barrier prevents the penetration of low pressure water but allows water vapour to defuse through it.

On medium pored stone the average penetration in field or site conditions is approximately 3mm for brush or spray application. The surface impermeability rapidly disappears upon exposure and the immediate surface being the top 2 or 3mm readily retains moisture. This can cause some failure due to the different moisture content of this thin skin.

Class C materials. Silicone may also be used on predominantly calcareous stones which includes all limestones. Here the silicone is in a water carrier and so can be applied to damp masonry. Penetration rarely exists beyond 2mm. Substantial damage has been caused to masonry in certain conditions by the application of water repellents and the treatment in this way of stones has become a specialist subject.

Silicone water repellents produce a surface which is not wholly impermeable to the passage of water. Treated stonework where exposed to normal driving rain will admit considerable quantities of water. Invariably this water will be acid charged on entering the stone but on egress by vaporisation it must leave its acidic content behind. While in the stone, water will absorb soluble salts and as the surface temperature of the stone increases on exposure to the sun the water containing the soluble salts and other dissolved chemicals migrates towards the surface until it reaches the thin artificial barrier where it deposits its soluble content. While the volume of water penetration on an untreated stone is not any greater it has freer access to the surface in the liquid form and the excess quantity of salt is then deposited relatively harmlessly on the face of the stone.

It is dangerous to use water repellents of a silicone nature on materials containing large quantities of soluble salts or where they are in contact with materials containing salts.

Cleaning natural stone buildings

Most moderately dirty buildings are cleaned for aesthetic reasons because the dirt disfigures and obscures the colour of the natural stone and may conceal a wealth of the detail that was originally evident.

The dirt may also conceal various structural defects and cause decay. Sulphur compounds in the atmosphere react with calcareous stones to form calcium sulphate and cause spalling and blistering. The reaction with sandstone and granite results in a thin hard dirt coated film which is virtually insoluble in water.

Cleaning

The various methods for cleaning stone buildings are:

Washing

Traditionally limestone was cleaned by softening the acculumated dirt with water sprays which washed the dirt away. This is the simplest and often best way to maintain the clean appearance.

However in certain circumstances the skin which has been formed on the surface of the stone is resistant to a gentle cleaning by water sprays and more drastic methods must be considered. In connection with limestone, dirt adhering to the porous surface of stone over a long period of time becomes attached to the stone with a binding matrix of calcium sulphate. Constant wetting of this skin over the years has dried out through the pores of the stone so that it is almost impossible to wash all the dirt out of the heavily soiled surface in one washing.

Often the removal of superficial dirt is followed by a staining of the surface by a ginger or tar solution coming from the pores of the stones. It is particularly common in the lighter coloured stones such as Portland and Kent rag, but is less of a problem on brown or yellow stone.

Sand blasting

In the case of sand blasting this is frequently followed by a browning of the stone as a result of a light tar solution being released by rain from the dust filled pores.

There is also a risk during the substantial water saturation of stone that moisture will go through and attack many of the fixings which are out of the reach of normal rainfall. These may become saturated and subsequently rust, stain or spall. Cracks and other open joints may be allowed to open up as a result of the build up of water within the stonework.

One must also consider the substantial amount of time which can be taken for the stonework to dry out following its saturation.

In many areas mains water is normally used, even when it is chlorinated. This is applied through fine to medium outlets arranged on booms which can be easily moved up and down the exterior of the building.

The best method of cleaning building exteriors is by using nebulous sprays; these produce a low volume of water by creating a wet mist. Their commercial use is limited because they take longer to produce the desired results.

Steam

Steam was used quite extensively before the last war but has fallen into disrepute partly because caustic soda added to the boiler water to avoid furring was deposited on the cleaned surface and remained there to cause decay. Hot water is no more effective than cold water in removing atmospheric dirt.

High pressure lances

The cutting action of the high pressure lance using cold water is used for removing stubborn patches of dirt. There is nothing to be gained by scrubbing or soaking with soapy water. Detergent powders containing sodium sulphate should be excluded.

Abrasive plastic cleaning techniques by using a compressed air and grit stream was developed initially to remove rust and scale from iron and steel sheets. During the 1960's it was applied to cleaning masonry. It was first used on sand and grit stones and then on limestone. It tends to blur a lot of the detail as a result of a loss of the surface. The main attraction of blasting is speed and the immediate and often dramatic overall result. The associated noise and dust is a disadvantage of this method. Irreparable harm will be caused by indiscriminate use.

Wet blasting

There are several types of wet head which allow water to be introduced into the air in an abrasive steam. The mixture of water and air tends to be less harsh than dry abrasives. It reduces to a minimum the free dust which can be such a nuisance to adjoining occupiers during the course of this work. On completion of the work the surface must be well washed preferably with a high pressure water lance to remove dried films of slurry.

Diagnosis of defects in the stonework

In much of the repair of any stonework attention must be given to cracks, bulges or signs of settlement. Encrusted air encourages decay in stonework and stone cleaning might be necessary for the sake of appearance prior to the repair of any damage which has occurred to the stonework.

Roots of plant life in the stonework will break down the mortar and force the stone to part allowing water to penetrate increasing root growth and causing rapid deterioration. Almost every defect that occurs in the stonework is caused by the presence of water.

Condition of joints

Faulty joints cause stones to be loose and allow rainwater penetration, this encourages vegetation growth and frost action, especially in exposed positions.

Condition of stones

Defects may be the result of the use of unsuitable stones or due to the stone being face bedded. In some cases expert advice may be needed. Accurate diagnosis is important. Any embellishments in stone may by their nature and position be more exposed than other stonework and are usually of a relatively soft stone to enable the mason to work them.

Repair and general defects

Structural defects which require stitching of cracks, the consolidation of the core of ancient walls, or underpinning should be dealt with before general stonework repairs. Careful diagnosis is crucial and may require accurate measurements over a long period of time. When the structural movement has finished some simple stitch or rebonding may be adequate.

Thick walls of ancient buildings probably comprise masonry skins with a rubble core which may be unstable. Cracking may be in one skin or through the wall or the core may settle causing the walls to bulge. Part of the outer skin may need to be removed to consolidate the core. In less serious cases grouting will provide sufficient stability.

Weak walls or walls leaning at the top may be held together with a reinforced concrete beam along the top within the core. The beam may be covered by rubble or a coping.

Bulging of one leaf of a wall may occur in buildings of the 17th or 18th century where walls were often built as two unbonded skins with no core. Higher load on one skin than the other will cause bulging in one leaf.

From the late 18th century onwards the stone facing was usually bonded to a brickwork backing. Sometimes this resulted in differential expansion where the materials were incompatible.

Creeper and vegetation should be removed carefully using weed killer which has been checked to make sure that it has no adverse effect on the stone itself.

All defective flashings and fillets should be

The Saxon church at Bradford on Avon. This is one of the oldest stone buildings in the United Kingdom.

STONE

repaired in order to prevent the penetration of excessive moisture into the stonework.

Adequate damp proof courses should be provided.

Repointing

Repointing is not only important in maintaining stonework but good pointing will unify and improve the appearance. Defective mortar should be raked out to a depth of one inch and washed out to remove dust and to reduce suction during repointing.

Modern cement pointing is usually too hard and too strong causing failures in the stone face. Mixes for pointing will vary with different stones, their hardness, and the degree of exposure. The mortar is usually lime and sand or lime sand and a little cement. Stone dust sometimes replaces the sand.

The hydraulic lime is best but is not readily available. Some hydraulic or semi-hydraulic limes are available as dry hydrates and occasionally as a lump for slaking. The sand which is used in the mortar will provide its basic colouring and some experimentation may be needed to make sure that it matches the existing pointing or is compatible with the colour of the stone. In medieval work crushed stone was used.

The use of a little Portland cement for speed hardening and increased strength of the cement will normally be needed for non-hydraulic or semi-hydraulic limes.

Lime sand mixes would be about 1 to 3 in proportion and the pointing should be kept moist for several days with sacking during dry weather. The choice of finish to pointing should be carefully considered. Weather struck pointing is usually out of character with the stonework and flush pointing results in feathered edges being formed which break away in time forming a lodgement for water.

Defective stones

If stones are defective, repointing should be deferred until repairs and replacement of the stones has been carried out. It is important that any replacement stones should be as close to a compatible stone as possible and if possible come from the same quarry. It may be possible to salvage stones from unwanted outbuildings or boundary walls.

Stones should be cut following the natural bed except for cornices which are better laid with an angled bed and arch stones which should be laid with the bed at right angles to the thrust. The natural bed is not easy to recognise and should be marked on each block in the quarry.

Once the stone is replaced it may be necessary to speed the weathering so that it blends in with surrounding stonework. The use of urine or cow dung or the painting of the surfaces with milk frequently obtains the right results.

Where metal anchorages have to be used they should be bronze or stainless steel. Galvanised iron or galvanised steel ties frequently have a short term life of up to 20 years and are not recommended as a permanent form of replacement. Where they have been used on existing stonework examination should be carried out preferably with the use of fibre optic probes which will enable examination to the rear of the stone face (see chapter on Equipment).

Anchor bolts

Anchor bolts are used to secure projecting cornices onto structural frame where there is insufficient masonry above.

Joggle joints

A joggle joint is a projection on top of one stone which fits into a matching recess on adjoining stones. Frequently the grooves are cut into stonework particularly where it is intended for use in cornice stones and a cement grout is poured into channels formed by adjoining cuts. This forms a 'y' shaped key and restricts the individual movement of the stone.

Re-dressing

Re-dressing of stonework may be possible where the original surface has eroded and the detail lost. This involves the recutting of the stone and can only be carried out where there is sufficient depth and the reduced sections do not cause other problems such as around openings in the main wall.

It is frequently used in repairs to cornices but naturally is expensive as the work is often carried out on site without removing the stones from the building.

Small repairs may be carried out using plasticised stone providing an adequate key is form by undercutting around.

Epoxy resin repair

Epoxy resin adhesive is very useful for fixing broken stones which have suffered accidental admage or spalling. It is also used to repair small local defects and this method has been used to repair gothic window traces in Ely and Carlisle cathedrals.

Salt removal

The movement of excessive salt from stonework may be carried out by the use of the laying of a poultice after the wetting of the stone. The poultice chosen is usually made up of absorbent materials such as clay of diatomacalius earth. This is left on the surface of the stone for periods up to one month which enables the salt-laden water to be drawn out from the stone. The poultice is then hacked off and the process repeated.

Dry stone walling in the Cotswolds.

STONE

Top: Stone used for roofing.
Above: Timber pegs fixing the stone slabs over timber battens.

Common defects associated with cleaning

Method	Defect	Cause and solution
Washing Wet blasting Pressure lancing	Tar stains	Tarry solution, formed during wetting, drying out of pores. Can be reduced by avoiding prolonged saturation and by further washing. Some tarry staining unavoidable.
	Dry rot Rust expansion Flooding	The results of penetration can be minimised by careful sealing of all open joints and cracks before cleaning, and by taping and sheeting all openings. Watersheds and catchment sheets on rigid supports with falls to gullies will avoid flooding risk.
Dry blasting Wet blasting	Pitting of surface Blurring of arrises	The result of wrong choice of method on stone which is too soft or a careless operative or the use of too harsh abrasive.
	Gun shading	Especially on wet blasting erratic movements with blasting gun leaves mottled effect. Slight damage often unavoidable. Pronounced damage the result of using inexperienced operative.
	Blasted glass	Careless use of gun and inadequate window protection. Glass should be coated with peelable protection.
Wet blasting	Slurry deposits and film	Unfinished job. All dust deposits and slurry should be hosed or jetted off.
Acid	Brown stains on sandstone	Stone with high iron content. Acid combined with rust inhibitor should be used.
	Etched glass	Lack of protection or if occurring after cleaning, due to acid vapour on scaffold. Peelable plastic coating should be used on glass and scaffold boards washed and lifted.
	Pavement staining	Splashes of acid not neutralised and washed away.
Alkali	Efflorescence	Excessive number of applications used, careless washing off or wrong use on too porous a material.
	Streak staining	First wetting and application carried out from top to bottom. Risk can be reduced by working upwards.
Mechanical	Scour marks	Lack of skill or wrong use of method on moulded stone. Can be improved with hand rubbing.
	Wavy arrises	Lack of skill or more probably wrong choice of method on carved or moulded work.

Cleaning of natural stones

Stone	Method	Remarks
Limestone Marble	Clean water spray to soften deposits followed by light brushing. Marbles and hard, polished limestones can be washed with water containing mild detergent, rinsed with clean water, dried with wash leather and polished with soft cloth	Relatively slow. Not suitable for heavy encrustations, in frosty weather, or where buried ferrous metals or timber might be adversely affected and the method may cause brown stains on limestones. Marble is sometimes difficult to clean. To maintain the colour of dark, particularly green, marbles externally, after cleaning them, beeswax and natural turpentine should be applied and polished several times each year.
Granites	Ammonium bifluoride	Acids, particularly hydrofluoric acid, are extremely dangerous in handling.
All stones	Hydroflouric acid (about 5 per cent concentration)*	Risk of severe damage to surrounding materials, particularly glass.
	Grit-blast:** dry	Rapid, even with heavy encrustations. No staining of stones. Requires skill to avoid damage to soft stones. Very dusty process, operatives must have independent breathing supply. Close screening is required but some dust escapes into atmosphere.
	wet	Generally as above. Water reduces visible dust, but may give rise to the objections to the water spray process.
	Mechanical: abrasive power and hand tools brushes	Rarely necessary for limestones.

*Steam sometimes helps to remove deep seated soiling after acid cleaning.
**Only non-siliceous grit should be used.

Building stones thermal and moisture movement

	Thermal movement mm/m per 90 deg C per cent approx.	Moisture movement mm/m for dry-wet change
Granites	0.93	none
Sandstones	1.0	approx. 0.7
Limestones	0.25 (porous limestone) 0.34 (dense limestone)	0.8 negligible
Slates	0.93	negligible
Marbles	0.34	negligible
Quartzites	0.90	none

Properties of some limestones

Type	Weathering	Resistance to pollution	Porosity per cent	Absorption per cent
Portland: Dorset				
Roach	Excellent	Excellent	21.4	4.3
Whitbed	Excellent	Excellent	12.0	3.7
Basebed	Excellent	Excellent	16.9	5.7
Bath Stones:				
Box Ground	Excellent	Excellent	23.2	6.8
Monk's Park	Good	Good	17.5	7.76
Clipsham: Rutland	Excellent*	Excellent*	14.0	4.7
Doulting: Somerset	Excellent	Good	17.0	8.6
Guiting: Gloucestershire	Excellent*	Good*	20.0	9.7

*These remarks relate to the better quality material. Information from Stone Firms Limited

Physical Characteristics of Commercial Building Stones

Stone	Relative Hardness	Durability to Weathering*	Cleavage or Fracture	Porosity per cent
Argellite	Soft	Poor	Excellent cleavage	0.29
Granite	Very hard	Excellent	Fracture	2.0
Limestone	Med. to soft	Good	Fracture	4.10
Obsidian	Very hard	Excellent	Fracture	4.0
Quartzite	Hard	Excellent	Fracture	0.60
Travertine	Soft	Poor	Fracture	1.9
Marble	Med. to soft	Fair to poor	Fracture	0.25
Schist	Medium	Poor	Fair cleavage	0.75
Serpentine	Soft	Excellent	Fracture	0.76
Sandstone	Medium	Poor to excellent	Fracture	5.26
Slate	Med. to soft	Fair	Excellent cleavage	0.30

(Southern Alberta Inst. of Technology)

*Resistance to weathering subject to type of climate and air quality conditions – extremes, moisture, industrial gases.

STONE

Top left and far left: Regular ashlar blocks.
Left: Face filled pointing to irregular ashlar.
Above: Irregular ashlar.

Masonry Protective Coatings

Type of Coating	Suitability for Masonry Walls				Comparative Coating Factors								
	Above Grade		Below Grade		Permeable Film	Alkali Resistant	Surface Condition Required	Usual Application	Moist Curing Required	Type of Thinner Used	Type of Finish	Coats Generally Required	Resultant Coating	Expected Service Life (Years)
	Int. Surface	Ext. Surface	Int. Surface	Ext. Surface										
Parging			x	x	Yes	Yes	Moist	Trowel	Yes	Water	Flat	2	Opaque	Indefinite
Bituminous coatings				x	No	Yes	Dry or moist	Brush or spray	No	Solvent or water	Flat	2	Opaque	Indefinite
Portland cement paints	x	x	x		Yes	Yes	Moist	Brush	Yes	Water	Flat	2	Opaque	5-8
Latex paints	x	x	x		Yes	Yes	Dry or moist	Brush or roller	No	Water	Flat	2-3	Opaque or clear	4
Oil-based paints	x				No	No	Dry	Brush or roller	No	Solvent	Flat or gloss	2	Opaque	4
Epoxy coatings	x				No	Yes	Dry	Spray	No	None	Flat or gloss	2	Opaque or clear	4-6
Polyester-epoxy coatings	x				No	Yes	Dry	Brush roller, or spray	No	None	Semi-gloss or gloss	2-3	Opaque or clear	8-10
Chlorinated rubber-based paints	x				No	Yes	Dry or moist	Brush or roller	No	Solvent	Semi-gloss	2	Opaque	5
Silicone-based coatings		x			Yes	Yes	Dry	Spray flood	No	Water or solvent	Flat	2	Clear	5-8
Metallic oxide waterproofing compounds			x		No	Yes	Moist	Trowel	Yes	Water	Flat	2	Opaque	Indefinite
Elastomeric sheet coatings				x	No	Yes	Dry	Brush	No	None	Flat	2-3	Opaque	Indefinite

Concrete

CONCRETE

Previous page
Top: The surveyor will rarely have seen the building framework. He has to guess from his experience and knowledge what goes on under the cladding or tiling that will conceal this frame.
Bottom: Fractures in this concrete column were caused by corrosion of the reinforcement.

This page
Below: Cable anchorage to post-tensioned concrete beams. These are under such tension that all buildings incorporating this type of reinforcement should have a clear warning plaque affixed to them warning anybody involved in alterations at a later date.
Right: Staining of a concrete column caused by impurities in the aggregate.

Concrete usually has to meet one or more of four functional requirements:

(1) strength;
(2) durability;
(3) fire protection;
(4) thermal insulation.

Its properties depend on the correct selection of materials as well as good practice on site.

Most concretes are made with ordinary or rapid hardening Portland cement.

Ordinary cement develops strength over 7 to 14 days, the rate of hardening being accelerated or retarded by the temperature. It has a low resistance to attack by acids and sulphates.

Rapid hardening cements develop their strength much more rapidly although they do not set any quicker than Portland cement. Formwork can usually be struck earlier, the sulphate resisting cements are useful where there is a risk of sulphate attack in certain ground conditions.

Extra rapid hardening cements usually contain 2% of anhydrous calcium chloride. These are not recommended for use in reinforced or pre-stressed concrete. Because of the speed of the accelerated hardening they have to be placed and compacted within 30 minutes of mixing.

Ultra high early strength cements contain a higher proportion of gypsum than ordinary Portland cement and the initial development of their strength is much more rapid than for ordinary or rapid hardening cements, but there is little increase in strength after 28 days. These are suitable for use in reinforced or pre-stressed concrete but the creep of the concrete may be higher than for concrete made with other Portland cements. These higher heat Portland cements, the heat resulting from the rapid hardening, tend to be prone to the risk of surface crazing or fractures on the cement after it has gone off.

Lower heat cements are used where large masses of concrete have to be laid and it is important that they develop their strength more slowly to avoid surface cracking.

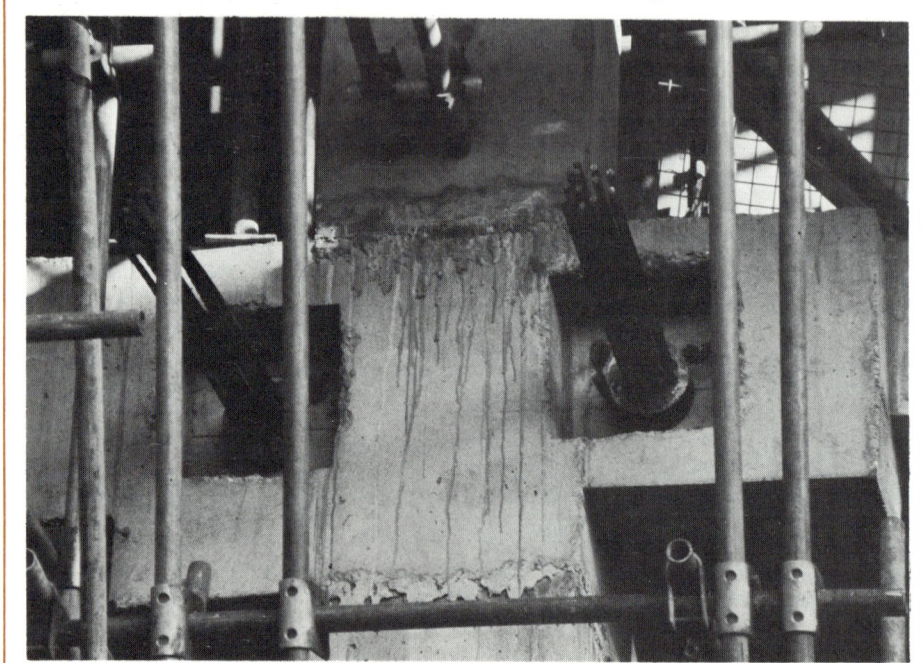

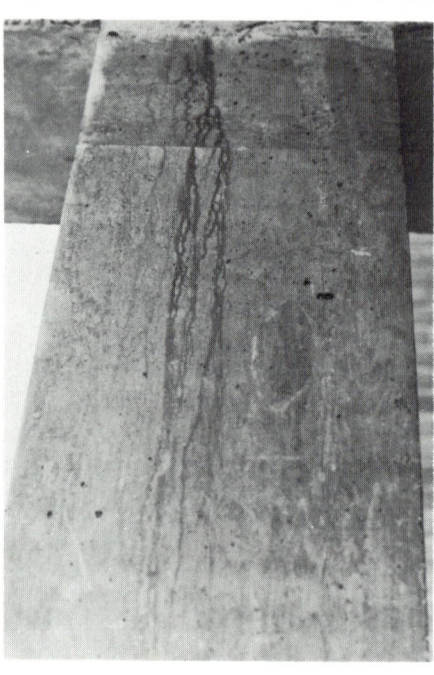

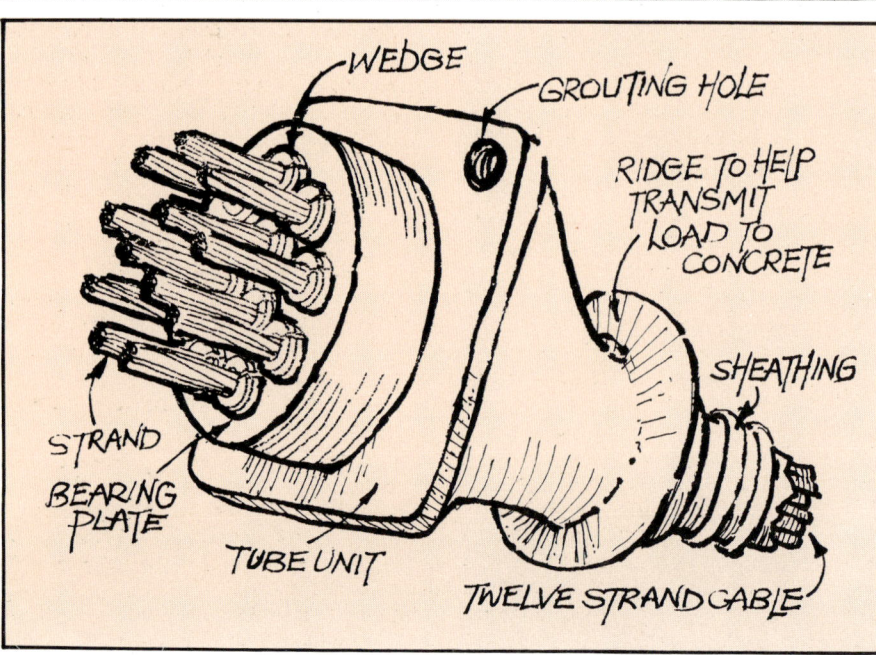

High alumina cement differs in its manufacturing characteristics from the Portland cements. Concrete made with it develops very high early strength which enables shuttering to be struck within 24 hours of the initial pouring. In factory manufacture it was used extensively to produce precast concrete sections such as flooring sections, beams, sill sections etc. If the temperature of the concrete during the mixing or curing exceeds 30°C there is a high risk that it will lose a substantial proportion of its eventual strength. High alumina cement has a higher resistance to attack by acids and sulphates but the water cement ratio needs to be extremely carefully controlled because it has a substantial influence upon the eventual strength of the cast material.

Concrete failures

Most concrete failures result from the following:
(1) Failures of the reinforcement resulting from inadequate protection caused by a failure to provide a sufficient cover of concrete.
(2) Advanced corrosion having already started prior to the installation of the reinforcement.
(3) Chemical action on the reinforcement causing fractures in the concrete which result in moisture penetration and accelerated corrosion of the reinforcement.
(4) The failure to select the correct type of concrete for the purpose required.

The majority of failures in concrete result from the corrosion of the reinforcing bars which are cast into the concrete. By an examination of the steel bars the surveyor may be able to distinguish between the different faults which are occurring and which have caused the corrosion. If there is no corrosion to the reinforcing steel there should be no problems with the concrete. If the corrosion forms an even layer of rust around the rod this will mean that it is being corroded by the by-product of infiltrating air. A pitted rod would indicate that water or chloride salts are attacking the steel. In order to diagnose the failures which are occurring within the concrete it is important to be able to identify the cause of the corrosion. The corrosion which is caused by air is usually less serious because the reinforcing rod erodes evenly and weakness will not occur for some considerable time. As the corrosion occurs the rod will expand and the damage will show up as unsightly fractures in the concrete.

A greater problem is being found with corrosion which is being caused by the presence of salts within the concrete. This can seriously weaken a rod at individual spots without necessarily causing visible signs of failure on the surface of the concrete. It is important that this type of cracking is rapidly diagnosed and dealt with with the minimum of delay.

In theory reinforced concrete should not fail. The Portland cement which bonds together the aggregate in the concrete is an alkaline material and this reacts with the steel to put a coating upon it which inhibits rusting even if some water does penetrate to the material.

Salts containing chloride ions destroy the protective surfaces of the steel rods. In some cases these salts were added to the concrete to speed up the setting of the concrete and also to speed up the building process.

Carbon dioxide and industrial pollutants from the air can also affect the steel. The carbon dioxide will combine with water in the cement to form carbonic acid. This will neutralise the alkalinity of the concrete and allow corrosion to begin. Carbonation is a natural process within concrete but usually only attacks the outer layers of the material. It is for this reason that reinforcing rods are usually placed 50 or 75 mm (2″ or 3″) below the face of the concrete so that they are remote from the carbonation zone.

While I was working in the Middle East in the mid-1970s I observed that a number of concrete buildings developed failures of somewhat alarming proportions. Chunks of concrete could be knocked off columns with a minimum of force. The fault in this case was traced to the use of sand swept directly out of the desert. The absence of sharp edges to the sand and the large presence of dust within the material made it act like graphite and prevented the concrete having any strength. The use of contaminated water or sand in the original concrete mix will affect reinforced concrete structures.

Most concretes provide an alkaline environment which tends to react with steel bars so that the steelwork will acquire a thin oxide film which protects the metal against further reaction. Concrete is permeable and the extent of this permeability will depend upon: (1) the aggregate which was used in the original mix; (2) the cement content; (3) the water/cement ratio.

Corrosion of steel in the concrete occurs as the result of micro galvanic cells. Where there is an addition of chloride of iron such as calcium chloride in the concrete mixture the risk of corrosion is progressively increased. It is important for the surveyor to be able to determine whether such irons were added and if possible whether they were mixed by being added as a water soluble solution, added as a fluid into the concrete mix, or in flake form to the dry materials. If they are added in flake form there is a risk of a build up of substantial quantities of calcium chloride which will result in rapid decomposition of the metal. If it is added in a water solution the chemical tends to be evenly distributed throughout the concrete and is less of a problem.

The normal requirement for cover for steel used in reinforced concrete varies dependent upon its location. A ground beam set below ground needs 100mm cover, whilst above ground this may be reduced to only 30mm. In much reinforced concrete construction the safer amount of concrete cover produced heavy frame components which had a poor aesthetic appeal. Because of this the concrete cover was reduced, and extra protection given to the steel reinforcement. This was achieved through the use of zinc coating on the steel which successfully reduced the risk of corrosion. The zinc coating is soluble where humidity is high and in these conditions it will not have a long life. The concrete coating should protect the zinc from the adverse effect of atmospheric contamination because this is usually high in alkalinity. This alkalinity causes corrosion to the zinc coating.

Chloride attack may have occurred as a result of salts which are aggressive towards steel being added to the concrete. This may have occurred because sea water or unwashed sand and aggregate from marine sources was used in the mix. The adding of calcium chloride in excess of 2% by weight of cement will increase the risk of corrosion in the steel reinforcement.

In some cases the exterior of the concrete is cleaned by the use of hydrochloric acid. This should be closely controlled because variation in the absorption of chloride leads to an increase in the risk of corrosion in the steel reinforcement. Hydrochloric acid may also be used on concrete flooring to make the concrete floors less slippery or to prepare the floor surfaces prior to the laying of epoxy resin surfaces.

Precast reinforced concrete mullions in many buildings constructed between 1950 and 1965 had an inadequate concrete cover over the reinforcement bars. Fractures in the concrete have resulted where rust has attacked the reinforcing bars. Eventually the concrete covering has spalled. These failures resulted from the slenderness of the mullion dictated by the aesthetic appearance required by the architects. The inadequate cover to the reinforcement caused many failures to occur within four or five years of construction.

In certain circumstances sulphate-bearing soils can cause damage to concrete. The sulphates are carried in water into the concrete. The chemical reaction which takes place depends upon the nature of the sulphate and the type of cement. Sulphates occur mainly in the strata of London clay, lower lias, Oxford clay, kimmeridge clay and keuper marl. The most abundant salts are calcium sulphate, magnesium sulphate and sodium sulphate.

The rate of attack on concrete is greatest where there is the risk of a substantial amount of movement of moisture through the material. This occurs when the concrete takes in moisture and then loses it by evaporation or leakage. There is hardly any deterioration of mass concrete foundations where the evaporation is reduced. Where there is an air presence to one side of the concrete there is a greater risk of fluid movement through the material. This may occur in basements, culverts, retaining walls and ground floor concrete piles.

When one has identified the problem and exposed the members, the extent of rust build-up on the reinforcement can be assessed. This rust build-up increases the volume of the steel. Provided there is sufficient sound steel remaining for the structural performance of the concrete, remedial work may be limited to the removal of the corrosion and reinstatement of the concrete. The qualitative analysis of the condition of the reinforcement may be undertaken by sonic testing.

Failures have also occurred within concrete construction where expansion joints have not been installed within the original framework.

The traditional methods of repair involve the exposing and mechanical cleaning of the corroded steel before resealing the metal surface. This was often done by painting the steel before replacing the concrete. Further failures often occurred as a result of the difficulty of cleaning all the surfaces of the steel.

New methods which are currently in use and for which few failures have been

CONCRETE

currently recorded include the treatment of the exposed metalwork with icosite plastic 256 dbh with a dry sand covering to the second coat application to provide a key for the subsequent repairwork.

The repair of the concrete has been achieved by the use of a combination of Icoment additive and Icomen repair mortar which uses acrylics mixed with cement and aggregate. This is applied as a mortar and should achieve a non-shrinkable repair.

Coloursite mortar 520 has been used as a finish on top of the additive repair mortar and may be used to fill small cracks and blemishes to the concrete surfaces.

Failures in high alumina cement

Since February 1974 when the roof at Sir John Cass School, Stepney, collapsed, high alumina cement has never been far from the headlines. Over recent years the background to the original failures has begun to drift into the mists of time and the surveyor may be unfamiliar with the research that was carried out nearly ten years ago.

The characteristics of high alumina cement concrete differ from those of ordinary Portland cement concrete.

After casting there is a very rapid gain in strength and after one day high alumina cement concrete may have attained a strength similar to that achieved by ordinary Portland cement concrete in 28 days.

High alumina cement when used in concrete undergoes a chemical change known as conversion. This is accompanied by loss of strength and reduction of resistance to chemical attack. The higher the temperature during the casting of the concrete the more quickly conversion takes place. Results show that most high alumina cement in concrete in pre-stressed beams in buildings more than a few years old is highly converted or is likely to become highly converted. The strength of highly converted concrete is very variable and is substantially less than its initial strength. Once a high level of conversion has been reached no further loss of strength due to conversion will occur in the remaining life of the material.

Highly converted concrete is vulnerable to acid, alkali and sulphate attack. For this to take place water as well as chemicals must have been present over a number of years at normal temperatures. Where failures have occurred there has been no significant sign of corrosion in the reinforcing steel.

The greatest risk lies in the use of precast and pre-stressed isolated roof beams. If it is known that these materials have been used the concrete should be appraised before any decisions are made as to the long term future of the components. The risk of structural failures in the smaller spans up to about 5 metres is small.

In July 1974 the Department of the Environment stated that all high rise buildings incorporating high alumina cement concrete must be regarded as suspect.

HAC continued to increase strength for some months but then its strength reduced over several years followed by a slight increase of strength. The reduction in strength is brought about by a change in mineralogical composition known as conversion. The greater reduction in strength occurred where a high water content was present during the period of mixing and high temperatures took place during curing.

The minimum strengths have been observed to occur between five years and ten years after mixing. It is reasonable to expect HAC to maintain its strength after the period of 10 years and some possible improvement may occur. In areas of warmth and moisture there is a possibility of chemical action occurring due to the types of aggregate that have been used. Further reductions in strength can occur in these circumstances. The process of conversion is not apparent in the surface appearance but the colour of the cement often changes from grey to chocolate. In assessing high alumina cement concrete it is necessary to investigate the concrete inside the members, rather than just make a surface examination.

An external examination for defects in HAC may reveal signs of failure such as excessive deflection, lateral bowing and cracking. The cracks may take the form of sheer fractures or flexing cracks. Special attention should be paid to any areas which have been subjected to high moisture penetration, such as areas of roof leaks or excessive condensation.

It must also be remembered that HAC is vulnerable to chemical attack from plasters

Assessment of high aluminous cement from its age and degree of conversion

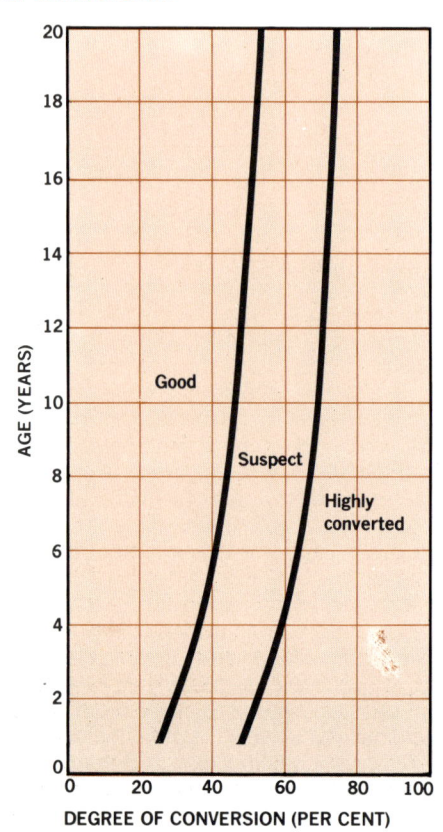

and wood wool slabs. Both white and black markings may be indicative of chemical attack. In order to carry out a chemical examination samples should be obtained from the suspect members. The method of carrying out the obtaining of samples is as follows:

Carefully remove all plaster and other surface particles, then drill using a number eight masonry drill a hole approximately ¼" (6mm) deep. Blow clear any of the surface particles that will have resulted from the first drilling.

The hole then should be drilled for a further ¾" (18mm) and the drillings caught on a card. Four such holes should be drilled over a distance of some 1'–2' (300-600mm). The mixture from the drillings from the four holes is placed in a sealed container and sent for analysis. (35mm film containers are invaluable for collecting samples and for storing 10p pieces for feeding avaricious parking meters.)

Initially such samples should be taken from at least two beams or 10% of the members. The samples should be clearly classified as to the position of the structure that they are taken from and the age of the concrete. Great care is required in taking samples so that you are clear about the areas of the building the analysis results will relate to.

(The degree of conversion which has occurred is found by using the following formula: the amount of gibbsite multiplied by 100, divided by the amount of gibbsite plus the amount of mono hydrate.)

For the addresses of laboratories which are able to carry out the testing of samples see Appendix on page 62.

For materials over ten years old the degree of conversion should not exceed 45%. Good quality HAC concrete is likely to give satisfactory service but should be reinspected at intervals. Where the material has been classified suspect it may give satisfactory service in the short term but tests should be carried out on an annual basis.

Where the material has become highly converted the concretes are vulnerable to chemical attack and there is a possibility that the minimum strength of such concrete will be equivalent to the strength of fully converted concrete.

The strength of the concrete can be determined by crushing small core samples cut from the structure. Care should be taken in the cutting of the core samples to ensure that one avoids the weakening of suspect areas.

The Building Research Establishment has come up with a method of relating the pull required to remove an expanding bolt from the concrete to the crushing strength of the concrete itself. This means it is now possible to assess the quality of the concrete without removing core samples which have caused substantial damage to buildings under test in the past.

Ultrasonic examination which measures the ultrasonic pulse of vibrations from one transducer to another through the concrete

CONCRETE

membrane has been described in the section on Equipment. The distance between the transducers and the ultrasonic pulse is measured.

This method can only give comparative test results because it will only identify the weaker and the stronger sections of the concrete. Lower pulse levels would indicate the weaker concretes but there are no pulse levels which can identify those sections which are of adequate strength.

Shrinkage in concrete

The addition of calcium chloride in reinforced concrete in amounts between 0.5 and 2% of the weight of the cement leads to a 50% increase in drying shrinkage.

Wetting and drying causes concrete to expand and contract. This movement is normally caused entirely by the expansion or contraction of the paste formed by the cement. Such movement is in the region of 0.04% but where the aggregate has a large wetting and drying movement such as occurs with some of the doleritic and basaltic aggregates from Scotland, such movement may be 0.1%. This will be further increased if calcium chloride was added to such a mix.

Such movement is difficult to accommodate in buildings and gives rise to cracking in reinforced concrete wall panels, the excessive deflection of balcony slabs and other structural defects. When these fractures are attacked by water penetration and subsequent fractures expanded by frost action substantial defects will result.

In certain circumstances brown stains have appeared on the face of concrete and this is often associated with the use of Thames gravel aggregate. These aggregates contain an iron sulphide mineral which, following a chemical reaction, forms brown iron hydroxide.

In other areas of the world outside the United Kingdom there are deposits of aggregate which react chemically with the sodium or potassium alkalis released during the hydration of the cement. These can cause a disintegration of the concrete. The surface fails in a series of interwoven fractures which often have dark staining.

The cracks are reminiscent of the surface failures that occur as a result of frost action. Frost and salt crystalisation can also cause failures in the surface of the concrete. The exposed aggregate may become loose as a result of further frost action.

System building

Many thousands of council flats built in the last 15 years have been so badly constructed that they are unfit for human habitation and will probably have to be demolished. These flats have been built using concrete slabs and panels in a method that is referred to as system building. It was introduced into the United Kingdom in the 1960's to meet the high demand for the development of a large number of homes in a very short space of time.

In many areas the present cost of carrying out repairs exceeds the value of the individual flats. In Hillingdon, flats constructed using the Byson wall frame method have recently been demolished because the repair cost ran to £20,000 per flat. Other buildings using the French Camouse method have suffered similar failures.

The system method of construction involved the erection of precast panels which were fixed and secured in a sort of house of cards assembly process. On a drawing board the system is sensible but it is only when you add the human element that problems occur. The installation and construction of properties using system methods require the highest levels of accuracy in the setting out of the site and in the installation of the individual panels. Minor errors which occur in the placing of panels at lower level compound and become major problems at higher levels, where substantial gaps may occur between panels.

In some cases rubber seals which protect the concrete joints were badly fitted, omitted, or became loose due to inaccurate panel location. In other cases sealants which have been used only had a life of two years.

The insulation within the panels was usually inadequate, leaving cold surfaces which later suffered from a build up of condensation on the inner face when the properties were occupied. Poor ventilation aggravated this situation.

Ill-fitting panels and window frames often resulted in water penetration. The design of windows in tower blocks of 15 and 20 storeys was frequently only adequate for exposure ratings compatible with properties four and five storeys high. Failures in the steel reinforcing rods which were placed around the joints resulted in the fracture of the concrete panels and the vulnerability of the structure.

The pressures which were exerted upon local authorities in the 1960's to erect houses at a super-human rate resulted in the adoption of systems which were not fully tested and used unskilled and untrained labour.

In 1969 the National Building Agency reported on defects which began to appear in system built properties and, in particular, Hillingdon's Byson Estates. The report criticises the open drain method used in the jointing of the precast concrete panels. A rubber seal is used in this building method but there is no way of checking whether the membrane has been properly fixed.

Adhesion failure

The failure in adhesion of the surface materials used to conceal the concrete framework. This has included the failure of mosaic tiling used on the face of concrete frames and the more serious failure of brick slips of cut bricks used to conceal edge beams or floor panels. The increasing problems which will occur as a result of this type of defect cannot be underestimated.

Whenever multi-storey properties are examined the drawings showing the form of construction which was used should be

Careful examination of the fixings of all external cladding, whether stone, mosaic or sheet material, must be made in order to advise upon the anticipated life of the construction.

The provision of expansion joints in modern multi-storey concrete frame buildings must be checked. The application of any cladding to a concrete building where there is inadequate provision for future movement will lead to failure, as can be seen in the deterioration of the mosaic tile finish to the concrete framework shown below.

Types of concrete

Type		Aggregate	Compressive strength at 28 days N/mm²	Thermal Modulus of elasticity N/mm²	Drying shrinkage per cent	conductivity (dry) W/m deg C	Main uses
Dense concretes		Iron shot	up to 69			—	Radiation shielding
		Gravel	14.0–70.0	20 700–34 500	0.03–0.04	1.4–1.8	Fire resistance Class 2
			41.4–69 (special purpose)	34 500 44 800			
		Crushed limestone	24.1–34.5				Fire resistance Class 1
		Crushed brick	13.8–27.6				
Lightweight concretes	No-fines	Gravel			0.016–0.028	0.08–0.94	Structural
		Clinker	2.76–6.89		0.033–0.040		
	Lightweight aggregate	Foamed slag, sintered PFA or expanded clay with some natural sand	2.0–62.0 (structural concrete minimum 15.0)	6 890–20 700	0.030–0.070	0.24–0.93	Superior thermal insulation and fire resistance Class 1
		Various	1.40–27.5		0.03–0.09	0.25–0.35 0.16–0.91	
	Aerated	—	1.38–10.35	1450–3120	0.22 air-cured 0.06 autoclaved	0.08–0.26	

(Information from BRE Digests 150 and 123 and other sources)

inspected. Where brick slips have not been secured by metal frames or ties they must be regarded as defective. A detailed examination will probably include the removal of a small section of the brick edge finishes. If they are inadequately fixed, the brick edge will have to be replaced with a correctly supported section.

At the present time there appears to be no successful and acceptable method of glueing and securing brick slips to the face of concrete framework. The use of non-ferrous support framework is the only acceptable alternative.

Joint Failures

The large size of the components used in industrialised building has placed a greater demand on the design and performance of a joint. The larger the components the more difficult it is to make them accurately. The joint must be able to cope with the movement that may occur, be it structural, thermal or moisture.

Rainwater will run off the face of the panels onto the joints. Because many of these panels are impermeable the volume of water will increase as it progresses down the face of the building. With brick construction the water is taken into the surface during rainfall and discharged in sunshine. This reduces the volume of water running down the face of the building.

Joints at present used for building precast concrete wall panel construction are of two main types:

Filled joints

These rely exclusively upon a sealant or gasket to seal the joint and keep out the weather. Mortars are unable to accommodate large movement and can only be used for small brick of block sized panels. Sealants are dealt with in the following chapter. Gaskets can be shaped to meet greater variations in joint width, but suffer from a number of limitations and should not be used exclusively as a weather protection.

Open drained joints

These joints involve a two stage defence against the weather. The joint relies on an air tight barrier at the back of the joint to prevent through flow of air and a joint layout which traps most of the rain in an outer zone.

An intermediate baffle usually divides the vertical joint into two zones. The outer zone deals with most of the water which is then drained away. The inner zone will drain away the small amount of water that will pass the baffle. The main seal is at the back, where it is least exposed. It is unfortunately concealed from the surveyor during his survey and the site supervisor during construction. Failures are not uncommon. If the rear seal fails to remain an air tight barrier, water will be blown past the baffle and into the building.

Concrete cancer

This is the nick-name for a defect which was first diagnosed in the United States in the 1940's. The technical name for the failure is Alkali-Silica-Reaction (ASR). It is caused when alkaline chemicals present in cement react with silica. (Silica is a substance occurring naturally in certain types of aggregate or stone used to make concrete). A jelly-like substance forms around the stones, this absorbs water and expands in a way that can crack the concrete. If water gets in and freezes, the cracks become wider and let in still more water. The steel reinforcement is then vulnerable to water penetration and corrosion takes place causing further defects. The Building Research Station warned of the dangers in a paper in 1958, but the first major failures were found in this country after 1976.

There are now buildings throughout England, Scotland and Wales where remedial work is required as a result of this defect. In Plymouth, Devon, a multi-storey council car park has developed severe cracks and is now being strengthened at a cost of 1.5 million pounds. Cracks have also been found in bridges on the A30 Exeter/Plymouth trunk road. There is a possibility that the Koss Farm bridge may have to be demolished and rebuilt, so serious are the defects which have been found. The cracks in the bridges are mainly in the concrete supports.

The cracks develop slowly and become noticeable long before there is any major danger but there is no known cure to this defect. The problem can be avoided in new building work by using a low alkali cement or by avoiding the use of chemically reactive aggregates.

If the concrete is not subject to water penetration the failures are unlikely to occur and it is probably for this reason that the defects have not been found in residential properties.

sealant failures

Since the introduction of flexible applied sealants in the 1950's, specifiers have been faced with a bewildering array of sealant materials and systems. These sealants have sometimes been unwisely used as the only defence against entry of wind and water. In those early days the life expectancy of many sealants was as little as two years and the maximum life of those available may have been no more than fifteen years. Recent development in sealants has extended the life expectancy of certain modern sealants to up to 50 years. Many buildings erected in the late 1950's used untried and untested materials and now have serious problems of water penetration due to sealant failures. A sealant should be used as a method of sealing a joint and should not be regarded as the sole method of preventing moisture penetration.

Sealants have been used to infill joints which have occurred due to poor setting out of the building. The assumption that sealants are capable of covering all gaps in a property has extended onto many drawings, where the phrase "mastic joint" is used without any description of the materials which are to be used, to cover any gap that is inherent in the construction.

Causes of failure

Approximately 80% of sealant failures occur for the following reasons:

1. The application of the sealant has been too thin.
2. The sealant has not been laid on a properly primed surface.
3. The sealant used was the incorrect material required for the joint.

The lack of knowledge of the materials and their method of application and poor workmanship on site contribute to these defects.

Application

Sealant should be applied to a joint so that the depth of the sealant is twice the width of the joint.

It is important to make sure that the sealant adheres to those surfaces that it is intended to stick to, and does not adhere to other surfaces which could place greater strain upon its flexibility.

If the sealant is used on an expansion joint it may be required to cover a gap that may become wider than the material can cope with. If the bond is not broken in the appropriate place, the sealant may be asked to deal with a movement which is beyond its capability. If the sealant is unable to act over the full width of the joint but has the movement concentrated on a small section of the sealant, failures will occur.

Where sealants are used as triangular pointing fillets the diagonal face should not be less than 10mm and the surface should be convex rather than concave. If there is a gap greater than 5mm where this bead is being installed it is important that a back up material be inserted and the sealant applied in a sufficiently large fillet to ensure adequate adhesion to both surfaces.

It must be remembered that sealants may react with certain materials and it may be necessary to provide a bond breaking tape or other materials to prevent the contact of the sealant with, for example, board joint fillers which may be incompatible with the sealants being used.

Some sealants require primers to be applied. These are usually applied half an hour before the application of the final coat of sealant. The primer tends to go yellow as it is exposed to daylight in this time span. If it is necessary to determine whether the primer coat was applied, the removal of a small section of the joint should reveal the yellowish section in the primer.

Types of sealant

The groups of sealants which are now in common use are oil based or oleo resinous; butyl, acrylic, polyurethane, polysulphide and silicone and bitumen. The sealant types are grouped in polymer classification for identification purposes on the table. The failures of various sealants have been listed on the table on page 125 setting out the various characteristics, life expectancy and joint performance of each of the various categories. This table should be used for guidance only and not for the selection of sealants.

Failures

There are two types of visible failure which exist with sealants. These failures are:

1. Adhesive
2. Cohesive

There are a larger number of potentially contributory factors that may lead to these situations occurring.

1. Adhesive failures

These may be recognised by the pulling away of the curved sealant bead from the joint interface and result from the sealant having lost adhesion.

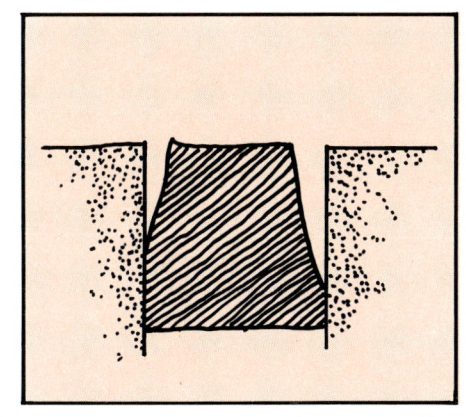

These failures may have been caused by one of the following reasons:

(a) Poor preparation of the joint interfaces prior to the application of the sealant. (Many types of sealants require primers to be applied to porous surfaces of the interface or to glass where the sealant may be light reactive.)
(b) Dusty joint interfaces not adequately cleaned prior to the application of the sealant.

(c) The incorrect primer used due to poor specification or negligence.

(d) No primer used.

(e) The sealant applied to damp surfaces.
(Some sealants cure upon contact with atmospheric moisture but may provide ineffective adhesion with the interfaces of the joint if the surfaces were damp at the time of application.)

(f) No joint backing caused an incorrect sealant configuration.
(The design of the joint will have determined the section of the sealant. The sealant will be required to carry out a specific duty within known tolerances of expansion and contraction. If the backing is inadequately prepared the joint movement may be concentrated on one part of the sealant and not spread over the full width of the joint.)

(g) Lack of tooling during application causing incorrect sealant configuration.

(h) Comprehensive thermal movement prior to the full cure of the sealant.
(Certain sealants take a substantial time to reach their full flexibility. Certain of the polysulphide sealants take up to four days to cure completely throughout the joint and any movement that takes place in that time may cause rupture to the sealant.)
Silicone sealants can take up to seven days to cure and are also vulnerable.

(i) Sealant cannot accept joint movement and the bond is strained.

(j) The depth of the sealant is inadequate and is unable to cope with the strain placed upon it.
(This may have occurred as a result of poor workmanship where the design depth was adequate, but the application was incorrect or as a result of the design having been incorrect and the application having met the requested standards.)

(k) The coating of the sealant being incompatible with the material.
(Some of the sealants improve with paint finishes being applied to the surface but others will react to certain finishes and this may result in breakdowns in the material.)

2. Cohesive failures

The cohesive failures occur when the material itself splits, as opposed to the adhesive failure which results from the sealant breaking away from the interface of the joint. The cohesive failures may have been caused by any of the following failures.

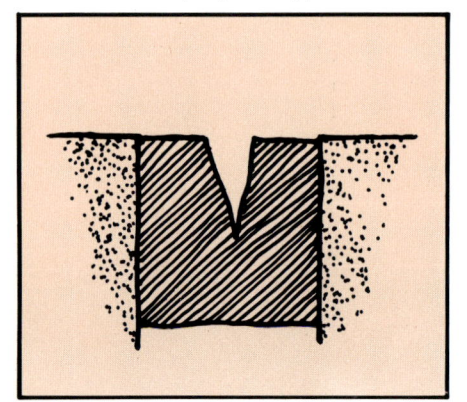

(a) The thermal movement is too high for the elastic value of the installed sealant.
(The table on page 125 shows the maximum width of joint for most of the sealants. In many cases this does not exceed 50mm. If the joint widens to a greater extent or if the range of movement is beyond the capacity of the materials, failures will occur.)

(b) No joint backing present causing incorrect sealant configuration.
(The design of the joint or failures in the construction have resulted in the sealant being unable to spread movement over the full width of the joint. If the stress is placed on too small a part of the sealant failure occurs.)

(c) The sealant is of inadequate central depth.
(This occurs where there is poor application resulting in the depth varying over the joint and weaknesses occur towards the centre of the joint.)

(d) High thermal movement occurring prior to full cure.
(Under adhesive failures, the vulnerability of polysulphide and silicone sealants to early movement which occurs prior to the joint having fully cured was noted. This can result in either an adhesive or cohesive failure.)

(e) The joint design is too small for the movement which occurs.
(Joint movement may be greater than that which was calculated in the design processes, or joint design may have allowed too high a value for movement for the particular materials.)

(f) Poor quality sealant that shows no inherent strength and ruptures under stress.
(Certain materials have got limited movement potential, with the oil based butyl and bitumen sealants having the minimum movement accommodation and the polysulphides and silicones having some of the greatest flexibility.)

(g) Incorrect use or inadequate bond-breakers being applied to the rear of the joint. This is due to inadequate design or defective installation.

(h) Entrapment of air behind the sealant.
(This can result in the expansion of the air which has been trapped and the sealant bubbling or being fractured by the pressures exerted.)

(i) Poor mix of component parts to produce an unstable sealant.
(Manufacturing errors in quality control do occur.)

Design

The correct design and selection of a sealant will require the following items to be given adequate consideration.

(a) Careful inspection of the sub-strata or faces of the joint.
(One will have to ensure that all coatings and finishes that are found on the base elements are known. For instance polyester paints on aluminium or preservatives on timber frames must be of chemicals that do not affect the sealant.)

(b) Calculate thermal movement; so that maximum movement to the joint can be assessed.

(c) Settlement.
(Depending on the speed of the loading of the structure, the condition of the sub-soil and speed of drying out, the joint detail will be substantially altered. At what time will these changes take place?)

(d) Manufacturing tolerances.
(One has to ensure that maximum and minimum component size is adhered to.)

(e) Construction tolerances.
(Does the joint take into account component alignment during construction?)

(f) Primary joint size.
(Having allowed for all these factors it is now possible to determine the probable movement of the joint.)

(g) Building location.
(A check must be made of the reaction to climatic conditions upon the joint. For sealant design windows, the dynamic pressures that the wind may apply, the effect of sunlight upon the joint, the flexing of glass as a result of wind, airborne pollution, salts and other chemicals which may be carried in the air must be considered in the design calculations.)

(h) Aesthetic requirements.
(Has the designer required a joint of a particular appearance, is the joint to be disguised or has it to be of a uniform width.)

(i) Joint configuration.
(Can the necessary breaks in the bond be applied or joint backing be easily incorporated. Is access to the joint possible so that sealant may be installed.)

(j) Primary sealant selection.
(Once the sealant has been selected its compatability with other materials that it will come into contact with will have to be checked. The movement capability will have to be analysed in respect of anticipated future movement the joint may face.
The life expectancy of the joint must be related to future maintenance cost or the cost in use of the building. The case of replacement of joints should be considered if sealant life expectancy is under 15 years.)

Repair work

Once one has isolated the problem it becomes necessary to select the treatment and the cost. The safest remedial treatment is to remove, wherever possible, the offending material, clean up the joint and replace the seal. Where there are heavy oil compounds in the original sealant it will not be practical to grind the interfaces of the joint because a smear will occur over the surface. The new sealant should be applied in strict accordance with the manufacturer's instructions.

Where there are original butyl seals some erroneous recommendations have suggested overcapping with one part polysulphide without primer. The butyl will migrate from the joint and contaminate the adjacent areas which will result in further failures. It must be remembered that it may not only be the sealant but the cleaners and primers that can react with any residues because of the chemical inter-reaction between any one or all of the components. It is essential that if any remedial work is to be carried out on the basis of the application of an overcapping that the onus for the responsibility for the sealant selection is placed upon a well insured specialist. If he is able to maintain that form of insurance he is highly unlikely to recommend that course of action.

Conclusions

Jeff Webb and John Palmer, of Joint Masters Ltd, one of the few companies in this country that deal with the remedial work caused by previous generations of defective installation and selection, have in a remarkably generous gesture suggested ways of putting themselves out of business.

To assist in the removal of whole areas of remedial work the following changes could be implemented.

(a) Removal of oil based pointing mastics from manufacturers' ranges to avoid serious compatibility problems in continual remedial or maintenance works. Upgrade to a more sophisticated sealant such as acrylic or one part polysulphide because the cost of application is similar and the increased cost of material is insignificant.

(b) Incorporate wider glazing rebates to enable adequate sections of compound to be employed. The cost of applications is similar, the increased cost of sealant insignificant. The extra aluminium or timber used would cost much less than potential remedial action in the future.

(c) Think about joints being at least as important as component parts. Sealants and mastics are advanced materials and require due consideration.

Vandalism

Even these actions will be unable to prevent vandalism because at low level the pliable surfaces are all too attractive to the curious young child and the stupid older one.

The action of birds and other animals are also difficult to control but there may be occasions where the actions of sea birds and squirrels should be considered in the selection of jointing materials.

BS 6213: 1982 *British Standard guide to selection of constructional sealants.*

British Standard documents:
Joints and their design
BS 5606: 1978 *Code for accuracy in building*
BS 6903: 1981 *Code of practice for design of joints and jointing in building construction*

Sealant selection
BS 6213: 1982 *Guide to selection of constructional sealants*

Sealant specification and tests
BS 544: 1969 *Specification for linseed oil putty for use in wooden frames*
BS 2499: 1973 *Specification for hot applied joint sealants for concrete pavements*
BS 3712 Parts 1-3: 1973-74 *Methods of test for building sealants*
BS 4254: 1969 *Specification for two-part polysulphide-based sealants for the building industry (& AMD 1749)*
BS 5215: 1975 *Specification for one-part gun-grade polysulphide-based sealants*
BS 5212: 1975 *Cold poured joint sealants for concrete pavements*
BS 5889: 1980 *Specification for silicone-based building sealants*

Sealant type	Characteristics	Life expectancy (years)	Form in which available	Max width of joint	Min depth (in ideal conditions)	Use	Remarks
OLEO RESINOUS *Not recommended for use*	May harden after a few years, gun applied mastics tend to remain pliable for longer periods, painting improves durability. Fibre board, rope or expanded plastic are some of few suitable backing materials.	2–10	Pre-extruded strip	50mm	5mm	Limited use.	Very small movement tolerance. Mainly black or stone grey but also coloured.
			Knife	25mm	10mm		
			Gun	15mm	15mm		
BUTYL	Tend to collect dirt as some don't skin. Cellular polythene most suitable backing material. Easy to place, may split through centre if subject to repeated slow movement.	10–20	Pre-extruded strip	50mm	5mm	Glazing, bedding, window and door frames.	Grey and black.
		2–10	Knife	25mm			
			Gun	15mm			
ACRYLIC	Needs warming for application (50°C). May collect dirt. Not recommended for exposed joints. Liable to breakdown if subjected to prolonged wetting.	15–30 can be more in certain cases	Gun	15mm	5mm	Cladding joints, window, door, curtain walls and glazing.	Any colour to your choice.
POLYSULPHIDE	Slow to cure. Cellular polyethylene or cellular polyurethane suitable as backing material. Primers required on porous or glass substrates. Cure is initiated by atmospheric moisture. Resistant to acids, alkalis, petroleum and most solvents but not chlorinated hydrocarbons. Difficult to replace. Not suitable for joints where early movement takes place before curing complete throughout joint (4 days).	20	Gun	15mm can be up to 50 with 2-part	5mm	Joints subject to traffic, joints in water retaining structures, cladding, window, door frames, curtain wall joints, glazing.	One part. White, black, grey and brown are available in 2-part.
BITUMEN	Overheating destroys properties, suitable for joints in mastic asphalt and tarmac paving. Surface tends to oxidise and craze, difficult to paint.	3–10	Strip	50mm	5mm	Joints subject to traffic, water retaining structures.	Black
			Knife	25mm	15mm		
			Gun	15mm	20mm		
			Hot poured	50mm	20mm		
SILICONE	Slow curing (7-10 days), cellular polyethylene or cellular polyurethane are suitable backing materials. Cure initiated by atmospheric moisture. Durable under water. Good adhesion, but elasticity places adhesion under stress. Preparation vital and primers required on porous surfaces. Picks up dust.	20–30	Gun	25mm	5mm	Cladding, pointing, curtain walls, glazing.	Translucent white and colours.
POLYURETHANE	Truly elastic, adhesion under stress is only fair, therefore surface preparation vital and primers may be required.	15	Knife	50mm	5mm		Cream.
			Gun				

SEALANT FAILURES

Typical sealant failures

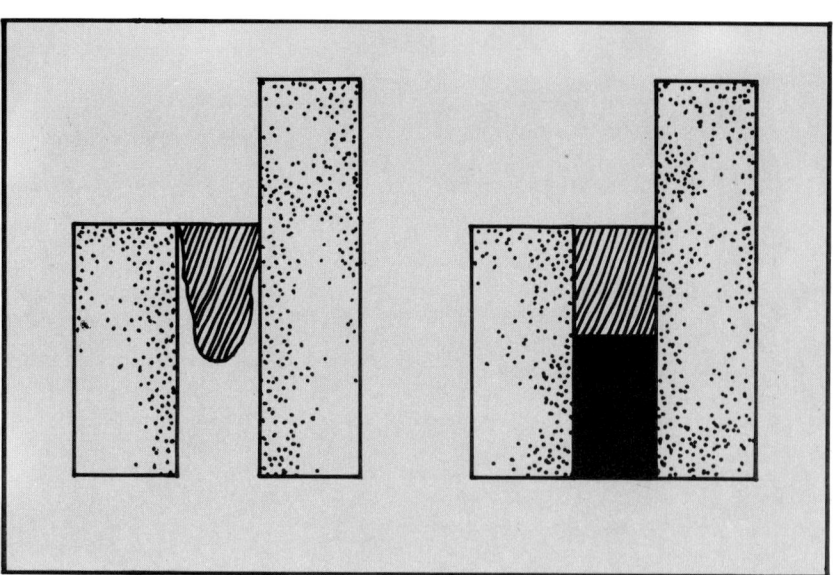

The need for joint backing to force the sealant against the joint interfaces to promote adhesion.

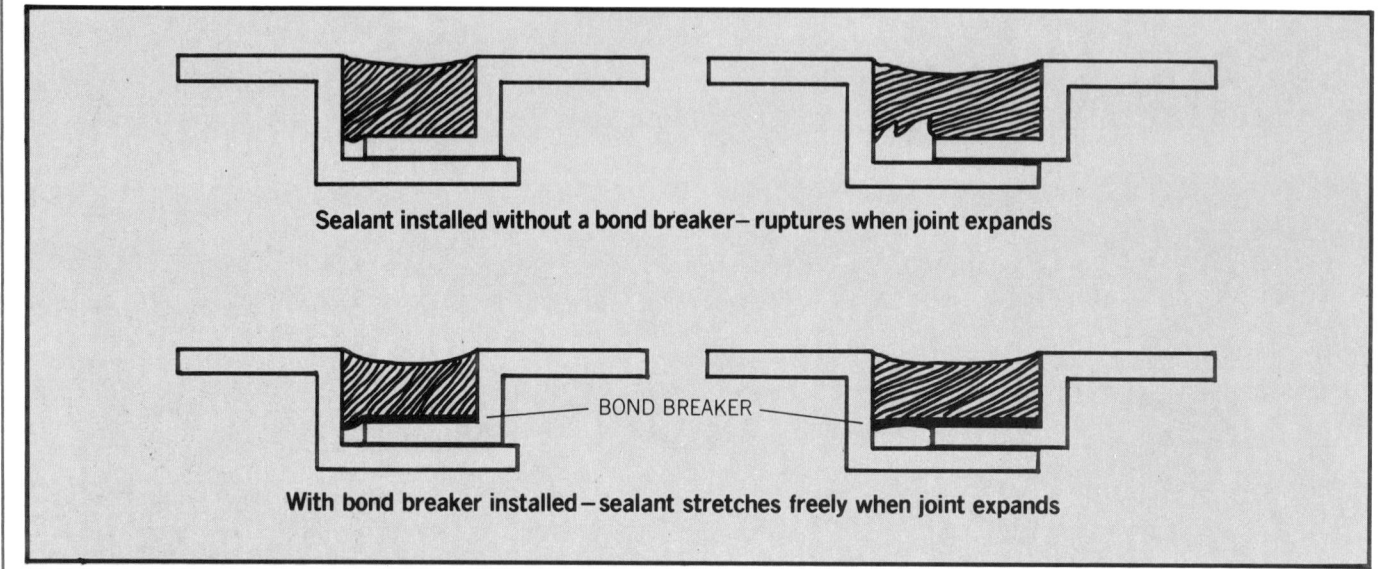

Sealant installed without a bond breaker — ruptures when joint expands

BOND BREAKER

With bond breaker installed — sealant stretches freely when joint expands

Where one cannot use the joint backing a bond breaker ideally should be installed.

Typical failure of oil based mastic pointing due to drying out from the surface and eventually failing to accept movement thereby failing cohesively.

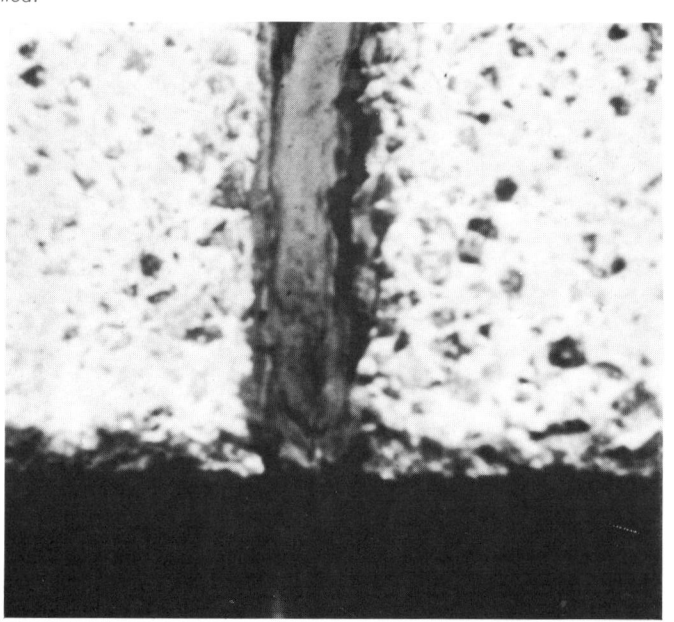

Parapet joint in "butyl enriched" sealant incapable of accepting rapid major movement has failed adhesively.

SEALANT FAILURES

Two-part polysulphide sealant is too thin and thermal movement has caused adhesive failure.

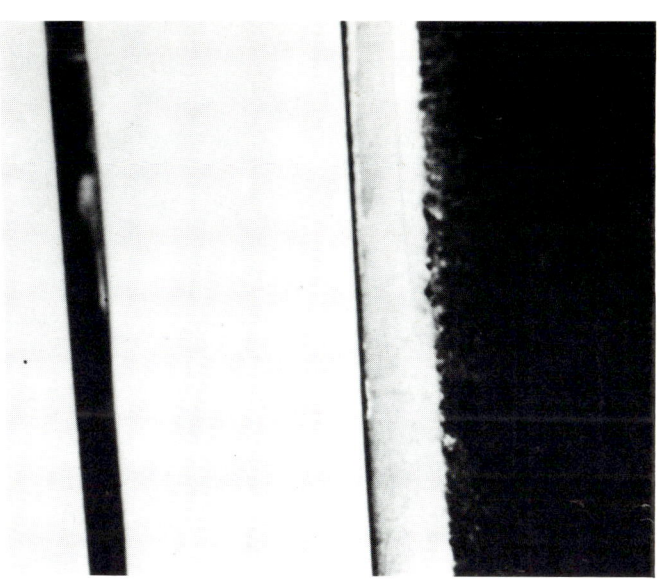

Perimeter pointing one part polysulphide has clearly failed adhesively with the timber frame due to incompatibility with the coating used.

Poorly shaped joint slot too small for movement being experienced. Polysulphide sealant is completely over-stressed causing cohesive failure.

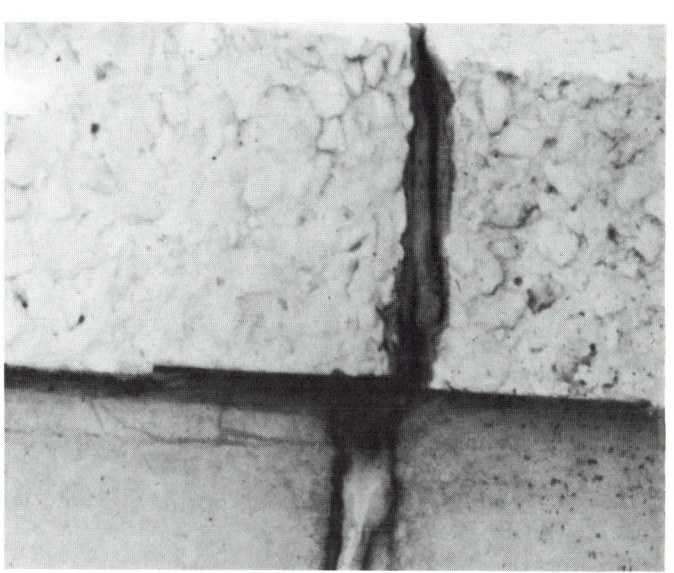

High movement is causing heavy compression and extension of this polysulphide seal. Deep creases are the first sign of impending cohesive failure.

Joint between GRP and timber window is subject to very rapid movement not allowing this acrylic sealant to relax and reform causing stress lines eventually leading to failure.

Lack of priming to anodised aluminium has meant adhesive failure of this acrylic sealant.

SEALANT FAILURES

Poor detailing coupled with poor sealant type selection has caused significant stress to this mullion bandage joint which will suffer cohesive failure in the near future.

This window head joint against a raised aggregate concrete panel has failed adhesively due to the uneven nature of the joint slot.

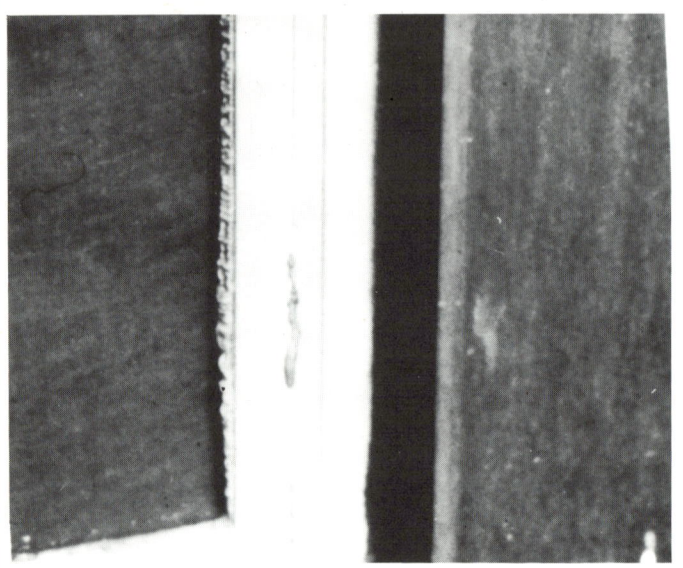

Glazing compound dried out and exuded from the glazing space/"creep" of butyl based sealants due to migration and dirt pick-up.

Overstressing of a two part polysulphide seal due to high thermal movement causing splitting and, although near to the bond line, cohesive failure.

Poor application of silicones or how not to seal a curtain wall. No provision for external seals.

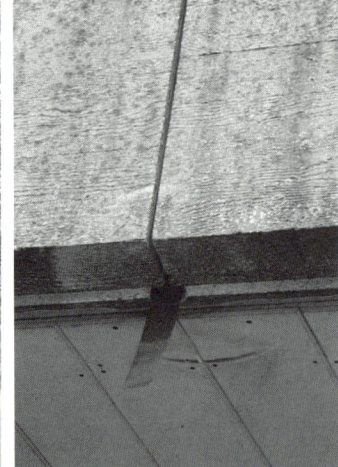

Damp penetration damage as a result of movement of sealant baffle.

Plastics

Numerous products available for use within the building industry are now made from many different plastic materials. Each of these materials has widely differing properties and it is important that the surveyor should be capable of identifying the relevant plastics where these properties may cause problems in the buildings' performance.

There are two basic types of plastic. Thermoplastics and thermosetting plastics. In thermoplastics the long chain-like molecules are not linked together and can move relative to one another. These can be moulded and will retain their shape on cooling but can be reformed again on heating. Thermosetting plastics also have a simple chain structure of molecules prior to the final shaping operation but in the moulding process cross-links are formed between the chain so that the material cannot be reformed on heating. The variation of the additives incorporated in the plastics as well as variations in the basic materials can produce profound changes in some or all of their properties. This permits the design of materials to suit particular applications and can make it more difficult to identify the individual material.

These additives can take on many forms: pigments may be added to give colour, fillers may be added to increase hardness and reduce cost. Plasticisers in general reduce mechanical strength but increase flexibility and ease of processing. Reinforcements, notably asbestos or glass fibre, increase the strength and rigidity. Flame retardants play an important part in diminishing the risk of ignition of plastics and determining the behaviour of the plastic in the case of fire.

Properties

In general plastics have moderately good tensile strength though their rigidity is generally low. Their mechanical properties are time-dependent and there is a tendency to creep under stress and show stress-relaxation. There is a fall-off of compressive and tensile strength on heating. Some plastics are very tough whereas others have a poor resistance to impact. Because of their basic carbon-hydrogen structure all plastics are combustible, although some burn more readily than others. They are generally resistant to the action of water and to corrosion by most of the water-based corrosive agents that affect metals. Some are resistant to a wide range of chemicals. For the most part they have good electrical resistance and are low in density. Some of the materials have high resistance to weathering whilst others are bio-degradable and break down quickly out of doors.

The surveyor's main requirement in his diagnosis of the defects of a building where plastics have been used is the performance of that particular plastic.

In the event of fire it is important to know how a particular plastic will react. For example, polystyrene will melt as it burns. The drips of the material can spread fire as well as cause serious burns to anyone who comes into contact with them.

The gas which is released during a fire is a problem with certain plastics.

The fire behaviour of most of the plastic components used in buildings does not have an immediate effect upon the rate of fire growth. The major hazards depend on the overall increase in fire load and production of smoke.

An exception must be made in the case of plastic ducts or a piped system. The updraughts caused by the failure of the plastics result in the creation of flues linking the various levels. The majority of pipes which are used within plumbing ducts are of the thermoplastic group and are liable to fail due to distortion caused by the heat. Where hot gases flow through the pipe failures will occur within a very short period of time.

Where decorative finishes are of a plastic material the identification of the material will enable one to indicate the performance of that material in the event of fire.

The recent requirement for improvement in thermal insulation has resulted in a number of materials being used, including expanded plastics. Some expanded plastics are used as a loose fill between the rafters at ceiling level or in cavity wall insulation. The materials which have been used include waste chips of expanded polystyrene, urea formaldehyde, and expanded polyethylene.

Certain of the expanded plastics inter-react with the plastics used for the insulation of electrical cables. It is important to ensure that the two materials do not come into contact. The insulation of buildings should not be carried out without checking the materials which are to be used and their properties.

PLASTICS

PLASTIC	CHARACTER-ISTICS	COLOUR	WEATHERING	FEEL	REACTION TO FLAME	USE
ACRYLONITRILE –BUTADIENE –STYRENE COPOLYMER ABS	Tough, rather hard material, like PVC in appearance, mouldings have a good finish, good impact resistance, suffers from creep	May be transparent or any colour	Poor surface weathering (may be coated or metallised)		Melts on warming to 100°C, burns readily, smell similar to styrene	Sink and bath wastes, cisterns (NOT for drinking water), vent pipes
POLYACETAL (POLYFORMALDE-HYDE) POM	Tough, rather hard material, good mechanical properties, resistant to fatigue or creep, can be machined	Translucent white or range of colours			Melts on warming, burns fairly readily with pale blue-ish flame, smell of formaldehyde	Taps, window furniture
POLYTETRA-FLUOROETHYLENE PTFE	Tough, very smooth, chemically and thermally inert, low friction, non-stick in use, expensive	Usually whitish translucent		Waxy	Does not melt on heating, very difficult to burn	Jointing tape
PHENOL-FORMALDEHYDE PF	Hard and difficult to cut, inexpensive	Dark colour, often brown	Reasonably good weathering		On heating does not soften, smell of phenol, very difficult to burn	Used as fillers or binder for paper reinforced laminates, wc seats, electrical plugs and sockets, door furniture
NYLON	Tough, smooth and hard material, resistant to creep, mechanically strong, low friction, expensive	Usually translucent, can be black, white or any colour	Adversely affected by sunlight	Waxy	Softens and melts on warming, can be pulled into threads, burns with blue flame, smell of burning hair	Taps, window furniture, canopies, roller blinds, hinges, sliding door runners, air supported buildings
POLYCARBONATE PC	Very tough and hard, does not cut easily, excellent impact resistance, expensive	Attractive transparent, or may be coloured	Reasonably good weathering		Softens on heating at 110°C, burns reluctantly with black sooty flame and smell of phenol	Light fittings, lampshades, window glazing
RIGID POLYVINYL CHLORIDE STRUCTURAL FOAM PVC	Microcellular structure.	Any colour		Hard, solid and dense	Softens on warming, burns only with difficulty, stops burning on removal from flame, acrid fumes	Solid material as core of composite board
POLYETHYLENE (POLYTHENE) PE	Fairly flexible, easily cut, floats on water, large thermal movement, may suffer creep, relatively cheap	Usually translucent, often colourless may be black	Weather resistant only when black, good chemical corrosion resistance	Smooth waxy feel	Melts on warming to 100°C, burns readily with smell of candle wax	Water pipes, cisterns, waste pipes, electrical cable covers, damp proof courses

PLASTICS

PLASTIC	CHARACTERISTICS	COLOUR	WEATHERING	FEEL	REACTION TO FLAME	USE
POLYPROPYLENE PP	Floats, large thermal movement, may suffer from creep, very similar to polythene but harder	May be translucent white, black or other colour	Not good in outdoor use	Smooth waxy feel	Burns readily with smell of candle wax	Hot water pipes, cisterns, wc cisterns, waste pipes, ventilation ducts
POLYMETHYL METHACRYLATE (ACRYLIC) PMMA	Hard, rather brittle material, attractive appearance	Transparent or any colour	Excellent weathering		Burns readily with black smuts of soot, melts	Taps, baths, basins, sinks, light fittings, dome and roof lights, glazing, door furniture, shop fascias
POLYSTYRENE PS	Brittle, hard material, metallic ring when tapped, relatively inexpensive, excellent impact resistance	Any opaque colour	Changes on weathering and becomes more brittle, affected by many solvents		Melts, low softening point, burns easily with black smuts of soot, smells of styrene	Cistern floats wc cisterns, light fittings, shades, wall tiles, lightweight aggregates in concrete
UNPLASTICISED POLYVINYL CHLORIDE UPVC	Easily jointed by solvent weld, good strength properties, can be suitable for outdoor use	Transparent, translucent, pigmented	Good	Hard, light, solid	Burns only with great difficulty, low softening temperatures	Mains and water service pipes, gutters, downpipes, soil pipes, drains, electrical conduit, translucent corrugated sheets, external weather boarding
POST-CHLORINATED POLYVINYL CHLORIDE CPVC	As above	As above	As above	As above	Higher (100°C) softening point than upvc	Hot water pipes and fittings, sink, and bath waste pipes
PLASTICISED POLYVINYL CHLORIDE PVC	Flexible, easily cut, good impact and abrasion resistance	Translucent or any colour			Low softening temperature, acrid fumes, may burn readily	Cable covering and insulating materials, wall coverings, wallpaper, suspended ceilings, canopies, roller blinds
POLYESTER UP resin which sets on curing glass reinforced	Hard, rather brittle in cast form, most commonly has glass fibre reinforcement. High tensile strength though low stiffness	May be transparent or any colour	Good weather resistance (adversely affected by incorporation of flame retardents)		Does not soften much on heating, burns readily with yellow sooty flame and smell of styrene	Resin binders for concrete, industrial flooring
EPOXY RESIN resin which sets on curing	Hard, sometimes reinforced with glass fibres, excellent adhesion, rather brittle	Translucent or may be filled and coloured, yellowish tinge			Does not soften on warming, burns readily with a fishy smell	Adhesive, or in repair work, industrial flooring, resin binder for concrete repairs

PLASTICS

PLASTIC	CHARACTERISTICS	COLOUR	WEATHERING	FEEL	REACTION TO FLAME	USE
UREA-FORMALDEHYDE UF available as resin for adhesives or as mouldings	Hard and difficult to cut fishy smell, inexpensive, rather brittle, can be a soft or low density foam	Light colours, white and off-white usually	Poor weathering behaviour		Does not soften on heating, very difficult to burn	WC seats, electrical plugs, sockets, switches, adhesive binder in chipboard or plywood
MELAMINE-FORMALDEHYDE MT	Hard, difficult to cut, relatively expensive	May be clear or coloured			Does not soften on heating, very difficult to burn, fishy smell when warmed	Surface laminates
EXPANDED POLYSTYRENE EPS	Bead or microcellular structure, may have a ringing sound on tapping	Usually white or pastel coloured			Melts, burns readily, black smuts of soot, smell of styrene	Underfloor insulation, flat roof insulation, wall cavity filling, impact sound absorption, ceiling tiles
FOAMED RIGID POLYURETHANE	Fairly soft microcellular material, not very flexible, partial cell collapse under compression	Usually creamy white, yellow, brown, grey			Does not soften on warming, burns readily with smokey flame, slight acrid fumes	
FOAMED UREA-FORMALDEHYDE UF	Very soft texture resembling cotton wool, friable	White or tinted			Does not soften on warming, burns only with great difficulty, smells fishy	Wall cavity insulation
PHENOLIC FOAM PF	Soft friable, broken cellular structure	White or tinted			Does not soften on warming, burns only with great difficulty, smells fishy	
FOAMED POLYISO-CYANURATE PIC	Fairly soft microcellular structure, not very flexible, partial cell wall collapse under compression	Brown, yellow			Does not soften on warming, burns only with difficulty, slightly acrid fumes	Ceiling boards
FOAMED FLEXIBLE POLYURETHANE PU	Soft, tough, resiliant microcellular structure, rigid or flexible, usually as a foam, can be supplied as liquids which foam and set on mixing	White, yellow, brown or pastel			Softens and shrivels on warming, burns fairly easily, slightly acrid fumes, produced in large amounts which is toxic in large volumes	Cavity wall inslulation
EXPANDED POLYVINYL CHLORIDE PVC	Even cellular structure, cells may be small or large, fairly rigid and tough	Yellow to brownish colour			Shrivels and decomposes on warming, does not burn readily, acrid fumes	
EXPANDED POLYETHYLENE	Microcellular structure, flexible and resilient, recovers if indented with finger nail	White, sometimes black		Soft, waxy	Melts on warming, burns readily with smell of candle wax	

Openings

OPENINGS

It may be said that man sees the world through a window but a surveyor can see much of the condition of a building in a window.

The window has developed from the narrow slits of medieval times to the modern curtain walls of glass which were made possible by the development of frame construction.

The development of glass towards the end of the Middle Ages, which enabled windows to throw light into the interior of the buildings, encouraged the advancement of interior design and decoration.

This development was restrained from 1695 to 1851 when windows were taxed. The early cottages of the Middle Ages had small openings in the walls protected by shutters on the inside and with a louvred finish to the outside of the opening to prevent water penetration. This system was developed into a rudimentary casement window in Tudor times and these were often of metal construction set between stone mullions. The sliding sash window was developed in the 17th century and was in universal use until the development of the various window forms that have become the vogue during the 20th century. The shape and appearance of windows has been an indicator of the age of buildings and has identified the development of various styles in architecture over the centuries.

The surveyor approaches the window on the basis that it is the weak spot in the external wall of the building. It often carries as much information as a tell-tale would in the area of a structural failure. In order to determine the failures in any building it is important that one has a working knowledge of the construction in use at the time the building was constructed.

The use of timber for the support over openings was common in brick built buildings from the 1730's until the 1940's and although the development of iron and steel occurred much earlier the use of metal lintels did not become standard until after the Second World War. Solid stone lintels were used extensively and these are usually clearly visible from the exterior as they pass through the complete thickness of the wall. The use of the brick arch started around the early 1720's and was later modified to the cut brick square-head which is still based on the arch form of structure by using a key stone. In many cases the external brick finish was not the main structural component, the major load being supported by an internal timber lintel.

Stone sills have been in common use right the way through to the present century. In the 18th and 19th centuries they were often formed of a very soft stone which deteriorated as a result of weathering.

An external examination through binoculars will soon reveal failures in the cut brick arches or the alignment of brickwork over the heads of the windows. The frontages of a number of other properties of a similar type within the same area should be examined to see whether there is a common defect in that age and style of property.

Slippage of brick arches is very common. It sometimes occurs because the splay cut brick arch utilised soft bricks, which were easy to work. The arches are also prone to failure following minor movement in the brickwork surrounding the opening. Whilst it is possible to repair an arch and repack the joints, on many occasions the remaining brickwork of the property has also taken up a new line making the re-alignment of the arch very difficult.

Movement within a building usually causes fractures to the stonework of the sill; the stone not being capable of taking stresses set up by the redistribution of weight in the surrounding brickwork. In some cases movement resulted in the re-alignment of the sill so that water flowed backwards off the sill and into the building. If this has occurred the defect must be stressed in the report, even if there is no sign of current failure.

It is important also to make sure that the sill section has a drip cut in the underside to prevent water running back on to the brick faces below the sill, particularly where that brickwork is part of a non-cavity wall. Without that drip the brickwork beneath the window becomes damp and that dampness is transmitted through the solid brickwork. Any window opening without an adequate sill in sound condition and with a substantial fall and a good drip should be commented upon within a report.

The design of the brickwork around a window varies depending on the type of window which is installed. In the case of the sliding sash window the brick opening has a

The gap between frame and sill will allow water behind the apron flashing below the sill. The decorations will not protect the wood either. At this stage stripping of the paint will be necessary – so read the paragraph on lead in paint.

set back so that a section of the front face of the timber window frame is concealed. The gap between the window frame and the brickwork was often covered by a render fillet or by the facing of the complete window reveal and head with a cement render. Any movement within the external wall was often revealed by the formation of a fracture between the timber frame and the brickwork or fracturing within the fillets or linings of the opening. It is an important part of the surveyor's inspection to thoroughly check each opening to each level so that he can determine whether any movement has taken place in the external walls. Where failures occur the surveyor must be sure to discover whether it is the frame itself which has failed through fungal decay or rot or if it is the brickwork which has moved.

Because the window forms a panel, all the rainwater which falls upon that section of the opening will discharge down its face and across the sill at the base. This means that there is a substantial pounding dished out to the wooden sill surfaces which are often exposed to a greater amount of water than the average weekend sailor's yacht.

In view of the substantial amount of water which falls across it, any minor failure may result in water penetration. Any window that was inserted more than 50 years ago is liable to have defective putty surfaces retaining the glass. A careful examination of the surface of the putty and the condition of the seal between the glass and putty should be made. On a lot of the older windows the timber mullions and transoms were of such minor dimensions as to give a minimal amount of rebate for any glazing and the failure of putty not only results in the absence of the securing of the glass but also

Above left: Defective pointing and poor render filler between frame and wall.
Above right: Always check the condition of the junction of the window frame and the brickwork. If the wall has moved there will be a crack at this point. Make sure it is not just the window having come loose.

Right: The timber sill has begun to decay as a result of the poor decorative finish and the absence of a drip groove on the underside of the sill. Check the putty on all windows. Failure in this allows water into the window frames and will cause decay.

in the exposure of the timber to deterioration caused by water retention to the back of the putty.

The surveyor must check the inner face of the wall immediately beneath the sill to see whether any dampness has occurred at that point. One should also use a sharp probe to determine whether the timbers of the window frame are sound or have begun to decay. If the lintel to the head is at all accessible it must be inspected to find if it is in sound order.

Although it may sound pedantic it is also worth mentioning when glass is fractured or cracked within sliding sash windows, the current replacement glass may be heavier than the glass which was originally inserted and the sash weights will be an inadequate counter-balance for the extra weight of the frame caused by new glass being fitted. Many surveyors are well aware of the guillotine effect that this produces. This becomes only too evident when they are carrying out a detailed examination of the sill section. It is only the absence of a woman knitting below the window which prevents a full re-enactment of a decapitation during the French Revolution.

The vertical hung sliding sash window is essentially formed from two frames of similar sizes which slide vertically between two boxes containing pullies and weights. There are two weights to each window frame. The sill sections were often of oak or similar hardwood and the box framework was also often of hardwood construction. The frames vary considerably with the type of construction and the quality of the

Top left: Mould growth on a softwood window frame is a clear indication of water penetration.
Top right: Rusting of the metal plaster bead and signs of water penetration. Could this be caused by condensation of water on the windows which has run onto the window sill and dampened the plaster?
Above: A vertically hung sliding sash window without a lock can easily be opened from the outside. Also note the failure of the stone sill. The repair with a cement fill has broken away and there is risk of water penetration into the building.

property. The sash weights which are solid cast iron tubular cylindrical objects were connected to the window panels by bits of string known as sash cords. These were haphazardly nailed to the side of the window frames and the whole contraption tended to work after a fashion for many years. The sash cords in the early days were waxed rope. On some occasions chain was used but nowadays wire reinforced rope, nylon reinforced rope or nylon ropes are often used.

It is fairly easy to spot whether a sash cord is in good condition, this is done by pulling out the cord and seeing if it breaks or bending it to see if it has dried out and is liable to fail. Even with the current nylon ropes which have a very long life it is difficult to know whether these have been adequately secured to the frames.

In order to carry out a replacement of the sash cord one has to remove the beading which forms the internal or external guide to the two frames. The beading which keeps these frames apart is known as a parting bead. A word that must have been selected by a surveyor because of the aptness of the description.

These beads usually suffered at the hands of the carpenters who carried out window overhauls throughout the ages.

The sill section of these windows, if they have not been maintained in good decorative condition, are often deeply ridged. These sills although they look terrible can often be of very sound timbers and consideration should be given to replaning the surface. This tends to be more satisfactory than trying to fill the cracks with proprietary external fillers.

The sash window has served well over the centuries but it must always be well maintained. Before the arrival of security locks they used to be the burglar's friend because it was relatively easy to slide a sharp object between the meeting stiles of the two windows and release the catch. The only problem with the arrival of Messrs. Banham and Chubb is that most surveyors now find that they are unable to carry out a detailed examination of the window or they are left with a six inch gap through which to push their head if they wish to look at the exterior of the timber work. Most surveyors acquire a set of various types of window lock keys but this may result in lengthy and somewhat embarrassing conversations with the members of the local constabulary. It is to be regretted that they seem unable to make a distinction between house breaking and a structural survey. The surveyor must request that all such keys are available when one makes the appointment to inspect.

The casement window which was introduced in the early Tudor buildings was a metal frame often set into grooves in the stone surrounds to the head and sill of the opening. The metal hinge was prone to failure usually by shearing of the pin of the hinge but the rusting of the metalwork is usually not nearly as advanced as that in the metal windows of this century. In those early days the diamond shaped leaded lights enabled some movement to take place within the frame without fracturing the glass. In later years the bulge or distortion in the leaded glazing had to be supported by metal bars across the window.

Where one comes across leaded lights it is important that one points out that any burglar can fold back the lead holding the glass panels and enter the property. This lead work is usually very weak and some water penetration will take place particularly in an exposed location. The replacement of the small panes has become a specialist's job not usually attempted by the door to door double glazing salesman.

The more modern metal windows, often called 'Crittalls' in honour of their early manufacturer, were also prone to failure around the hinge. In this day and age it is becoming very difficult to find people who can carry out repairs to these windows, but they do exist.

Minor movement of the property often results in distortion of the main frame and the failure to obtain a good fit when one opens a casement window can indicate current or historic movement in the building. Distortions in the frames in modern buildings can be caused by the builder trying to find out if the window frames bounce when they unload them.

Rusting of the metal casement can lead to fracturing in the glass panels. In many cases it is necessary to remove the glazing to adequately de-rust the frame. It is then possible to prepare and prime the metal prior to redecoration. The maintenance of the putty in good order as well as the

external decoration in good condition is essential for the long life of this type of frame.

The timber casement which returned to favour in the 1930's has gone through many developments and refinements. The early massive timber mullions having gone out of fashion and been replaced by slender and therefore more vulnerable timbers. In the early 1960's the English Joinery Manufacturers Association developed a timber cross section for use in the construction of windows. Window failures in the property constructed around that time may be attributed to defects in this design. One should look very carefully at any window frames in properties constructed during the period 1960-1970 as they usually show indications of failures within the casement timbers as well as deterioration within the sill sections. The damage was caused by the reduction in timber sizes which made them far more vulnerable to minor deterioration as well as the use of more vulnerable soft woods. The formation of grooves between casement and frame was

Below left: The horrifying result of water penetration due to putty failure.
Below right: Deterioration in a modern door. The frame is built to a minimum size so any failures will weaken it. A poor seal between glass and the door and an inadequate gap between the door and sill have resulted in water getting into the wood.

OPENINGS

intended to enable moisture which had penetrated the external rebate to be carried away. In many cases these cuts seem to provide far more chance of the water entering the timber framework because the wood was not protected by a high standard of paint covering in the rebates. The vulnerable parts lay at the rebate and frame edge. The paintwork was often skimped or worn through tight fitting casements. Deterioration took place which resulted in many of these windows needing replacement.

Because the number of vertical and horizontal bars restricted the view, many of these window frames have been replaced by householders who have a desire to gaze upon their carefully nurtured football pitches through expensive sections of plate glass. It is not the vulnerability of these new windows to Johnny's football that is a concern but the quality of the workmanship many of the companies who have been responsible for the reinstatement work. All too often the original solid timber frame has been cut away and a new highly glossed timber sill inserted. Although this is stained and varnished to look like hardwood on the "never have to paint that again in your life, guv" principle, in many cases these turn out to be soft wood of a square section with no external drip. They become vulnerable to deterioration after a few years. The securing of the frame to the structural opening either directly or via a sub-frame is often carried out to a poor standard. Where a new opening has been made the lintel used is often incorrectly installed and designed.

The universal use of mastic as a cure-all for any gap of less than 6 inches between window frame and brickwork has also resulted in some curious failures. On many modern buildings there is a cavity closing damp proof course immediately above the window which follows the line of the metal lintel. The spreading of mastic across the joint between the head of the frame and the brickwork may seal this point of discharge so that the liquid which had run into the cavity is returned into the inside of the property to each side of the window. There are also those people who seem to assume that mastic is also a glue and the need to provide adequate securings for the frame are obviated by its use.

The dimensions of the clear glazed openings must be checked because the glazed area must be not less than 10% of the floor area and the area of ventilation or area of open window should be not less than 5% of the floor area. A door does not count as a window and therefore does not form part of the calculation for ventilation.

Top: Putty and paint failures on a door panel will result in future failure.
Centre: A rotten bottom door panel.
Bottom: Putty and paint failures to a window frame.

Left: Chipboard used as a water tank cover has disintegrated due to absorption of condensation. Apart from rendering this useless as a cover, pieces of chipboard can drop into the tank and will cause problems in the water system. Chipboard should never be used in any situation where it is likely to be exposed to water, i.e. panels and floors in bathrooms.

condensation

With very rare exceptions the air always carries a moisture content. Warm air is capable of carrying a larger amount of moisture than cool air. When the air is carrying all the moisture that it is capable of at that temperature it is at saturation point and the temperature when this occurs is referred to as being the dew point. If this air meets a surface which is at a lower temperature than the air temperature, the moisture in the air will condense into water upon the cooler surface, thus forming condensation.

There is always a degree of movement of air within a building and a number of air changes per day take place through the wall surfaces alone. This means that there is a risk of the build up of moisture within a wall when the warm moist air meets the cooler external part of a wall and deposits the moisture that it had been carrying. This occurs when the internal temperature of the air is above its dew point and the external temperature is below the dew point. If the temperature of the internal face of the wall was at the level of the dew point condensation would occur on the face of the wall. It is important to identify the cause of the moisture content in the air within the building before one attempts to try and reduce the levels of condensation where problems have occurred. If the internal moisture content is caused by water penetration causing dampness to wall or floor surfaces it is this presence of moisture which is contributing to the condensiation. In most domestic properties the source of moisture is cooking, washing and body discharge.

It is important that there should be a minimum range in the temperature of the components within the interior of the property. Where this is unavoidable such as around windows adequate precaution should be taken to allow the discharge of water which will build up.

In the current economic climate great strides have been made in improving the level of insulation to buildings. This improvement in insulation usually reduces the ventilation which takes place. Little condensation would occur in an open barn with no external walls because the extent of ventilation would prevent warm damp air being able to condense against cooler surfaces. It is appreciated that this form of construction would be expensive to heat but the opposite example of concrete walled, floored and roofed units with no heating and no ventilation would mean that substantial condensation would occur even with only the moisture source of a human body. The substantial discharge of moisture from the body each night was recognised by our ancestors, and that is why they always used to air the bedding every morning.

With modern gas fired heating systems there is a tendency for the heat to be provided in limited periods throughout the 24 hours of a day. This means that there is a considerable temperature range throughout the day which brings with it the risk of condensation. These fluctuations also cause air movement which helps to reduce the problem.

The main problems which occur with condensation are where damage can occur to the surface or the construction of a property by the build up of moisture.

With pitched roof construction condensation will occur within an inadequately ventilated loft space on the underside of the roof surface. Where there is a roof pitch in excess of 30 degrees this should present few problems because the moisture will be discharged at eaves level.

CONDENSATION

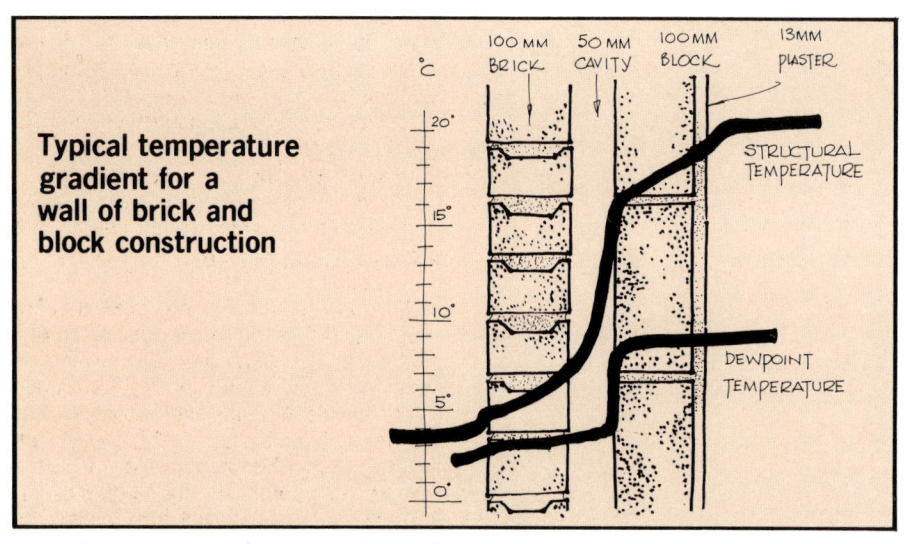

Typical temperature gradient for a wall of brick and block construction

In shallow pitched roofs of less than 20 degrees there is a possibility that this moisture may be allowed to build up against the timber components of the roof. The lower the slope, the greater the obstruction to through ventilation and the adequate discharge of the moisture. A roof space of the lower pitch with inadequate ventilation will be vulnerable to decay in the event of the build up of condensation within the roof void.

Where flat roofs are constructed the moisture build up due to interstitial condensation may cause embarrassing faults. Most timber flat roofs have no provision for ventilation.

Where insulation has been placed at ceiling level there is a likelihood that condensation will occur around the surface of the ceiling.

Where the insulation is placed beneath the roof covering condensation is likely to build up in the boarding beneath the insulation and to the head of the joist within the void. There is a lower likelihood of moisture occurring in the form of condensation within the void because of the absence of a cold surface so that ventilation is less important.

Where the flat roof is of concrete construction the colder surface of the concrete usually causes condensation to occur close to the face of the inner section. The same set of circumstances will occur where there is solid floor construction at ground floor level, and similar faults will occur in situations where the underside of the floor is exposed such as where there are balconies or garages or open areas beneath the property.

The majority of problems these days which are placed upon the desk of the surveyor relate to mould growth on clothing and shoes within cupboards and wardrobes. In property which is not centrally heated these storage areas should preferably be sited against internal walls which are subject to lower temperature variations. Where such problems occur additional ventilation should be provided to try and reduce the damaging effect of the settlement of moist air upon the contents of the wardrobe.

Where condensation occurs within a property an examination of the method of space heating should be undertaken. It should be remembered that non flued gas or paraffin heaters discharge a substantial amount of moisture into the air. Part of this moisture must fall as condensation. One of the symptoms of surface condensation is mould growth.

This occurs initially as spots increasing to larger patches and occurs in varying colours, usually green, black or brown. Mould growth is highly probable where there is a moisture content or humidity value of 70% for half a day. In most cases the mould can be removed by using diluted household bleach followed by washing with clean water and weak mixes of fungicide may also be used.

Where vinyl papers have been used in an area where condensation can occur to the rear of the paper, this mould growth cannot be dealt with until the surface of vinyl has been removed. Many of the pastes which this type of material is fixed with now include a fungicide to reduce the risk of mould growth to the rear.

The use of infra red film (see chapter on Equipment) will highlight the presence of mould growth and enable the full area affected to be treated.

When one is considering the enclosure of buildings and their performance relating to the movement of water vapour it is interesting to consider the situation in extremes of temperatures.

The average January temperature in Vancouver, Canada is 38°F while that in Ottawa is 10°F, the corresponding outdoor relative humidities in each location is the same, about 88%. When the outdoor air in these conditions is warmed to an internal temperature of 75°F the resulting humidity for Ottawa is 5½% while that for Vancouver is 26%. This is because although the external humidity is the same the different temperatures control the volume of water the air carries. In many public buildings having high ventilation rates and low rates of moisture supply, the relative humidities within the buildings will be close to these values.

When the ventilation rate is at the accepted minimum of 10 cubic feet per person the moisture provided by the occupant becomes quite significant and will result in humidities inside the buildings of 60% for Vancouver and 45% for Ottawa. This means that with no variation to the input of water into the air within the building by methods of air conditioning the internal humidity of buildings within two parts of Canada vary considerably.

The average Canadian house, as opposed to public buildings, has a transmission heat loss of about 0.04 BTU per cubic foot of building per degree temperature difference between indoors and outdoors. Ventilation will add to this to the extent of about 50% for each hourly air change. The tendency is to assume that the minimum acceptable air change per hour is approximately 0.3 but this is increased by loosely fitting windows and doors and it is normal for air changes to occur once per hour.

The average human being will produce between ¾ and 2 lbs of water per hour. The higher figures are based on water discharge into the air from activities such as washing as well as from the human body. If one is to maintain a relative humidity of 40% the air changes per hour for Canadian properties would vary so that a minimum rate of 0.4 would be required in Vancouver which would have to be increased to approximately one air change per hour on wash days.

In Ottawa the air changes would be required at the rate of 0.25 and 0.65. With the lower external temperatures providing a minimum weight in water for each cubic foot of air which is introduced into the properties it can be seen that there is a need to add water at the rate of approximately 10 gallons per day for each 10,000 cubic foot of building volume or per thousand sq. ft. of floor area.

In the United Kingdom where there are lower standards of insulation the introduction of water into the air within properties is not usually a problem which has to be faced but the increasing use of central heating within residential and public buildings has seen an increase in complaints from the occupants of excessively dry air. The external temperatures throughout most of the year mean that such claims are nonsense unless the level of seal to prevent air movement is extremely high.

If one is to avoid the build up of surface condensation on single glazed windows in the UK the humidity of the air should be about 60%, but in cold weather when the external temperature drops to freezing it is probable that the moisture content of the air within a building should not exceed 50%.

It has long been accepted that thermal sensations of comfort are reflected by relative humidity and that for the same comfort higher temperature is required to offset the decreased relative humidity. There is little evidence to show that either high or low humidities are in themselves detrimental to the health of normal people. It is generally assumed that extremes of humidity are undesirable and that it is better to keep within the range of 30% to 70%.

The survey and inspection of modern timber framed property

Check list

Is it timber framed?
gable
ventilators
windows
doors

Roof/Loft (pitched)
Is it ventilated? (3000mm²/m)
Where is membrane?
What type insulation?
Is roof frame braced?
Is there an expansion gap?
(brick panel increase, timber shrink)

Flat roof
Is there provision for cross ventilation? (50mm clear over insulation)

Cavity
Firestops
Vertical if cavity exceeds 8m long
Horizontal at ceiling levels
Fire stops at party walls
All around openings
50mm cavity essential to outside of frame and around chimneys (25mm)

Notching
Not permitted in trussed rafters
No hole to exceed ¼ depth or to be within 50mm of edge of timber

Membrane
Breather paper
Vapour barrier

Wall ties
400mm minimum vertical spacing
800mm minimum horizontal spacing
Holding-down bolts

How do you recognise a timber framed building? The exterior may resemble the traditional brick and block construction. On many estates timber framed and conventional buildings are situated side by side and are indistinguishable.

In the inspection of any property it should become second nature to tap wall surfaces. This would indicate where there are defects in the adhesion of plaster to the wall surfaces and would reveal whether there is a timber framework, stud or other form of less conventional construction. This does not automatically mean that the property must be timber frame but should be an early warning of an unconventional construction.

The external detailing of window frames may indicate the type of construction that has been used and the seals and finishes around windows and door frames should always be closely examined.

If, at this stage, you are suspicious of the type of construction and wish to be certain whether it is or is not a timber framed building you should carefully inspect any gable or party walls from within the loft. If it is timber framed construction the face of the party wall in terrace or semi-detached formats will clearly indicate the form of construction which has been used.

If the layout of the roof is of hipped construction and there are no party walls or gable walls on view you will have to crawl into the eaves to obtain a good view of the head of the wall construction. With the use of a flashlight and mirror you should be able to examine the construction at the wall head. If there are indications that the wall plate is not fixed to blockwork but to a timber framework you may well be carrying out the inspection of a timber framed building.

Timber framed construction and conventional block or brick cavity wall construction have one major difference, other than the obvious variations in the use of materials. A timber framed building has to have cavity closers located at 8m centres vertically and horizontally at each ceiling level. These cavity closers are essential to prevent the spread of fire between components of the building, and the condition of these closers should be inspected wherever possible. Unfortunately the inspection of a completed building will rarely give you a clear view of the cavity but this is a point which should be specifically noted where one is inspecting during the course of construction.

There are a number of checks that we need to make. Is the roof space ventilated? If the construction of the building is correctly carried out the building is wrapped on the inside with a vapour barrier. This means that moist air which builds up through normal human activities or, in the case of children, inhuman activities, is unable to escape through the walls. The absence of open fireplaces and the current trends for reduced ventilation and the installation of double glazing have reduced air changes within residential properties to an unacceptable level. (Most double glazing appears to reduce the available open window space to less than one twentieth of the floor area.)

The damp air will move up through the building and find its last refuge within the loft space. Unless there is adequate ventilation, condensation will occur at this point, resulting in deterioration of the woodwork of the loft. The requirement is that there should be 3,000 square millimetres of ventilation per square metre of the ceiling, as measured on plan. (A Barclaycard is 4,500 square millimetres so the ventilation is required at the rate of one Barclaycard of opening per 1½ square metres of ceiling plan).

The framework within the roof should be examined to make sure that it is properly braced. This will apply to all forms of buildings which have a timber trussed roof. Diagonal bracing is essential to prevent what is known as racking. This is the collapse of roof trusses often caused by wind pressure and is similar to the collapse of books on a bookshelf where the bookends are insubstantial.

Common faults in trussed rafter roofs include structural stress due to bad storage, manufacturing faults, unauthorised cutting of the timbers, inadequate support for cold water storage tanks, missing diagonal and horizontal bracing and lack of support at the change in the roof line.

Detailed inspection of the support for a cold water storage tank should be made and with a lightweight form of trussed roof it is important that the load should be evenly spread. (Make sure that there is a cover to the tank because there is little point in pumping warm moist air into the roof space on summer days.)

It is important to check that the timbers used within the roof are correctly sized. Where gang nailed joints have been used those should be closely examined. Many are of galvanised covered metal and failures are beginning to occur on the crease or fold marks on the plate in some examples of this type of joint. As they are the sole structural component of the joint the failure of these plates will have serious consequences when related to the life of the building.

Check that there is a reasonable expansion gap left above the brick external panel. This brick panel is mainly decorative but as it will have been built from new bricks there is a probability that they will increase in size as they take up moisture. They were naturally burnt dry during their manufacturing process. Timber, on the other hand, will have started moist and will have had its moisture content reduced over the first few months of the building's existence. This will result in the timber frame shrinking. If no gap is allowed the brickwork will lift the timber roof trusses clear of the wall plate and will set up strain which will result in deflection on the roof surfaces and probably water penetration.

The location of the membrane within the timber wall construction is of vital importance. It is essential that moisture in the form of damp air should not be able to pass out through the timber wall and condense within the insulation. Such an occurrence would result in damage occurring to the timber frame.

Most timber framed constructions are formed of timber studwork in-filled with insulation with a ply or similar board finish to the outer face and a polythene or similar membrane to the inner surface. This is then overcovered with plasterboard. The outer face of the ply is frequently covered with a breather paper. The intention of this breather paper is to restrict water penetrating into the timber frame during construction and also to prevent damage from any water which may bridge the cavity after completion. It is essential that any water that does enter the timber frame should be able to move out through the breather paper and evaporate into the cavity which must be ventilated. The examination of these membranes is occasionally possible in that narrow space within the loft at eaves level.

The insulation in most cases will not be visible and pattern staining on the surface of the interior would probably indicate any major failures. Often there is a slump in the fabric leaving a colder surface at high level. Pattern staining should indicate if this failure has occurred.

A detailed examination of the external face, particularly if it is brick should be carried out to make sure that no careless cavity insulation salesman has persuaded the houseowner to accept the benefit of cavity wall insulation. The cavity in timber frame construction must be kept free as the ventilation of this cavity is an essential part of the construction. There have been occasions where construction has been damaged by cavity insulation.

The cavity, which is an essential form and part of the construction, should be 50 millimetres wide and at least 25 millimetres all round any chimney brickwork that may be part of the construction. This cavity will be bridged by wall ties. These are required at 400mm centres vertically and 800mm centres horizontally. The framework should also be secured to the ground slab by holding down bolts.

The cavity must be ventilated and bearing in mind the location of fire stops, the provision of cavity ventilators is greater than the ventilation provision in conventional construction and may be yet another indication of the property being timber framed.

Because timber framed construction is not as forgiving as conventional construction concern must be voiced on the quality of work which has been carried out where the building has been altered.

The do-it-yourself home owner may be unaware of the importance of the polythene wrapping around the building components and may have removed sections of the building from their polythene bags. There may be constructors who are unfamiliar with the specific details of this form of construction and who may be responsible for future failures by poorly designed extensions.

It is evident that the surveyor who is inspecting a building which is timber framed will be able to express an opinion about those sections of the construction which are visible to him. A structural survey should be a surface inspection of a property with advice given on the information which can be obtained from the surface.

Because many of these buildings are erected to order upon prepared sites, the ground slab is frequently left exposed for many months. Where this slab incorporated a horizontal membrane it is possible that this may have been retained in position by the placing of bricks or blocks upon its surface or by the installation of the floor screed. Polythene is light reactive and may become brittle if it is left exposed to sunlight for substantial periods. Some failures have already been recorded in membranes of this type where this has been allowed to happen.

The examination of a timber frame building will not be able to reveal the presence of fire checks or fire stops, nor the location of insulation vapour barriers or vapour checks or breather membranes. It is important that the surveyor does check the moisture content of the timbers forming the base of the stud framework and to the floor plate. If there is a fault in the location of the various vapour checks and breather membranes the moisture content of the timbers will be in excess of 15 percent. Of the buildings that I have inspected over the last five years, I have yet to find a moisture content in excess of 12 percent. In order to carry out an examination of the moisture content of these timbers it will be necessary to use either a moisture meter which is calibrated to take readings of moisture content behind plaster board or to use a hammer probe where insulated pins are driven through the plater board and into the timber frame behind. It is important that the reading be accurate and isolated from any material which may influence the reading. The insulated prongs would prevent a foil-backed plaster board interfering with the reading, or condensation dampness on the face of the plaster board within the building influencing the moisture content that one is taking of the timbers behind. If you obtain high readings from an electrical resistance moisture meter, check, if you are able, other timbers because it is possible that they may have been treated with a copper chrome arsenic wood preservative. The copper element within this preservative means that there is a natural conductivity to the woodwork which can falsify the readings you obtained from this type of moisture meter.

It is important that the surveyor has a knowledge, not only of the method of construction used and its limitations or failures, if any, but also of the marketability of this type of building construction. The surveyor must point out in his report the type of construction which has been used, and may also desire to comment upon the continuing debate over the long-term qualities of this type of construction.

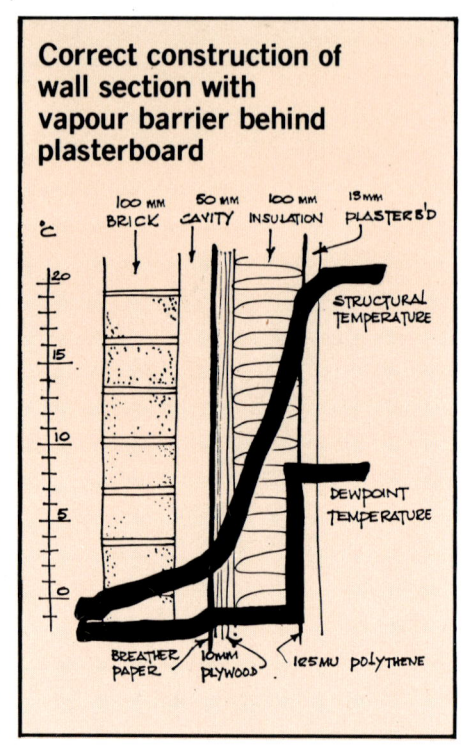

Correct construction of wall section with vapour barrier behind plasterboard

Modern timber framed buildings have been in existence for over fifty years. The majority of those few failures that have been found are the result of the design or construction deviating from the ventilated cavity. Water penetration into the frame has occurred in exposed locations, where the door and window frames were poorly fitted. Damage occurred where the exterior was rendered onto expanded metal fixed directly onto the frame.

The danger of fire is no greater than that to brick and block buildings. Fires in the cavity are few, although often well publicised.

Timber framed buildings are an accepted and reliable form of construction which demands knowledge and thought from the surveyor during his inspection.

Rust on a cast iron rainwater pipe has split the collar to the lower section.

Corrosion of metals

Corrosion usually results from a complex electrical action where dissimilar metals are electrically connected via a conducting liquid and a short circuit galvanic cell is formed. Metal in the form of positively charged ions is removed from the anode but the cathode does not corrode. Thus zinc protects steel because it develops a lower potential and becomes the anode in the cell. The rate at which this electrical corrosion proceeds depends on the potential difference between metals in a particular environment. Within buildings the most important relationships to bear in mind are those between copper (and the copper alloys of bronze and brass) and zinc (including galvanised coating and aluminium).

A table shows the relationship of various metals. The speed of deterioration will increase the further the metals are apart in the table.

Copper and the copper alloys will attack cast iron, mild steel, plated steels including stainless steel, galvanised steel, zinc and

143

CORROSION OF METALS

aluminium. Aluminium is also vulnerable to nickel and mild steel.

The rate of corrosion increases with temperature and in the case of the combination of zinc and steel the reversion of polarity occurs above 70°C. This makes water pipework mixing these materials in a hot water system particularly vulnerable.

These electrolitic actions may occur where particles are deposited upon a vulnerable material. For instance if hot water flows through copper pipework there is a probability that it will pick up small sections of copper and these could be deposited on any galvanised pipework or sections of the water cylinder.

Aluminium and cast iron gutters may also be vulnerable to rainwater which has passed off a copper covered roof surface.

Corrosion can also take place if pipework comes into contact with chemicals which are injurious to it. Sometimes this may be the result of metal components being in different environments such as steel pipes passing through a damp plastered wall. Such pipes should always be wrapped to protect them from the dampness and the rusting of the surfaces.

Copper pipework which is embedded in concrete or cement screeds without protectional wrapping is vulnerable to corrosion. The deterioration in the copper leaves it perforate, but the failure often goes undetected for months and maybe years and is very expensive to repair.

Keen sailors will be aware of the advantages of having what is known as a sacrificial anode under a boat. This is a lump of metal placed to the underside of the boat which will corrode instead of the bronze surfaces of the propellor and propellor shaft. In a building an immersed magnesium anode connected electrically to a galvanised steel cylinder or cistern protects the zinc coating while a protective scale is forming. This is particularly effective in soft waters and the scale deposited on the inside may become an effective protective.

Examples of corrosion of copper, aluminium and galvanised pipe components.

Table of metals and their interaction

Anodic	Magnesium
	Zinc, including galvanised coatings
	Aluminium
	Cadmium
	Copper-aluminium alloys
	Iron and mild steel
	Chromium
	Lead
	Tin
	Nickel
	Brass
	Bronze
	Copper
	Stainless steel
Cathodic	Silver

CORROSION OF METALS

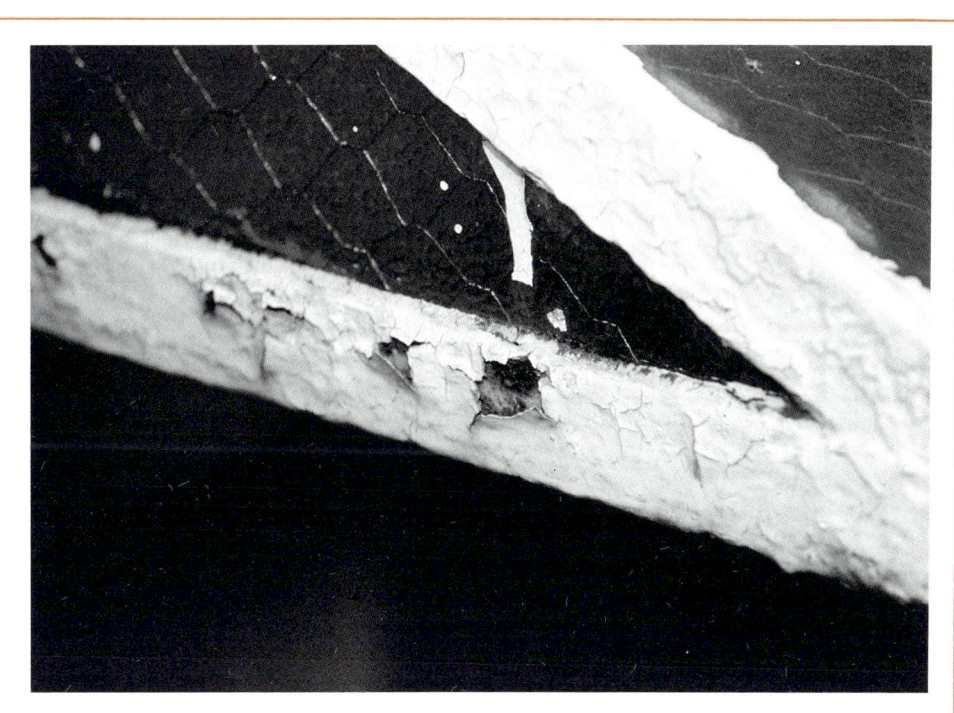

Right: Corrosion of metal window frame.
Below: Corrosion of metal rooflight.

CORROSION OF METALS

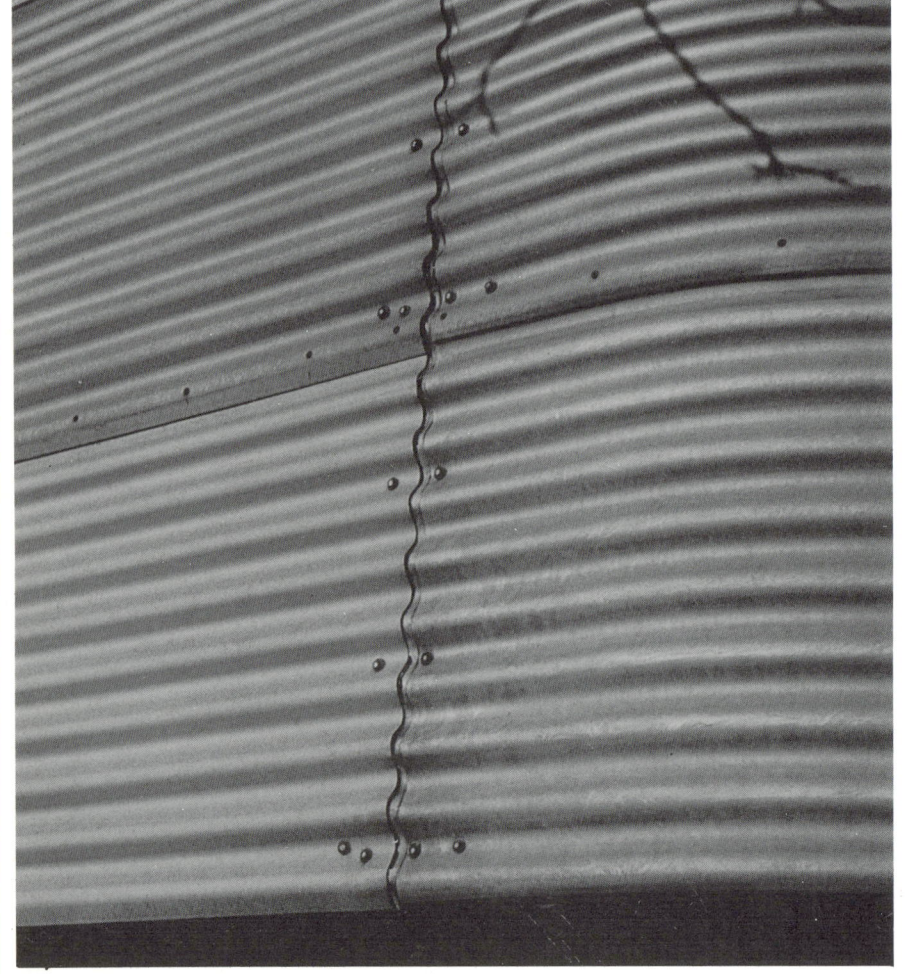

Above: Typical example of coated steel cladding.
Left: Corrugated galvanised iron used as cladding.
It is important that the surveyor should familiarise himself with the processes involved in the construction and installation of this type of structure so that he may be aware of faults and corrosion that might be hidden.

Asbestos

Asbestos has long been suspected as being injurious to health but it was not until 1920 that the airborne fibrous dust was first recognised as being the cause. Asbestos is a generic term for the fibrous forms of several mineral silicates that occur naturally in seams or veins generally between 1 and 20mm in width in many igneous or metamorphic rocks, and belong to one of two large groups of rock forming minerals: the serpentines and the amphiboles.

The Serpentine group contains the type of asbestos known as chrysotile (white asbestos) which is the only asbestiform member of this group of minerals and by far the commonest and commercially the most important type of asbestos.

The Amphibole group contains crocidolite (blue asbestos), amosite (brown asbestos) and anthophyllite, actinolite and tremolite. Amosite is mineralogically known as gunnerite asbestos. Tremolite may occur as a contaminant with chrysotile and with other minerals such as talc.

All these materials have been exploited commercially but the chemical composition of each is different. Chrysotile fibres tend to be whitish in colour and have a smooth silky texture, while crocidolite fibres are usually blue and shorter, straighter and less silky than chrysotile.

Amosite tends to be brown with fibres more brittle than either chrysotile or crocidolite. The colour is not a dependable guide for identification. The main properties which give asbestos commercial value is its incombustibility, strength and effectiveness as a reinforcing or binding agent when combined with other materials such as cement or plastic.

Chrysotile or crocidolite fibres when broken down and separated are strong enough to be spun and woven into cloth. The largest world producer of asbestos fibre is believed to be the USSR but the main sources of asbestos imported into the UK are Canada and Southern Africa.

Raw asbestos used in Britain is usually milled in the country of origin. The milling process separates the asbestos fibre from the useless non-fibrous material. In 1975 chrysotile accounted for about 85% of the total import of fibre into the UK. Almost the whole of the remainder was composed of amosite. It should be remembered that at this time the dangers in the use of crocidolite were known, the importation of crocidolite ceased after 1965. Crocidolite was used in yarn, cloth, rope lagging, webbing and packings, in the production of asbestos mill boards and insulation mattresses, in preformed thermal insulation, sprayed asbestos, thermal and acoustic insulations, battery cases, filtration materials, asbestos cement flat sheets, corrugated sheets, pipes and moulded products, asbestos cement pressure pipes and the reinforced plastics.

Crocidolite or blue asbestos is no longer used in the manufacture of asbestos products in this country nor has any import taken place since 1970. Although imports of raw crocidolite ceased in 1970 substantial quantities of material remained as insulation materials in building, ships locomotives and railway carriages. Most of these are likely to cause problems only when they are broken up or damaged so as to release fibres into the air. The risk of this is greatest with some types of insulation products.

At the present time a high percentage of the asbestos fibres used in the production of asbestos based products in this country is the chrysotile type. This is extensively used in the production of asbestos cement for building works, friction material, floor tiles and fillers and reinforcements in mastics and adhesives.

Asbestos cement is cement reinforced with asbestos fibre of approximately 10% to 15% of the total by weight. Although it is a non-combustible material it will not withstand exposure to direct flame or temperatures above 260°C.

The asbestos dust levels generated in asbestos cement manufacture generally are low in comparison with most other manufacturing processes involving asbestos. Most asbestos cement does not undergo any further manufacturing operation after leaving the factory. They do undergo an intermediate stage during which they are cut, drilled or otherwise processed by specialist companies to meet the requirements of the contractor before reaching him. When this work is carried out there is a likelihood of contact with dust from sawing or drilling operations. Asbestos manufacturers have recently tried to reduce the amount of on-site cutting and drilling which is necessary by increasing the range of sheet size available.

Roofing felt is often made of asbestos paper impregnated with bitumen and coated with fine sand or chalk. For damp proof courses bitumen impregnated felt is further laminated to metallic sheeting. Site operations using these products are likely to generate small amounts of asbestos dust.

Asbestos fibres are often added to paint to give it not only the ability to produce various finishes but also strength. One of the most common products that was on the market was Artex compound which is manufactured from water soluble resin and combined with fillers and pigments.

As the paint has a very low asbestos fibre content which is effectively sealed *in situ*, particularly with each further application of paint, the risk is minimal. There is however a small potentional risk despite the lower asbestos fibre content that fibres could be released when the surfaces are rubbed down, for example before recovering with new paint. The manufacturers have now replaced asbestos and the revised formula is available from 1983.

Many plastic flooring or floor covering materials contain asbestos. One type of material in this category is the thermo-plastic tiles, pvc, vinyl and asbestos tiles. The tiles are made from a blended composition of thermo-plastic binder, asbestos fibre fillers and pigments.

Dust generated from cutting tiles is minimal and from analysis is shown to be nearly all limestone, the other major non-polymeric product in flooring.

Asbestos is also used in the nosing of staircases. The composition of this product is very similar to that on a moulded brake lining and was intended to give a non-slip surface to the edges of stair treads. Ferodo have discontinued this use of asbestos because high dust levels were found to arise during cutting and fitting.

Insulation board is a major use of asbestos. Insulation board is used for partitions and suspended ceilings that are required to provide a fire barrier, or to protect structural steelwork. Potential risk to operatives is great during installation and dust respirators must be used during all cutting and sanding operations.

Sprayed (Limpet) asbestos

This was extensively used to produce fire resistance for structural steelwork and was also used to provide sound absorption. The application process is relatively risky for the operative and sprayed asbestos tends to deteriorate fairly quickly and give off fibres in the surrounding area unless it is sealed or coated. The process has now been discontinued but many buildings are insulated or the steel protected in this way.

Asbestos lagging for thermal insulations has in the past taken the form of rigid pipe sections or slabs as well as sprayed coating.

The asbestos content of such sprays could approach 100% and crocidolite was one material used in them. For this reason its use was discontinued.

Boiler lagging and engineering works often consisted in the past of layers of 85% magnesia reinforced with 15% asbestos fibre, which in many cases was crocidolite with a surface coat of asbestos composition about 1cm thick. The fact that magnesia is a white substance is often misleading as it suggests that no crocidolite is present. Similar materials have also been used in locomotives and railway carriages built before 1967.

An important use of crocidolite is in the manufacture of asbestos cement pipes. Crocidolite may be used there, but not in excess of 20% to 25% of the total asbestos content, because its special properties make it more suitable than other types of asbestos for the autoclave process of pipe manufacture. This process is not used in the United Kingdom.

Sprayed asbestos coating and asbestos containing thermo-insulation material should be thoroughly soaked or sprayed with water before any stripping may take place.

With thorough soaking the dust levels can be brought below five fibres per ml. Even with wetting the operatives must wear protective clothing and must use protective equipment.

It must be remembered that health injuries occur when there is prolonged exposure to an intake of asbestos fibres and it is unlikely to occur where there is an intermittent or isolated exposure.

Identification of crocidolite in insulated materials and sprayed asbestos can only be achieved by one of the following methods:

> x-ray diffraction;
> optical microscopy;
> electron microscopy.

The Asbestos (Licensing) Regulations 1983 came into force on 1 August 1984. They impose strict controls on work with asbestos insulation or coating. Only a contractor who holds a licence granted by the Health and Safety Executive may carry out work involving the use of asbestos products. The exceptions to this are short-term repair or maintenance, or an individual carrying out work to his own premises. The licensing system does not extend to such asbestos based materials as cement, insulating board, roofing felt or vinyl floor tiles.

There is no regulation which requires that any asbestos-containing material be removed from the building. It is frequently advisable not to disturb the material which contains asbestos, but to seal it in its existing location. Where this is not possible or where the asbestos-based material has already been damaged, it should be removed in carefully controlled circumstances.

Testing

Corrugated asbestos roofs

It is now believed that about 1% of the asbestos cement roofs currently installed in the United Kingdom may contain crocidolite. If a surveyor is inspecting such a building, he may recommend that a section of the asbestos be sent away for analysis. A list of laboratories that carry out this work is set out on page 149. The majority of laboratories will respond within days for a fee in the region of £50-70.

Airborne fibres

Where a surveyor is inspecting a building which has sprayed asbestos coatings or damaged asbestos insulation, he may feel it necessary to recommend that a test of the fibre content of the air be carried out. This will enable a decision to be made as to whether the asbestos may be retained by sealing it or whether it should be removed. The laboratories set out in Appendix will carry out this function or the assistance of the local Health and Safety Executive may be sought.

Control limits

The Control Limits for asbestos dust levels in the working environment are laid down by the Health and Safety Executive in guidance note EH10 (revised July 1984). The United Kingdom limits are the most stringent in the world. United Kingdom standards:

Chrysotile (white)
1970 — 1982
2 fibres per millilitre of air
1983 — August 1984
1 fibre per millilitre of air
1 August 1984
0.5 fibres per millilitre of air

Ammersite (brown)
1970 — 1982
2 fibres per millilitre of air
1983 — August 1984
0.5 fibres per millilitre of air
1 August 1984 0.2 fibres per millilitre of air

Crocidolite (blue)
1970 — 1982
0.2 fibres per millilitre of air
1983 — August 1984
0.2 fibres per millilitre of air
August 1984
0.2 fibres per millilitre of air

These control limits represent the maximum levels of permitted exposure. If there are more fibres per millilitre of air in the tested sections, that part of the building will be in need of urgent attention and no work should be carried out within that area until remedial work has been completed and the fibre count reduced.

ASBESTOS

Organisations undertaking asbestos analysis for industry on a repayment basis

	Organisation	Contact		Organisation	Contact
1	Air Sample Counts Analysis 2 Hereford Way Stalybridge, Cheshire	061 338 5645	17	Central Electricity Generating Board Scientific Services Department Ratcliffe-on-Soar Nottingham NG11 0EE	0602 830591
2	Air Sample Counts Analysis 7a London Road Staines, Middlesex	078 61302	18	Chemical Defence Establishment Porton Down Salisbury, Wiltshire SP4 0JQ	0980 610211
3	Albright and Wilson Analytical Services PO Box 80 Oldbury Warley, West Midlands	021-552 3333	19	City of Birmingham Council Engineer's Department Industrial Research Laboratories Brasshouse Passage Broad Street Birmingham B1 2HR	021-643 6987
4	A H Allen and Partners Public Analysts Laboratory 67 Surrey Street Sheffield S1 2LH	0742 21687	20	City of Birmingham Council Environmental Department 120 Edmund Street Birmingham B3 2EZ	021-235 9944
5	Analytical Laboratory 7 Offham Road Lewes, Sussex BN7 2QR	079 16 4534	21	City Laboratories 126 Mount Pleasant Liverpool L3 5SR	051-709 3932
6	Analytical and Consulting Chemists Metallurgists and Engineers The Laboratories Fortune Lane Elstree, Hertfordshire WD6 3HQ	01-953 1306	22	City of London Polytechnic Department of Metallurgy and Materials Whitechapel High Street London E1	
7	Atek (Pollution Control) Ltd Barclays Bank Chambers Hick Lane, Batley West Yorkshire WF17 5SQ	0924 471272	23	Commercial and Forensic Laboratories 220 Elgar Road Reading RG2 0DG	0734 82428
8	Atomic Research Establishment Environmental and Medical Services Division Harwell, Oxfordshire	0235 24141	24	County of Avon Scientific Services County Laboratory Canynge Hall, Whatley Road Bristol BS8 2PR	0272 312371
9	Baine's Environmental Consultants Ltd 140 Bolton Street Ramsbottom Lancashire BL0 9JA	070-682 2156	25	County offices Matlock Derbyshire DE4 3AG	0629 3411
10	Bath University South West Industrial Research Unit Claverton Down Bath Somerset	0225 6941			
11	Birbeck College Department of Crystallography Male Street, London WC1	01-580 6622			
12	Bostock Hill and Rigby 35-39 Birchfield Road Birmingham B19 1SU	021-359 5951			
13	British Belting and Asbestos Group Ltd Industrial Health Unit PO Box 18 Cleckheaton, Yorkshire	0274 875711			
14	British Ceramic Research Organisation Queens Road Stoke-on-Trent, Staffordshire	0782 45431			
15	British Rail Research Department Engineering Division Derby	0332 49203			
16	Cape Asbestos Ltd Fibre Research Laboratory Iver Lane, Cowley Uxbridge, Middlesex UB8 2JQ	89 37111			

DECORATION AND EXTERNAL SURFACE FAILURES

Frost action

Deterioration of materials by frost action is caused by the expansion of water when it changes from a liquid to a solid upon freezing.

The performance of certain materials vary when subjected to repeated freezing and thawing.

Materials vary in the amount of pore space that will be filled up with water. If nearly all the available space is filled then the frost action will result in the creation of enormous pressure within the material. If only part of the pore space is filled, there may be room for expansion, and less damage caused. The saturation coefficient of a material often gives some guide to its survival in the event of frost.

A laminar structure is usually more liable to deterioration. The photograph showing a failure in a render surface illustrates the vulnerability of this form of construction where water is trapped to the rear of the render finish.

The brick failure has occurred in faced bricks, and the surface coating has been pushed away from the brick body.

The exposure of reinforcement in the concrete column may have resulted from corrosion of the steel. Did the steel corrode due to water gaining access as a result of fractures in the surface being opened up by frost action?

The metal handrail failure was caused by water becoming trapped inside a hollow section of steel. No drainage hole was provided at the bottom of the slope. Severe winter conditions resulted in this water freezing and 're-shaping' the steel section.

A new problem has recently been reported. Asbestos tiles or slates are much smoother than the clay or slate products they imitate. When snow settles upon an asbestos slate-covered roof it is unable to achieve the same friction grip that it could with natural materials. Snowslides from the roof tend to occur in much larger and heavier quantities.

The gutter bracket or support for plastic guttering tends to be uanble to cope with this shifting snow load. A surveyor should comment upon the risk, and may recommend that snow-guards should be fitted to the roof surface.

Decoration and external surface failures

In the preparation of this chapter I am indebted to the assistance and advice of the technical services section of BLUE CIRCLE CEMENT. The photographs which have been used in this chapter have also been provided by them.

The Surveyor should be able to identify failures in the external finishes of buildings.

This chapter sets out examples of failures for easy recognition of the defect and sets out the remedial work which would be required.

As a general rule a paint coating combines the functions of a waterproofing and a decorative medium. In the examination of the decorative failure it is important that the waterproofing function is considered, because a failure in this function will require a close examination of the component which may have deteriorated as a result. Most of the materials available fall into four groups: cement paints, emulsion paints, normal oil paints and silicate paints.

These coatings will all require periodic replacement. Where paint failures have been caused by building defects it may be possible to reduce future rapid failure of the repainted surfaces. Where this is possible the recommended repair is set out below.

Gloss paint on woodwork

Oil paint is usually impermeable to water, and is likely to fail if it is applied to a damp surface or if the surface becomes damp at some stage. The trapped moisture may force the paint film away from the surface. Plastic paints tend to blister and fail, or peel in flexible strips, whereas lead paint comes away in brittle flakes. The success of this type of paint depends on satisfactory maintenance. As soon as the surface has cracked or split it is necessary to repair the 'hole' in the protective membrane. Water seepage into the wood surface will cause further paint failure and probably result in damage to the woodwork occuring.

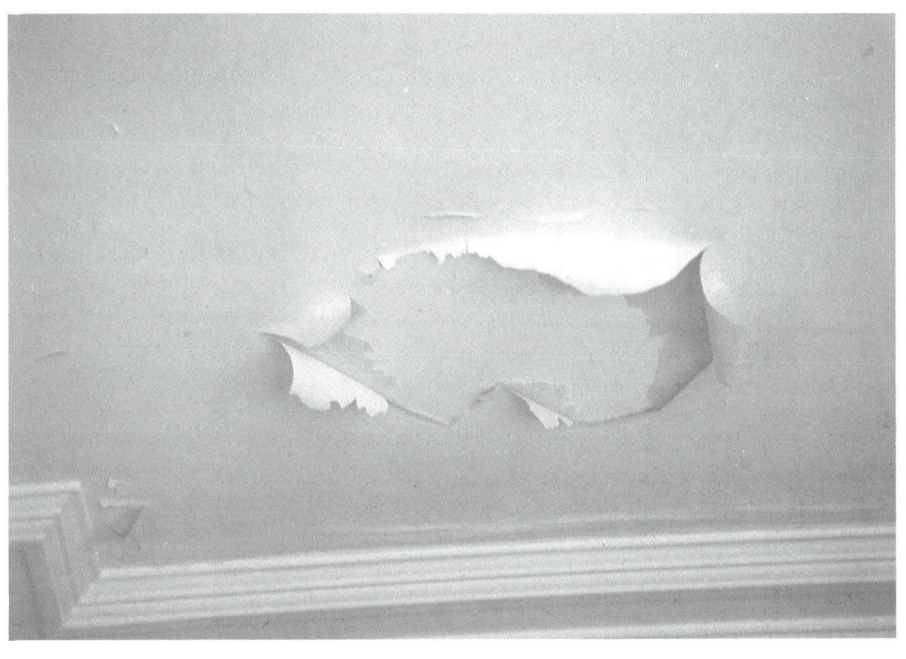

DECORATION AND EXTERNAL SURFACE FAILURES

Rendering

1. **Paint failures caused by decorations having been carried out on unstabled surfaces.**

   The loose and flaking material will have to be removed by scraping, or by the use of a stiff bristled brush to produce a sound edge. The use of wire brushes is not recommended. In cases of a heavy build-up of decorative coatings it may be necessary to use high pressure water jet equipment grit blasting, needle gunning or burning off. The surfaces must be clean, dry and free from anything which will interfere with the adhesion of the materials which are to be applied.

2. **Organic growth on friable undecorated rendering.**

   All visible signs must be removed by scraping, or by stiff bristle brushing. The spores must then be killed by the application of a fungicide in accordance with the manufacturer's instructions. In extreme cases of organic growth it may be necessary to use high pressure water jet equipment or further applications of fungicide.

3. **Failures due to insufficient cement content in the original cement/lime/sand rendering.**

   Areas of defective rendering must be removed and the surface stiff bristle brushed. The surface must be then damped with clean water and made good with a render coat suitable in type and strength. It is important that the patch material relates to the existing finishes so as to minimise the fracturing between the patched and original surfaces.

   Horizontally key each of the render undercoats with a wire scratcher.

DECORATION AND EXTERNAL SURFACE FAILURES

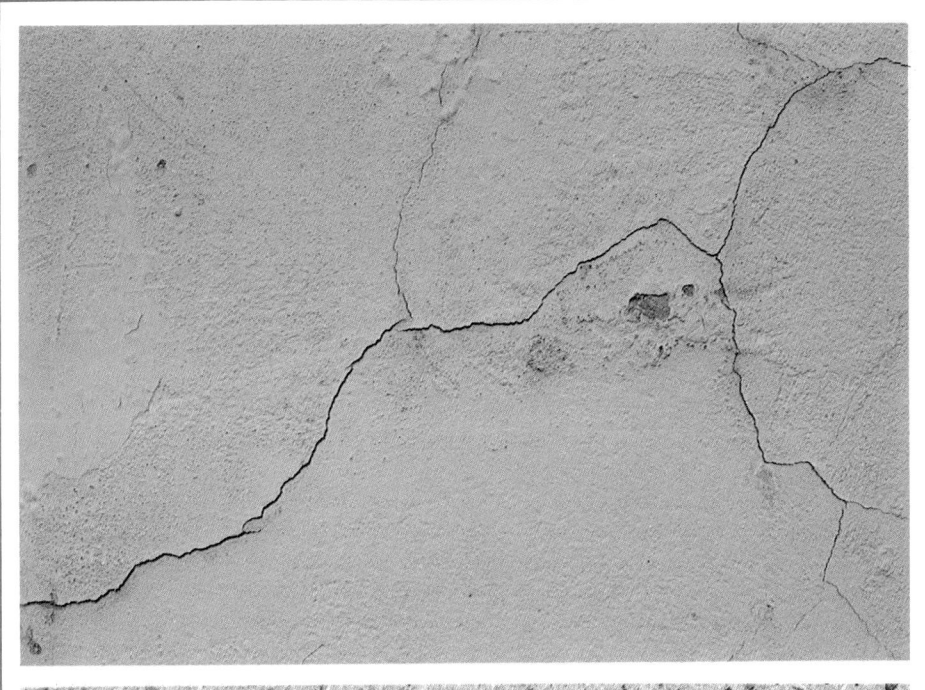

4. Typical random cracking.
Rake out and make good with a proprietary external filler in accordance with the manufacturer's instructions.

5. Friable rendering often has a sandy appearance and is usually found on weathered elevations.
Surfaces to be decorated should be treated with a stabilising solution until the suction from the undecorated surfaces is controlled. Some of the stabilising solutions react with bitumen and extreme care must be excerised if there is any trace of bitumen in the surfaces to be treated.

6. Undecorated surfaces with suction.
Test for moisture absorption with a wet finger. Surfaces to be decorated should be treated liberally with a stabilising solution until the suction is controlled.

DECORATION AND EXTERNAL SURFACE FAILURES

7. Severely chalky surfaces.

These may be tested by wiping the hand over the surface. If there is a substantial build up of deposits on the hand, the surface must be fully stabilised.

8. Cracks caused by moisture ingress through an open coping stone joint with no damp proof course.

The infilling of the fracture in this case will not be the sole cure, and ideally the coping should be raised and a damp proof course laid on the underside. Treatment may be successful where the coping stones are pointed as required. After the pointing has been allowed to harden, any stabilising of the rendered surfaces following the infill of the fracture should take place. Once the stabilising solution is dry, a bitumen base coat can be applied over the top of the coping stones and down the face of the render.

Whilst the bitumen is still wet, the surface should be treated with an open weave glass fibre membrane. When the surfaces have dried, a second coat of bitumen should be applied.

9. Parapets without coping stones.

Paint failures are caused by moisture ingress through the unprotected parapet. The treatment of the faces of the brickwork or render surfaces may be dealt with in the same way as the parapets with coping stones without an effective damp proof course. It is important that any drip which is incorporated within the mouldings of a coping or the parapet, should not be obstructed by the application of bitumen.

DECORATION AND EXTERNAL SURFACE FAILURES

10. Paint failures caused by moisture ingress through cornices.
The remedial works can be carried out in much the same way as that dealt with in the comments under Parapets with coping stones not having an effective damp proof course.

11. Decorative surfaces that are lightly chalky.
The test for this is when the surface of the wall is brushed with one's hand and the build up of chalk upon the surface is noticeable, but does not fill all the pores of one's hand. In this case it will be necessary to apply a stabilising solution but this may form part of the priming coat of the future redecorations. In other cases it may be necessary to treat the repairs and preparation in much the same way as one would do if the surfaces were very chalky.

Cement based or calcium silicate boards or oil tempered hardboards

1. Loose or flaking material.
Paint failure due to decorating on an unstabilised surface. All surfaces must be clean, suitably dry and free from anything that would interfere with the adhesion of the materials to be applied.

DECORATION AND EXTERNAL SURFACE FAILURES

Tyrolean, pebble dash or rough cast

1. **Paint failures due to decorating on an unstabilised surface.**
   This leaves loose or flaking material and all this material must be cleaned off, prior to any decorative applications. In the case of a heavy build up of decorative coatings, it may be necessary to use high pressure water jet equipment to remove the loose and flaking material.

2. **Organic growth on undecorated pebble dash.**
   All visible signs must be removed by stiff bristle brushing. Wire brushes should not be used.
   Spores must then be killed by the application of a fungicide in accordance with the manufacturer's instructions. In extreme cases of organic growth, it may be necessary to use high pressure water jet equipment or a second application of fungicide.

3. **Defective scratch coat**
   This usually occurs in small areas on exposed corners of a property, which are vulnerable to extremes of weather conditions.
   The areas of defective scratch coat must be removed to a sound edge and the surface stiff bristle brushed. The surfaces of the render will then be dampened with clean water and re-rendering carried out with a mix of a suitable type of strength to bond into and be compatible with the existing surface. The surfaces will be stabilised using an appropriate stabilising solution after the render has dried.

DECORATION AND EXTERNAL SURFACE FAILURES

4. **Loose blistering or balding of butter coat from scratch coat.**

Blistering and balding of weak butter coat. The areas of defective butter coat must be removed to a sound edge and the surface stiff bristle brushed. The surface can then be stabilised using an appropriate solution. After this has dried, the surface can be re-textured using a sanded cement paint, which is scrubbed into the surface with a well-loaded brush.

5. **Tyrolean finishes lifting from a weak render coat.**

Lifting Tyrolean occurs when the original application is too thin or the render is deficient in cement. Areas of defective Tyrolean must be removed to a sound edge and the surface stiff bristle brushed. A stabilising solution should be applied, until the surface is fully stabilised. To retexture the stabilised surface, apply a proprietary scrub coat in accordance with the manufacturer's instructions. Retexture the surface using a Tyrolean machine and blend into the surrounding areas.

6. **Undecorated surfaces made friable by weathering.**

Friable pebbledash is usually found on weather elevations. Surfaces to be decorated should be treated with a liberal application of a stabilising solution until the surface is fully stabilised.

DECORATION AND EXTERNAL SURFACE FAILURES

7. **Decorating surfaces that are chalky.**
   A heavily chalky surface. Apply a stabilising solution until the surface is fully stable.

8. **Decorating surfaces that are lightly chalky.**
   Light chalking surface. Apply a stabilising solution as a priming coat.

9. **Localised decorative failures on otherwise sound surfaces.**
   An example of localised failure. Remove loose or flaky material by stiff bristle brushing to a sound edge. All visible signs of organic growth must be removed by stiff bristle brushing. The spores must then be killed by an application of fungicide in accordance with the manufacturer's instructions. Prime the localised areas with a stabilising solution and allow to dry. Prime those areas which are defective and have been treated so that they are raised to the standard of the other sound surfaces in the area.

DECORATION AND EXTERNAL SURFACE FAILURES

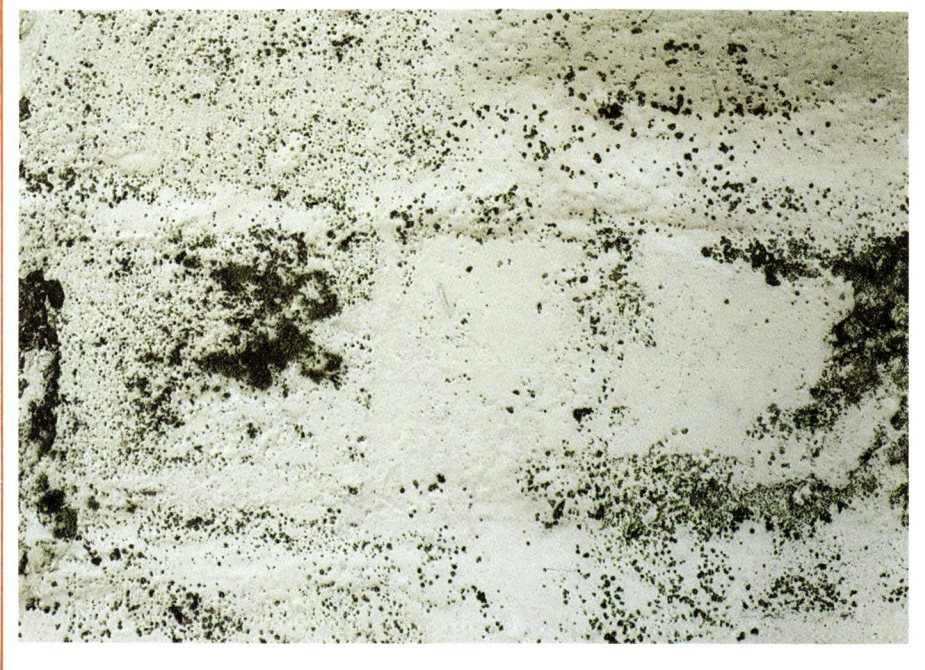

10. Localised repairs to previously decorated Tyrolean pebble dash or rough cast.

Localised repairs, prior to priming with sanded cement paint. Prime repair areas with one coat of sanded cement paint. Cure with water if necessary, then allow to harden and dry. Patch prime the areas affected with a coat of resin based masonry paint.

Brick block stonework and concrete

1. Loose or flaking material. Paint failure due to decorating on an unstabilised surface.

All surfaces must be clean, suitably dry and free from anything that will interfere with the adhesion of the materials to be applied. Remove loose or flaking material by scraping, or stiff bristle brushing to a sound edge. In the case of a heavy build up of decorative coatings, it may be necessary to use high pressure water jet equipment, grit blasting, needle gunning or burning off.

2. Organic growth

Organic growth on decorated brickwork. All visible signs must be removed by scraping and/or stiff bristle brushing. The spores must then be killed by an application of a fungicide in accordance with the manufacturer's instructions. In extreme cases of organic growth, it may be necessary to use high pressure water jet equipment or a second application of fungicide.

DECORATION AND EXTERNAL SURFACE FAILURES

3. **Defective mortar joints.**
Defective mortar joints causing localised paint failure. Rake out to a depth of at least 12mm (½"). Brush away all loose material, damp down with clean water and repoint with a suitable mortar mix. Cure with water if necessary, then allow to dry.

4. **Undecorated surfaces made friable by weathering.**
Eroded surfaces are usually found on weather elevations. Surfaces to be decorated should be treated with a liberal application of a stabilising solution until the surface is fully stabilised.
Undecorated surface with suction. Moisture absorption can be tested with the use of a wet finger. Surfaces to be decorated should be treated with a liberal application of a stabilising solution until the suction is controlled.

5. **Decorated surfaces that are chalking.**
A severely chalking surface. Apply a stabilising solution until the surface is fully stabilised.

DECORATION AND EXTERNAL SURFACE FAILURES

6. Shallow spalling bricks.
An unburnt brick causing paint failure. Areas that have been treated with a stabilising solution must be left for 24 hours to dry. Treat the individual bricks in the immediate surrounding area with one full-bodied coat of undiluted bitumen base coat. Allow to dry until black and shiny.

7. Parapets with coping stones not having effective dpc.
Moisture ingress through coping stone joints causing paint and mortar failure. Repeat as before.

8. Parapets without coping stones.
Paint failure due to moisture ingress into brick on edge capping. Areas that have been treated with a stabilising solution must be left for 24 hours. Liberally apply an undiluted bitumen base coat to cover the top horizontal surfaces and extend down the vertical surfaces for 150mm (6") on both sides. Allow to dry until black and shiny. Apply a second liberal coat of undiluted bitumen base coat and allow to dry.

DECORATION AND EXTERNAL SURFACE FAILURES

9. Cornices or ledges subject to moisture ingress.
Paint failure due to moisture ingress in the brick projection. Areas that have been treated with a stabilising solution have to be left for 24 hours. Then apply liberally an undiluted bitumen base coat to cover the top horizontal surface and extend up the vertical face for 150mm (6″) and down the projected surface to the nearest drip point of 150mm (6″). Do not brush out. Allow to dry until black and shiny. Apply a second liberal coat of bitumen base coat and allow to dry.

10. New concrete surfaces.
Mould oil must be removed by washing the surfaces with a detergent solution. Rinse with clean water and allow to dry. Remove any fins or thin concrete projections and fill any holes with a proprietary external filler in accordance with the manufacturer's instructions, or with a suitable mix of cement and sand.

11. Lightly chalking decorative surfaces.
Light chalking on decorative concrete. Apply a stabilising solution as a priming coat.

DECORATION AND EXTERNAL SURFACE FAILURES

12. Localised decorative failure on otherwise sound surfaces. An example of a localised failure.
Remove loose or flaking material by scraping or stiff bristle brushing to a sound edge. All visible signs of organic growth would have to be removed by scraping or stiff bristle brushing. Spores must then be killed by the application of a fungicide in accordance with the manufacturer's instructions. Patch prime the localised areas of damage with a liberal application of a stabilising solution, allow to dry for at least 12 hours, then patch prime the stabilised area with one coat of resin based masonry paint.

Woods

1. Loose or flaking material.
Paint failure caused by moisture movement in the surface laminate. All surfaces must be clean, suitably dry and free from anything which would interfere with the adhesion of the material to be applied. Remove the defective decorative coating by scraping or burning off. Rub down raised wood grain and remove all dust.

2. Organic growth.
Organic growth at the base of decorative plywood panel. All visible signs must be removed by scraping or stiff bristle brushing. Spores must then be killed by an application of fungicide in accordance with the manufacturer's instructions.

163

DECORATION AND EXTERNAL SURFACE FAILURES

Solar reflective treatments for mastic asphalts and felts

Clean the entire area by brushing and preferably hosing down and allow to dry before the application of any paint.

1. Organic growth.

An example of organic growth. All visible signs must be removed by stiff bristle brushing. The spores must then be killed by an application of fungicide in accordance with the manufacturer's instructions. Wash the surface by hosing thoroughly with clean water and stiff bristle brush to remove dead spores and all traces of fungicide. It must be remembered that the application of solar reflective paint prolongs the life of asphalt or felt surfaces, but is not a treatment which should be applied to any roof which is leaking, without that failure having been located and dealt with.

Plastic external paint finishes

In the early 1970's, as a result of nationwide advertising through national newspapers, many people carried out decoration to the exterior of their property using a plastic form of paint which was claimed to have a 14-20 year life. Unfortunately water penetration to the rear of this paint has caused failures where lifting of the surface has taken place. In these instances it will be necessary to carry out remedial work to the building to prevent further water penetration.

On many occasions the long life paint is found to have been sufficiently damaged by defects in the building for its complete removal to be necessary. This can be carried out by burning off where the paint shrivels as a skin and can be lifted off, but if for reasons of the risks involved this is not practical, wallpaper strippers can be used where it can be steamed off. Grit blasting and high pressure hoses can also achieve the same result.

Absorption of Solar Radiation

The table below shows the relative heat absorption of a selection of building materials and indicates the amount of linear expansion and contraction they are likely to undergo with variations in temperature.

Material	Percentage absorption of radiated heat	Mm per metre linear movement per 1°C
Aluminium (when clean and shiny)	2	0.024
Zinc (when clean)	3	0.031
(when coated with bitumen)	8	0.031
Sand lime bricks (pale colouring)	4	0.014
Limestone chippings (clean, without moss)	4	0.007
Concrete roofing tiles	65	0.006
Fletton bricks (dark colouring)	65	0.005
Concrete	8	0.011
Asphalt	9	0.014
(after treatment with reflective paint)	16	0.014
Lead	4	0.029

Lead in paint

Several recent studies have indicated the potential hazard to young children from concentrations of lead in paint and the dust caused by the stripping of paint during redecoration of buildings.

Since the early 1970's the paint industry has been replacing lead with other materials and the lead content has been progressively reduced in paint.

External paint degrades by chalking but where water or moisture gets behind the paint film flaking occurs. Where lead paint has been used this can cause an increase in the lead in the dust surrounding the building.

A recent report produced by the Inner London Education Authority has suggested where external stripping of lead paint surfaces is necessary buildings should be closed and remain closed until the work is completed. In many cases it may be necessary to contemplate the closing of a building such as a school for periods of up to 12 weeks.

In order to combat the health risk involved with lead paints the report suggests dry rubbing down of lead paint be forbidden.

Any debris which may occur during the stripping off of the paintwork must be dealt with carefully. When burning off external paint the danger of inhaling fumes is not great under normal circumstances and a mask will reduce the hazards to the operative. Where debris from stripping areas has been allowed to fall on the paving or floors and has not been cleared up or stored properly it has caused an unacceptably high concentration of lead in the dust.

When stripping is taking place dust sheets should be placed beneath the work areas to catch all stripped material whether by burning, scraping, rubbing down or chemical means. The dust from rubbed down areas should be damped down and swept up at the end of each working day or removed by industrial vacuum cleaner. The dust sheets used to protect areas must not be used elsewhere on site before they have been thoroughly cleaned. After cleaning up all the debris from the stripping should be collected and placed in sealed receptacles and disposed of by the contractor.

It may be felt that these are excessive precautions but the risk to the lives and development of young people mean that no surveyor should be ignorant of the responsibility involved.

Cracks in buildings

1976 saw an eight-fold increase in claims notified to insurance companies. Many of these claims followed the appearance of fractures in the walls of property. Ground movement had been caused by the drying shrinkage of the subsoil, particularly where clay was present, after a number of years with below average rainfall.

The shrinkage resulted in movement occurring to foundations which were not very deep.

Some claims were due to an over reaction by householders who attributed old cracks to recent events. The publicity the problems received was also responsible for an increase in awareness of the cover afforded by most insurance companies.

The number of claims far exceeded the problem because they represented a new phenomenon which had not been adequately researched at that time. Minor movement to properties often resulted in an over reaction and many instructions to underpin were based on movement which would now be regarded as of little consequence.

The building professions were not adequately prepared for this new type of problem. They often decided to recommend underpinning as this was the safe course of action.

The increase in claims may have been due to the insurance companies extending their cover in 1971 to include damage that resulted from ground movement. Damage that could be adduced to the failure of the ground beneath the foundations was covered, and repairs were carried out at the insurer's cost. Many of those repairs have failed however, because of the reduction of the repairs specification by the representatives of the insurers who were trying to reduce costs. The dry summer of 1989 has seen a similar expansion of claims.

There are often inadequate records of the work that has been carried out to remedy old cracking. Where records have been maintained, make sure that there is no deviation between the work that was recommended and that which was completed.

It is not just the insurers who reduced the extent of the work. Some ill-advised householders took the cash and carried out only cosmetic repairs. 1989 has revealed the deficiencies in old repairs, and the cracks that have appeared have required some explanations to be given.

If you are examining a building with cracks, consider that most domestic policies have an automatic cover for subsidence, except where the movement may have been caused by the variation in the level of the water table. This cover is not included in commercial policies, for example, cover which would have been taken out on properties such as a shop and upper part.

When one locates cracks in a building it is important that you carefully consider the implication of the cracking, both now and for the future. The surveyor has to be able to reach a decision, on the basis of what can be seen, on the consequences of the crack and the unremedied cause of the cracking. He must make a judgement. If the surveyor finds a crack there will be several options available to him when he comes to setting out his recommendations for his client.

The crack is either static or dynamic.

The crack is either a representation of the full extent of the movement that will take place, or is a representation of the movement that has already taken place, but no guide to the extent of the future movement that will occur.

If, in the surveyor's view, the cracking occurred some time ago and there is no possibility that it will get any worse, then he must be able to substantiate that opinion. Having stated that there will be no further movement, the surveyor will be at great risk if the crack later develops and expensive repairs are carried out. A judgement that the crack is old, and an indication of past movement and therefore not serious is a judgement that must only be made after the most searching of inspections. The notes of the information upon which this diagnosis is based must be retained.

If the surveyor thinks that the minute crack is the first indication of movement that will develop, and will lead to major failure, then he must advise the client of the future risk. This view will leave the surveyor with a series of options which can be put to the client.

Surveyors must not be frightened of advising their clients. They must be careful to ensure that the client is aware of the information and the level of inspection that has been carried out to enable the advice to be given.

Once a crack has been found, the surveyor will have to investigate the defect in the following way.

1. The correct approach

Set out any further examination, possibly outside the scope of the existing instructions, that would help to determine the exact cause and future prospect for the crack, together with approximate costs and the time scale of the tests. In this way the client knows what options are available to him to more precisely determine how serious the crack may be.

A further investigation that might be suggested should include;

Monitoring.

This will be over at least an 18 month period so that seasonal movement may be taken into account and eliminated from the concept of progressive movement. Measurement of a crack from September to March will usually show a high reduction because of the swelling of the building components caused by damp, timber expansion due to the climate, or the heave in the ground due to it being damper than in the summer.

The checking of cracks between March and September often shows them to become larger, for the reverse of the reasons outlined above. In order to monitor cracks, it will be necessary to measure each crack, and re-inspect regularly to re-measure. The most accurate way of doing this is with a Laser, where the volume of the building can be measured between certain reference points.

Monitoring a crack is done, not to see if there is any movement, but to plot the direction and amount of any movement that does take place. The use of tell-tales, however, will not give a sufficiently precise measurement, the accuracy of which is required to be within one tenth of a millimetre. Accordingly, they should not be used. Using Vernier callipers to measure between fixed points is an adequate level of monitoring dimensional movement for most buildings.

Ground investigation

In order to decide what might have caused the movement in the building, a thorough investigation of the ground below it may be necessary. This may involve digging trial holes to inspect the foundations.

Bore or auger tests. This involves taking a core from the ground at various points around the building. The cores will show the composition of the ground, and the presence of compressible materials, which may be contributing to the problem.

Drains test. These tests are fully discussed on page 110. It is important that any failure in the drain pipes below ground is identified, so that the possibility of the ground being washed away from below the property can be taken into account.

Geological maps. These maps are available for the United Kingdom and give information about the ground composition at a given point.

Old maps. Old maps may help you to understand the alterations that have taken place to the property that you are investigating, or the rearrangement of any local rivers or roads. Those changes may have set in motion the movement that you are trying to assess.

There is a book of old maps of London, called *The Village London Atlas 1822 – 1903* which duplicates the four stages of the development of Greater London. With this book you can date most buildings within 309 years, and see the changes brought about by the introduction of railways and tube lines.

Aerial photographs. Most of the United Kingdom is photographed from the air every ten years. A close examination of these photographs may show alterations, the removal of vegetation or other contributing factors that may have caused the cracking. The inspection of the aerial photographs taken after the hurricane of 1987, and the major storm in the south west in December 1989 may give some indication of tree removal or damage that occurred at that time.

2. The inspection of the crack

The surveyor must carefully examine the crack and note or record all information that is available to him. This will include the following:

The construction of the building. It is essential that the construction is carefully identified. The significance of the cracking cannot be determined without the method of building having been taken into account. The crack in a brick panel in a frame building may be of little import, whereas a similar crack in conventional construction may indicate a very serious fault in the foundations requiring very expensive repairs to be carried out.

Material which has cracked. Is it plaster, wood, brick, concrete or other component of the building? A certain size crack in concrete may be serious, whilst a similar sized crack in brick or timber may be of no significance.

Size of the crack. Measure the crack, and note the positions where the measurements were taken. The dimensions of the crack can then be referred to the guides for the implications of cracks of that size. This subject is dealt with later in this chapter.

Appearance of the crack. Is the crack clean or dirty? Is the colour uniform over its full length? Is there any indication of the crack being cleaner at one end than the other? Where is the end of the crack? Which is the wider end, and which is the narrower end? Are there variations in the dimension of the crack within its length? Is the plane of the cracked surface distorted? Has the crack distorted the face of the building? Is there lateral movement?

Direction. What is the overall direction of the crack?

Age of the crack. Is there any evidence that may help to date the fracture? For instance, the wider parts of the crack may be filled with old newspapers. The age of the crack may be deduced if paint is splattered around it and the date of decoration is known. Dirt build up on the interior face of the crack may give a guide as to its age.

Direction of movement in the building. The surveyor must identify the direction of the movement that has taken place. Hogging, heave, settlement, shear or horizontal movement must be identified.

Once this inspection has been completed the surveyor will have all the information that is available to enable a diagnosis to be made. This inspection must be made with great care, because if there is to be any claim of negligence it will be on the information that could have been noted in this stage.

3. Schedule the possible causes of cracking

The possible causes will have been suggested by the information that was gained during the inspection and assessment that you have carried out. The causes could include:

Loading. Variations in the loading on the building. For instance, a residential property which is used as offices where the floor loading for office use is greater than that for residential use.

The creation of new openings in a wall may result in the load being transferred from a uniformly distributed load to two point loads.

These variations may have influenced the building and set up a once and for all movement, which will stop when the building achieves a new equilibrium.

Partial support. Some buildings are not uniformly supported. The foundations are at different depths, and of varying qualities. This may occur beneath projections, such as bays or extensions, or may result from partial underpinning having been carried out at some time in the past.

Ground movement. The failure of the ground beneath a building may be due to trees in a clay area, drains failures, water mains leaking, or ground slip or erosion.

Structural failure. Failures in this category may result from corrosion of steel support, rusting of wall ties, decay of timbers built into walls as lintels or spreader plates, or chemical action between the various components of the building.

Thermal movement. Movement caused by the differing rates of expansion or contraction between the components of the property. Thermal movement may be caused by other defects within the building and can require remedial work to be carried out.

4. The advice

The client must be told upon what basis the advice is being given. The surveyor must also inform the client of further investigations that might enable a more accurate assessment of the faults and their implications to be made.

In advising on a building's faults and their implications, the alternatives from which the surveyor must select an option are:
- Serious
- Not Serious
- Repair Required
- Further investigation needed.

The last option should not be selected unless the result of the investigation gives the surveyor no option.

Classification of cracking

Building Research Digest 251 recommended various classifications of cracks in order to bring some uniformity to the interpretation of their implications based upon their Classification in dimensions.

One

Failures which affect only the appearance of the property and fall under the heading of aesthetic.

Two

Fractures and cracks which cause damage to the functions of the wall, such as its weatherproofing and insulation, and which will result in fracturing to service pipes, the jamming of doors and windows, etc. This category relates to the injury to the services of the building and is under the heading of serviceability.

Three

The third category relates to the situation where there is an unacceptable risk of failure in parts of the property, unless work is carried out to repair the structure of the building and improve foundations. The cause of the failure must be located. The fault may be outside the building, for instance, it may have been caused by the redirection of ground water, local trees or instability in the sub-soil caused by mining.

The classification is extended by reference to the ease with which repairs can be carried out. This catalogues the damage into six groups on a scale of 0 to 5, and relates closely to the three groupings already mentioned.

Categories 0 and 1 relate to minor damage and incorporate hairline fractures of less than 1mm in width and deal with circumstances which can usually be dealt with in normal redecoration.

Categories 2 and 3 deal with fractures where they are visible from the exterior of the building. Usually the fractures can easily be filled and dealt with during redecoration.

CRACKS IN BUILDINGS

Category of damage	Degree of damage	Crack width mm	Description
0	Negligible	Up to 0.1	Hairline cracks of less than 0.1mm width – would have no effect on structure or building use.
1	Very slight	Up to 1	Fine cracks which can easily be treated in normal decoration or redecoration. These are found in isolated slight fracturing in a building and are rarely visible in external brickwork. The damage is very slight and would have no effect on the building use.
2	Slight	Up to 5	Cracks can easily be filled, redecoration will probably be required. Recurrent cracks may have to be masked with a suitable lining. Cracks become noticeable to a minor extent on the outside and external repointing may be required to ensure weathertightness. Doors and windows may stick slightly.
3	Moderate	5-15	The cracks require some opening up and can be patched by a bricklayer or stone mason. Repointing will be required if fractures are a series of smaller failures or replacement of bricks if failure has split individual bricks. Doors and windows now stick and there is a risk of service pipes fracturing. Weathertightness is often impaired.
4	Severe	15-25	Extensive repair work involving breaking and replacing sections of walls, especially around doors and windows. Window and door panels are distorted, floors slope, walls lean or bulge. N.B. You can feel floor deviations in the region of 1:200 and can see variations in vertical and horizontal surfaces of 1:100 – variations in excess of 1:150 are undesirable. The bearing of beams may be affected and service pipes are disrupted. Stability is now at risk.
5	Very severe	25 plus	This requires a major repair job involving partial or complete rebuilding. Beams lose bearings, walls lean badly and require shoring, windows broken with distortion. Danger of instability.
6	Critical	I thought that was a window!	Run like hell!

The lines below have been drawn to provide an easy width reference.

0.1mm

1mm

5mm

10mm

15mm

20mm

25mm

Where they occur to the exterior of the property, repointing to the external brickwork and minor replacement would deal with the problem.

Categories 4 and 5 deal with extensive repair work. Damage placed in this category occurs when fractures are in excess of 15mm (½") in width. It should be remembered that in some cases a large number of minor fractures over a small area may not exceed 2 or 3mm each but may nevertheless be an indication of severe failure.

Major failures are usually indicated by substantial fractures. The severe and very severe categories of sections 4 and 5 deal with circumstances where rebuilding of part of the property would probably be required after the elimination of the initial problem.

The survey of the property which has been affected would include not only the sizing and location of the fractures but also a careful measurement of the variation in the level of the damp proof course all the way around the property. This would indicate any variations in level which could have resulted from ground movement or foundation failures.

Such a survey would record variations in the verticality of the external wall of the property. Once these have been plotted the location and points of movement will be more clearly determined. If this information tends to reveal that there is a problem, research should be carried out to try and locate or determine the type of construction which was used when the property was originally constructed.

A drawing prepared from this information will indicate the type of failure and the width of cracks. The direction of movement may be resolved.

In some cases it may be found that the cause of the fractures is related to material shrinkage, to steel corrosion, to decay or to differential thermal movements. It would be rare for such failures to exceed the damage in categories 0, 1 and 2.

Where the failures are more closely associated with ground movement or are caused by settlement or heave or failures in foundations or chemical attack on concrete or inadequate design, the resulting failures could fall into any of the categories shown on the tables below.

Where failures exceed category 2 it is probable that they have been the result of progressive ground movement. Two questions frequently confront one in examining such a property; is the damage due to a foundation fault or ground movement, and will the damage get progressively worse.

It may be necessary to carry out the excavation of trial holes to reveal foundations, if the research on the method of construction has not come up with an answer.

Before reaching a definite decision relating to the repairs that may be required, bore holes will have to be drilled as close to the location of the foundations as possible.

The failure of one property in Chislehurst was caused by the water mains entry to the property having failed and eroded a substantial amount of ground immediately below the party wall foundations. This failure had left a void or chasm over two metres deep below the house.

It would have been possible to have excavated and dug holes to carry out an examination of the foundations of the property every two metres around the building and totally missed the chasm which lay below. The testing of the ground with bore holes at approximately 1m centres revealed the failure and prevented an embarrassing error. The ground at that particular point was of a chalk material and it was found that the properties had been built in the area of a disused quarry. It is believed that the area of ground which was washed out was back-filled around the chalk workings and it was for this reason that there was such a sharp variation of levels.

If it is believed that the failures may be attributed to foundation movement it will become necessary to assess the stability of the failure. These failures may be the result of movement, shrinkage or heave in clay sub soils.

Floor slabs may fail due to movement caused by an inappropriate or poorly compacted under floor fill expanding or failing to give adequate support. Damage is generally confined to the slab at its junction with the external walls and with internal partitions carried on the slab. This type of problem usually occurs early in the life of the building. Failures may take place when a chemical reaction has occurred within the fill used beneath the solid floor constructions.

Chemical attack by sulphates can also cause failures but these rarely cause damage to residential properties as their foundations are usually sited above ground water level.

On sloping sites instability of the ground is frequently seen in the fracturing of roads, garden walls and services.

The major problem is to assess the risk of continuing movement.

Where made up ground occurs or there are defects in the ground in a particular locality the failures usually occur within the first ten years of a property's life.

Just outside Southport, in Lancashire, there is a collection of properties built before the last war where settlement occurred because the ground upon which they were built was formerly an inland lake. The ground was poor for conventional buildings and problems were experienced. The local grammar school which originally had four steps up to the front door eventually had to be rearranged so that there was a direct entry into the ground floor level from the front drive. These failures have continued over the life of the buildings and after some fifty years they have had to be abandoned because of continuing ground subsidence.

The use of geological maps to examine the ground in a particular area is invaluable. Aerial photographs are taken on a ten year cycle and may help to identify variations in the direction of drainage or water courses and variations in the use of a property or site over a period of time. This evidence together with discussions with local residents may help decide whether there is a continuing risk of progressive settlement.

In some areas mining subsidence is anticipated and the help of the National Coal Board should be sought. In an area where they extract salt or other chemicals from the ground the companies involved should be contacted so that their assistance may be given in assessing the probable causes.

Certain areas of the West Midlands have recently become blighted because of the ancient limestone workings that honeycomb much of the ground on which Walsall and the surrounding Black Country stands. Many of the workings were abandoned over the last century and now there are fears that the underground cavern roofs could collapse as water erodes the supporting pillars. Failures have occurred in the centre of Dudley.

Where there are indications of shrinkage or heave the property's location should be placed in one of the following categories:
1. In open ground away from major vegetation;
2. Near existing trees;
3. On sites where trees have recently been removed.

The chapter on trees deals with the damage individual trees may cause to property, the identification of the trees, and the safety zones for each tree.

In the event of damage to a property a full examination should be made. The damage should be carefully recorded and the sub soil should be examined by means of trial pits. These would reveal the type of foundations and the presence of any tree roots as well as their location in relation to the foundations.

The history of the site should be researched and the current or former position of trees and shrubs located. Where the failures occur in open ground and the foundations are of a depth of less than 1m, foundation movement may give rise to minor fractures which open and close seasonally.

Where the building is near existing trees the type of tree should be examined so that one can have a reasonable chance of assessing the likelihood of progressive movement. If the trees have reached or are close to maturity, minor shrinkage and swelling movements can be anticipated. Larger movements are only likely to occur in exceptional spells of dry weather. The felling of such trees in clay areas will result in the swelling of the clay. Tree pruning may offer an acceptable way of reducing the trees' influence.

If the trees are a long way from maturity there is a possibility that progressive foundation movement will take place giving rise to increasingly severe damage. It may be necessary to consider pruning or felling the trees in order to arrest the future and increasing failures that will occur.

Small alterations such as the repair of defective drains, the alteration of surface water where it is redirected from a soak away to the main drainage system, could increase the spread of roots as a result of the reduction in the provision of moisture to the tree.

Continuous movement in the property caused by movement in clay sub-soil may result in a hinge developing within the foundations thus concentrating all the movement on one point of fracture.

This is unlikely to occur where the damage is within categories 0, 1 or 2.

Where trees have recently been cleared from a site the clay will tend to absorb moisture and ground swell will occur. Sometimes this swelling may continue for up to eight years.

If trial pits have been excavated and desiccated clay has been found containing fine roots beneath the foundation level extreme care should be taken before carrying out the removal of trees. If these circumstances occur and the trees have been removed it may be a case of leaving the property for some eight years before the ground will achieve a degree of stability. At this point consideration should be given to carrying out full remedial works.

It is now believed that progressive subsidence damage is most uncommon

CRACKS IN BUILDINGS

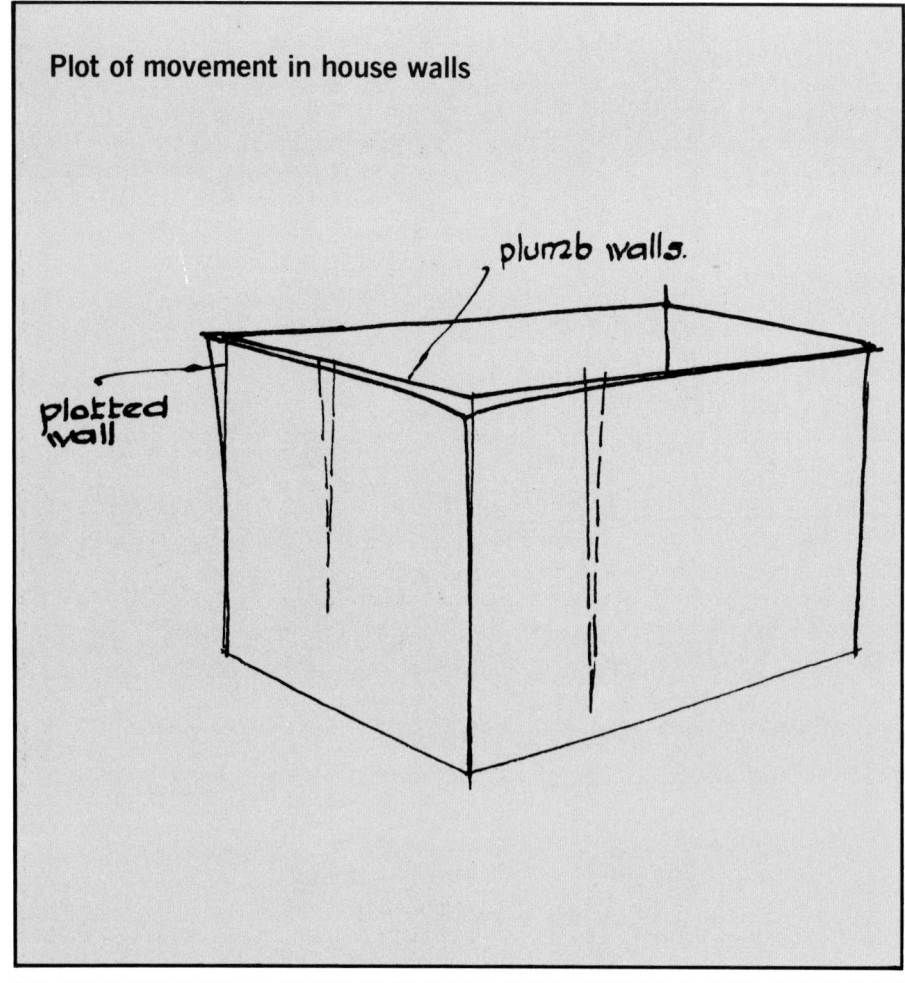

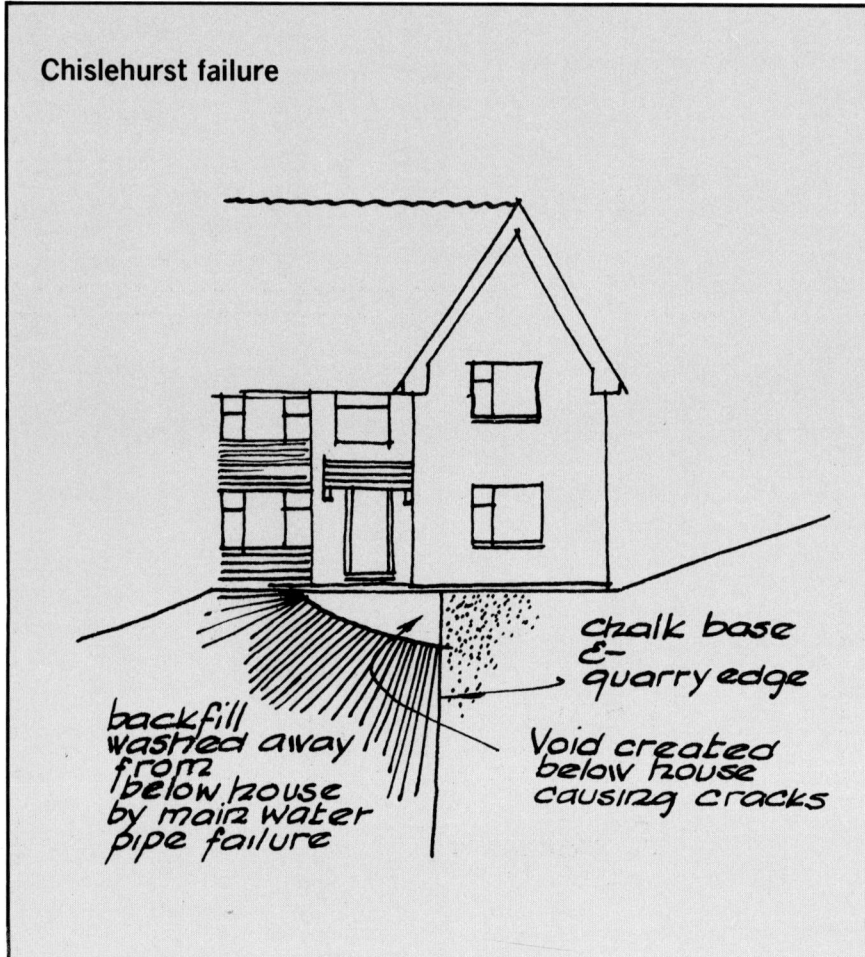

Plot of movement in house walls

plumb walls.
plotted wall

Chislehurst failure

chalk base & quarry edge
backfill washed away from below house by main water pipe failure
Void created below house causing cracks

Simple tests for the evaluation of ground conditions and their implications on foundation design

1. Rock
The rock should not be inferior to firm chalk. It should require at least a pick to make an impression upon the ground. Foundations should be at least 0.3 m (1′) deep to internal and external walls.

2. Gravel or Sand
Compact, this would require a pick. A 50 mm × 50 mm (2″ × 2″) peg can only be driven in a few inches. The foundation should be at least 600 mm (2′) wide and to a depth of at least 300 m (12″) for internal and external walls.

3. Clay, sandy
Stiff, this requires a pick to cut into it and it cannot be moulded between the figures when sections are cut out. Width of foundations minimum of 600 mm (24″), minimum depth of all foundations 1.2 m (4′).

Firm, can just be moulded between fingers and can be excavated with a spade. Mimimum foundation width 750 mm (2′6″), depth 1.2 m (4′).

4. Sand, silty or clay
Loose, can be excavated with a spade, a 50 mm (2″) peg is easily driven into it. Special foundation design will be required in these circumstances.

Soft, this is easily moulded in the fingers and is readily excavated. Special foundation design will be required in these circumstances.

Very soft, a winter sample would exude between the fingers when squeezed. Special foundation design would be required in these circumstances.

5. Peat or made-up ground
Loose, can be excavated with a spade, a 50 mm (2″) peg is easily driven into it. Special foundation design will be required in these circumstances.

Soft, this is easily moulded in the fingers and is readily excavated. Special foundation design will be required in these circumstances.

Very soft, a winter sample would exude between the fingers when squeezed. Special foundation design would be required in these circumstances.

CRACKS IN BUILDINGS

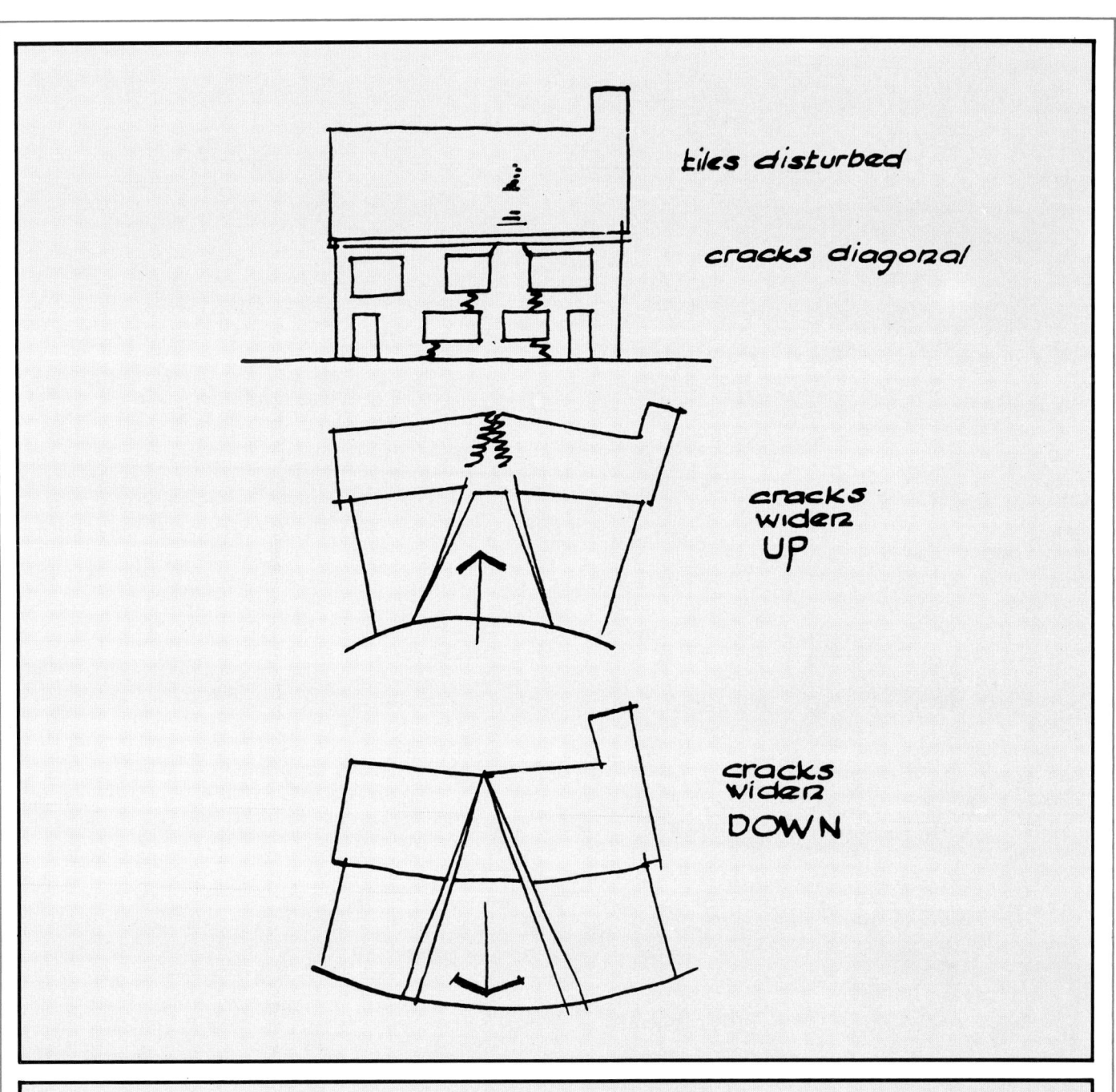

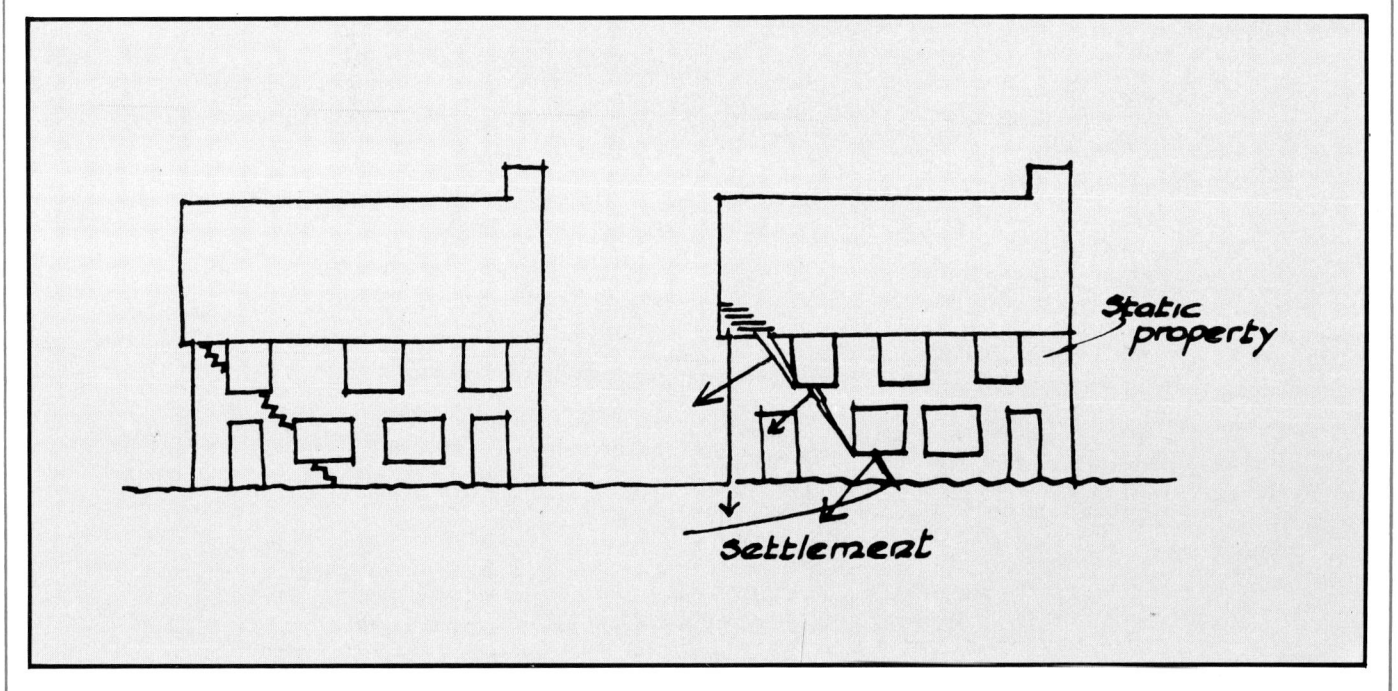

and is only likely to occur where major change has taken place in the ground such as the removal of trees. The removal of trees is unwise unless the level of damage is more than category 2.

If one has to inspect a property in order to record the degree of movement which has taken place it is advisable to provide and instal tell-tales to the interior and exterior of the property.

These should be of a form which will record progressive movement and not the type such as glass tell-tales which merely indicate whether or not movement has taken place.

Make sure that any tell-tale used has a date on it.

On some occasions the provision of a graph type of tell-tale to the exterior of the property has been unsatisfactory. Vandalism or casual curiosity has resulted in the equipment being repositioned or damaged. On one occasion where the equipment showed considerable movement over a short period of time it was later found that it had been knocked off a wall by the youngest member of a household and been repositioned by his elder brother in what they thought was the correct location.

The provision of small studs drilled into the wall surface is less likely to be the subject of such damage or be repositioned so easily. It is important to ensure that the measurements which are taken and checked are from the identical point on the stud, being either the clear gap between studs or from the centre of each stud, so that variations in the measurement do not cause an inaccurate diagnosis of the failures.

It may take more than a year before the pattern of the measurements taken will enable the surveyor to report on the extent of the continuous movement.

A recent review by Kew Gardens of the effect of tree roots on ground movement revealed that the dry summers leading up to 1976 have forced trees to expand their root systems in search of water. In areas with shrinkable clay soils these expanded root systems accelerated the pace of shrinkage and consequent house damage. In some cases the problem has been exaggerated by the success of environmental groups opposing the pruning of trees by local authorities. The trees that cause the most damage to property are poplars and willows.

Clay Soils

Clay is a soil which contains a large proportion of very small mineral particles. Those particles have to have a diameter of less than 0.002mm (very small in inches) and as such are invisible to the naked eye. The soils are plastic and smooth and slightly greasy to the touch. The more clay that there is in the soil relative to silt or other coarser grain materials the more pronounced these characteristics are. When wet, clays are soft and sticky and are likely to stick to footwear if one walks across them. When they have dried out they shrink and show fractures in the surface, and are hard to break up. A section of clay immersed in water will soften up slowly without disintegration. If it does disintegrate it contains silt and other coarser grained materials.

It is the mineral content and variation within a clay that contributes to its behaviour. Where there are minerals that hold water within their molecular construction the clay is often referred to as being fat. Conversely, where the mineral composition holds less water, the clay is known as lean.

In order to test the type of clay, the dry clay is mixed progressively with water until it becomes liquid. The amount of water expressed as a percentage of the dry weight of the clay is called the liquid limit. The higher the liquid limit is, the more the clay will influence the movement of the ground.

The plastic limit of clay relates to the amount of water within a clay soil below which it is no longer plastic and pasty but breaks up when worked in the hand. Below the plastic limit there is insufficient water to fill the spaces between the solid particles and increasingly the voids become filled with air. At and around the plastic limit the volume shrinkage of the clay becomes less than the amount of water that is removed.

Many of the clays encountered in the South East of England are termed "over consolidated" because the clay has become consolidated by loads far greater than they now experience in the ground. These clays often exist in their fully saturated condition at moisture contents close to the plastic limit. For London clay this is in the region of 25 to 30%. In the top 1 to 1½ metres of the ground in the London area the moisture content varies from as little as 15% in dry summer weather to 40% in wet winters. This results in substantial movement in the ground clay in this zone of 1½ metres (5ft).

As the water content of clay soils is reduced below its natural state the soil will shrink. The drier clay exerts more and more suction and if water is then brought into contact with such a soil it is absorbed and the soil swells.

The seasonal volume changes in clay attributable to the presence of vegetation other than hedges and trees extend only to a depth of 1 to 1½ metres into the clay and the major significant shrinkage is confined to the top 1 metre.

Large trees can create deep zones of dry and shrunken clay beneath them. In dry weather these zones will extend but will not be reinstated by a period of wet weather during winter months. The slower rate of penetration of water into the more impermeable dry clays will leave a permanent dry zone. The prediction of the shrinkage that will occur in certain clays is often difficult to determine. One can test the possible movement by cutting a regular shaped section of clay from the site and measure the extent of shrinkage caused by the gradual drying that will occur if it is placed at room temperature. This movement will usually exceed the degree of movement which will occur on the site.

Clays have been classified by their shrinkage and swelling and these properties are referred to as the "plasticity index" and the "percentage clay content" of the soil. The plasticity of a particular clay is obtained by subtracting the "plastic limit" from the "liquid limit". The "clay fraction" is the percentage of clay content in the soil.

For the London area the London clays have a "plasticity index" of between 28 and 52%. This is the difference between the amount of water expressed as a percentage of dry weight of a section of clay which is required to make the clay a liquid, and the percentage of water which is within a clay at the point when the clay becomes no longer plastic and pasty but begins to break up.

These are balanced with the percentage of clay that is in a sample. For the London area this percentage is between 60 and 65%. Soils which have this percentage of clay as a clay fraction have a high shrinkage potential.

CRACKS IN BUILDINGS

Settlement and Subsidence

Although the dictionary does not differentiate in any major way between these two words it has been attempted by certain insurance companies. Settlement is the subsidence of a wall, house or the soil, and subsidence is the sinking to a lower level.[1]

An insurance company whose policy covers the risk of subsidence in a property would not be bound to reimburse the property owner for the cost of repairing damage caused by ground heave following the removal of trees. This and any other upward movement is neither settlement nor subsidence. It has been suggested that the ground subsides and the building settles. The dictionary does not recognise this distinction.

The claims that were made following the dry weather culminating in the summer of 1976 were mainly for damage to property built before 1940. These houses had been subjected to continuous movement over their life time, mainly because of the shallow depth of their foundations. The most vulnerable areas of the country, based upon insurance claims made for damage caused by subsidence, were London, part of the Home Counties with clay subsoils and the county of Avon.

Settlement of foundations is not necessarily confined to very large or heavy structures. In soft and compressible gaults and clays appreciable settlement can occur under very light loadings.

If a structure settles evenly there is not going to be any detrimental effect upon the superstructure. The peat areas of central England and in Ireland have compressed, and settlement of up to 10 metres has occurred, sometimes without serious property damage. The effect on drains and services is a different matter and may cause more serious differential settlement due to the escape of fluid into the ground.

Differential settlement will distort a property, and the extent of the distortion will be a measure of the damage caused. Movement in a brick wall which has resulted in a bulge may often be tolerated provided the projection of the bulge does not exceed ⅓ of the wall thickness.

Most movement will occur in property with shallow foundations. In clay soils moisture variations were noted in 1976 to a depth of 3 metres.

Tree roots can vary the moisture content of the ground for a depth up to 4 metres. Those properties with basements are unlikely to be affected by the variations in ground moisture because their foundations will be between 3 and 4 metres below the ground level. Many houses, however, have a part basement level. The variation in the depth of the foundations for the various parts of the house frequently means that there is a greater risk of movement occurring in those parts with shallow foundations. The part with the basement will be more stable. Fractures will occur in the property if movement of the shallow foundations occurs due to variations in the moisture content of the ground.

Old houses with bay window projections or rear extensions or wings often show signs of movement. The foundations of the main house and the various projections are not built to the same level. The bay foundations are usually quite shallow and will be more likely to react to ground movement.

Over 99% of insurance claims made between 1971 and 1978 related to buildings on clay soils. The effect of tree root damage in other soils should be related to this information. The Tree chapter in this book deals with those trees which have some history of causing damage to buildings. All the trees are common in clay soils, and no tree which cannot grow in clay soils is listed.

Trees should be their mature height from a building. The heights of the most common trees are shown in the Tree chapter. The Building Research Establishment also points out that for trees in rows the distance should be 1½ times their mature height.

The "factor of safety" for trees, after

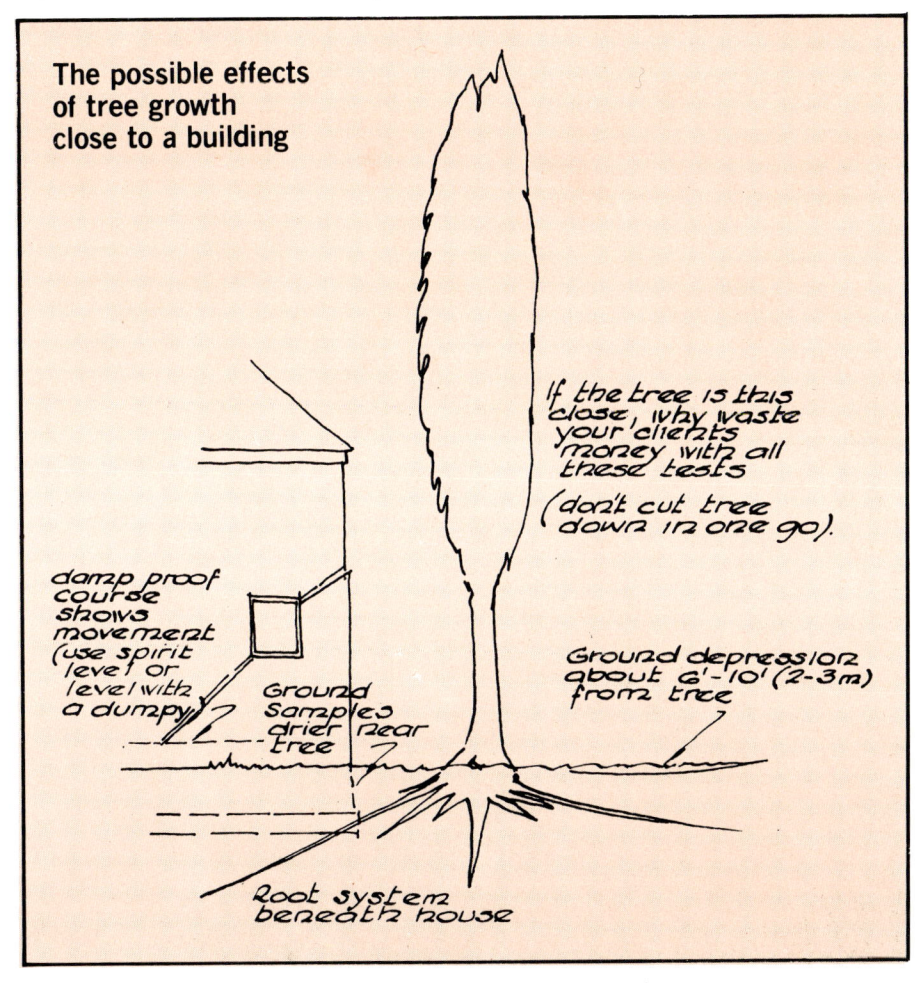

The possible effects of tree growth close to a building

If the tree is this close, why waste your clients money with all these tests (don't cut tree down in one go).

damp proof course shows movement (use spirit level or level with a dumpy)

Ground samples drier near tree

Ground depression about 6'-10' (2-3m) from tree

Root system beneath house

---

1. Oxford English Dictionary.

research by a number of botanical laboratories, rarely exceed their mature height. (The factor of safety is discussed fully in the Tree section.) The surveyor must be able to recognise all the common trees, know their characteristics, and also know the maximum recorded distance that their roots have spread to damage a building. The "factor of safety" is about 1 metre more than the longest recorded distance that an individual tree's roots have damaged a building.

Not all cracking which occurs in buildings is due to ground movement. Certain materials have a history of shrinkage, such as calcinated flint bricks. These often cause a vertical rather than a sloping crack. Sand lime bricks, with their shrinkage, cause similar failures.

The various types of movement are shown diagrammatically on the sketches on the right. Rotation is usually the result of differential settlement. The recording of the variation in crack sizes over the length of the crack helps to discover the part of the building which has moved and the part which is stable.

Horizontal movement on its own may be due to slippage on a sloping site. With trench fill strip foundations, horizontal movement or outward movement may have resulted from differential pressures being set up by the shrinkage of clays to one side of the foundation. Pressure on the other side, possibly due to expansion caused by variations in ground moisture content will result in horizontal movement.

The lime mortar joints used in older residential property was more tolerant and flexible where movement took place. It is possible that the owners of those buildings were also more tolerant and less litigious than the owners of modern property which is often built with more brittle materials and joints.

The sight of fractures in a building is not always a cause for concern and a balanced appraisal will only be achieved by a greater awareness of the causes of settlement and further research into the effects of trees and their roots on buildings.

Ground movement in areas near a deep foundation construction

Very little research has been carried out on the effect on compressible ground of the installation of extra loading by the construction of tall buildings which have deep foundations. The construction of the car parks to Palace Yard at the Houses of Parliament was carefully monitored and movement in the ground some distance from the construction of the car parks was noted. The construction of deep foundations for a new office complex situated over Fenchurch Street Station has also resulted in the movement of ground at some distance from the location of the piles.

Both of the building sites where monitoring has taken place have compressible clay bases but it is probable that similar variations in the ground around the building site would occur in different ground formations.

As a general guide it is reasonable to suppose that the area of surrounding ground which will be affected by deep foundations will approximate to a 50% extension of the area covered by the new building.

At the extreme edges of this area the maximum ground movement is likely to be in the region of 5% whilst the extent of ground movement close to the edges of the site area of the property may be in the region of 15%. In areas where the ground is cut away to form deep basements the surrounding ground may lift by up to 15 mm (0·6").

The major movement takes place within approximately 1 year and most of the movement will have taken place within the first 3 years of the construction of a building. These dimensions and areas are very much a generalisation but the surveyor should be aware in the inspection of large buildings of the effect upon the ground upon which they sit of major building operations which are being carried out, may be carried out, or which have just been completed.

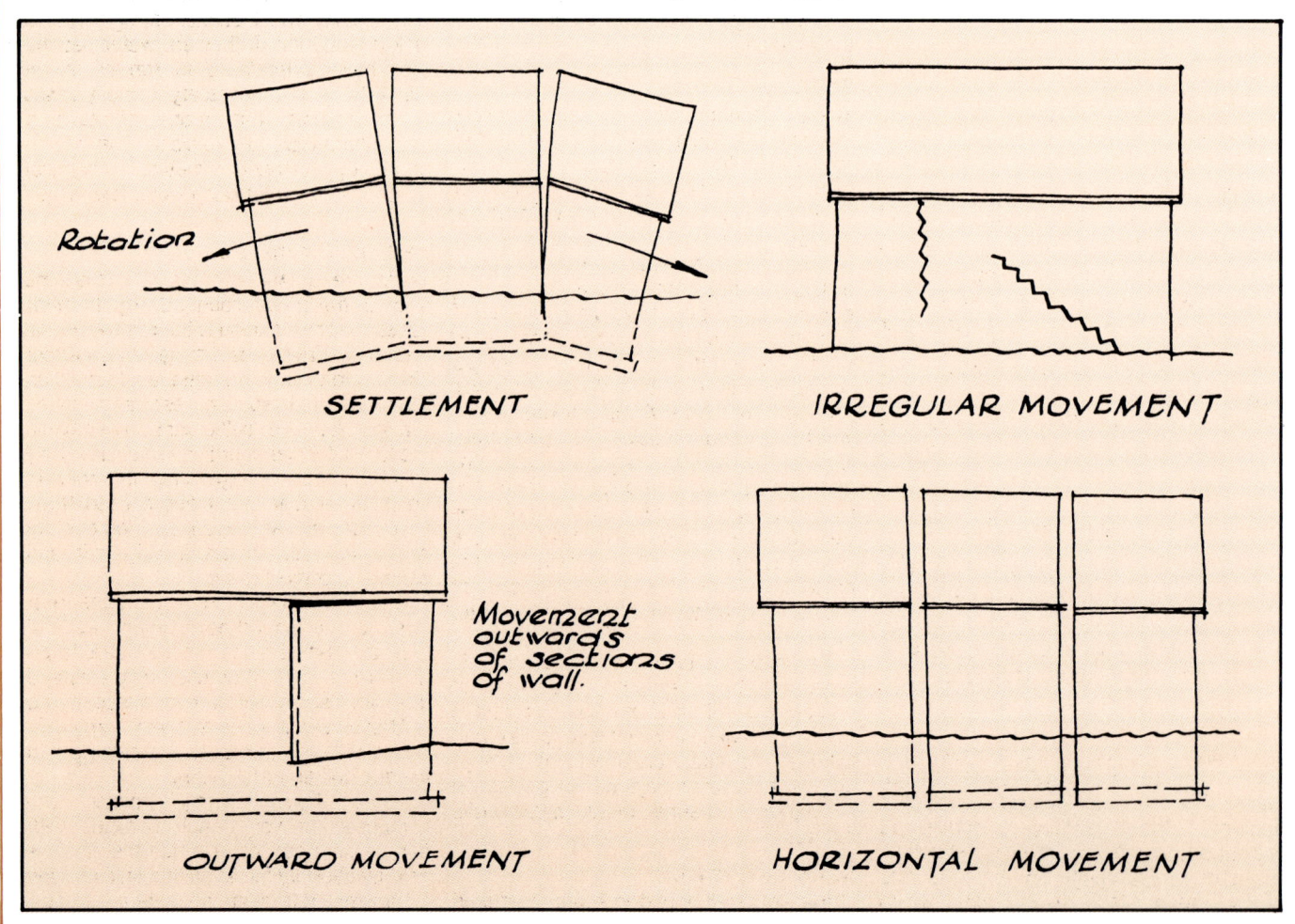

SETTLEMENT

IRREGULAR MOVEMENT

OUTWARD MOVEMENT
Movement outwards of sections of wall.

HORIZONTAL MOVEMENT

CRACKS IN BUILDINGS

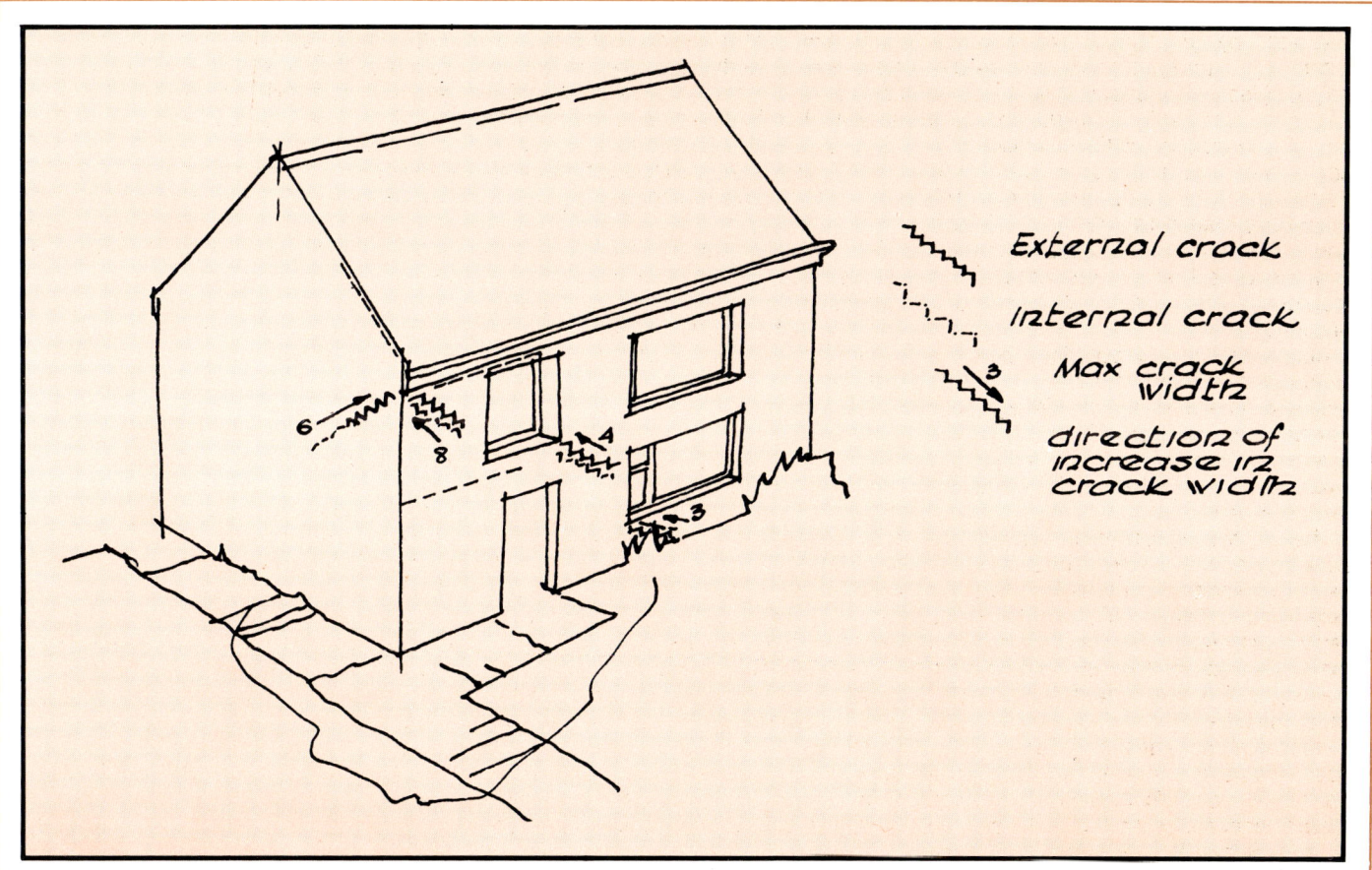

[Illustration of a house showing crack locations labeled 3, 4, 6, 8, with legend:
- External crack
- Internal crack
- Max crack width
- direction of increase in crack width]

Soil

The uncemented sediments such as sands, clays and silts are known as soils. Owing to their low strength, they often cause trouble in the design and carrying out of constructional works and may contribute to the failure in buildings at a later date. For this reason some knowledge of their behaviour may be of value. The development of the science of soil mechanics has only developed since the 1930s.

Soil Types

In discussing the various properties of soils it is convenient to divide them into four main types. There is a continual gradation from one type to the other so that the characteristics of certain soils may have to be interpolated between the characteristics attributed to the main classifications.

Soft Clays and Silts

These are of recent geological origin and have not been consolidated under a large overburden of subsequently deposited sediments. They are being formed at the present day in valleys, estuaries and deltas. The post-glacial up-lift on the land has exposed as dry land many regions which were tidal mud flats and estuaries at an earlier age. Soils of this type have very low strength which results in low bearing capacities. If excavations have to be made in this type of soil, the banks have to be cut to a low angle because of the instability of the material.

Stiff Clays

These belong, principally, to the tertiary and mesozoic formations. They owe their stiffness to the great depth of sediment which has been deposited over them at some period in their history and which has consolidated them from a soft mud into an almost shale material. The greater part of this overburden has been removed by erosion but the clays have not expanded. A London clay may have been consolidated under an overburden of about 30 tons per sq. ft. The water content of the clay, which is the ratio of the weight of pore water to the weight of solid particles, will have been reduced from 80% to about 22%. The water content has probably not increased by more than 4% or 5% since these clays were exposed. This means that they are very susceptible to expansion in the presence of an increased water content caused by changes in the level of the water table.

An important characteristic of the stiff clays is the frequent occurrence of numerous small fissures. These constitute planes of weakness which lead to slips and failures in the edges of excavations, cuttings and cliffs. Boulder clays do not possess in so marked a manner these characteristic fissures.

Sands and Gravels

These are rarely found in strata older than the cretaceous period and are commonly encountered in association with modern rivers and with glacial deposits both as outwash terraces and as englacial material. Their outstanding characteristics are high permeability and frictional strength with a lack of cohesion. If they are water bearing they have a tendency to run or fail in excavations. It is necessary in many cases to consolidate this group of soils if excavations have to take place. This may be done by cement grouting, chemical consolidation and ground water lowering.

Peat and Organic Soils

Peats are among the most troublesome soils for the construction industry. In many cases the only solution is the removal of the peat or to drive piles through it to a firmer stratum to provide an adequate foundation for a building. In some cases this will not be practicable. The settlement of embankments in the land drainage scheme may require periodic heightening in order to keep the embankment above flood level. Peat contains an exceptional amount of water and this makes it very soft and extremely compressible.

These soils, which form the uppermost layers of the ground have varying levels of porosity. Some of the water which is taken up from rainfall will be used by plants and trees and some will be evaporated. If the soil becomes saturated or is very permeable water will then percolate downwards into the sub-soil and rock below.

Clays, though highly porous, are composed of very small particles with minute pore spaces in which the water is held by capillary forces so that little or none is transmitted through the material. It has a very low permeability. Chalk is also highly porous but has a very low level of permeability. There are many small joints and fissures in the rock and as a result of this, although the rock itself may be impermeable, in the mass water does transmit through it. Compact rocks, such as quartzite, igneous rocks like granite and dolerites and finer grained rocks have a very low porosity. Limestones vary greatly in the void spaces they contain. A number of oolitic limestones are in excess of 20% porosity which is similar to that which would be achieved by loose sand and gravel.

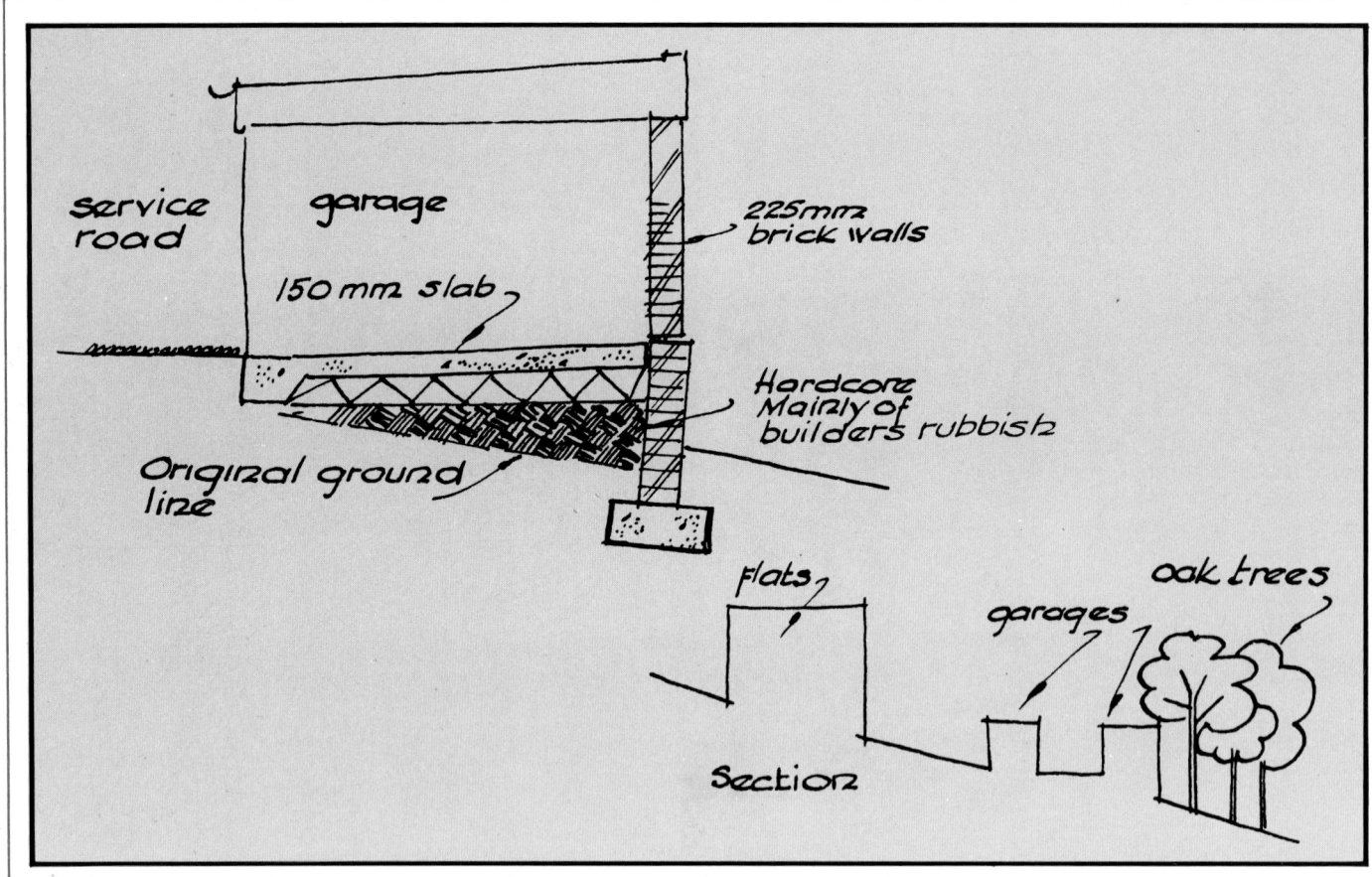

Sands and Sandstone

The porosity of sand depends on several factors.
1. The grade size of the grains. The sand containing mainly grains of one size will possess a higher porosity than one consisting of a mixture of grades. This also applies to sandstones.
2. The amount and kinds of packing which the grains have undergone.
3. The amount of cement present. The gaps between the grains may be filled to varying extents with mineral matter which binds the grains together. This will have a considerable effect on the porosity of the material. Example of the levels of porosity of varying materials the porosity is based on the relationship between voids and solids expressed as a percentage. Soils 55%, clays 46%, chalk 48%, loose sand and gravel 40%, sandstones 10%, limestone 12%, dolermitic limestones 5%, granites less than 1%, quartzite ½%.

The varying types of soils have corrosive effects upon building components within which they may come in contact. The light sandy soils and chalks are not generally aggressive to any material which will be commonly used in construction. The heavy clays which tend to be anaerobic, can be corrosive to ferrous metals. These clays prevent the air from getting to the metal and can support the bacteria known as the anaerobe which survives without air. In the excavations of the Mary Rose and its ultimate raising off Portsmouth, the absence of air from the timbers and artefacts contained within the vessel preserved their life and much of the wood, leather, bone and pottery was maintained in a condition which was remarkable bearing in mind that many of the items were over 450 years old. The cannon which were on the boat had extensively corroded, as had most of the cast iron and ferrous metals. The survivors were the bronzes or non-ferrous materials.

Saline soils can cause severe corrosion to aluminium and galvanised steel. Lead will also corrode in these conditions if it is connected to copper. The corrosive effects of the salts are well known to most week-end sailors.

Sulphate bearing soils can corrode cast iron if there is an absence of oxygen. The sulphates react with the iron to produce ferrous sulphide.

Clinker, ash, cinders and other forms of builders' rubble can often be corrosive to steel, copper and aluminium. The careful selection of materials used in hardcore and the blinding of hardcore is important where these metals may be present.

Soil effect on buildings, materials and pipework

Soil type	Effect
Light, sandy soils, chalk	Not generally aggressive.
Cinders, builders' rubble	Very corrosive to steel, copper and aluminium.
Heavy, anaerobic clay (air is prevented from getting into the clay and anaerobic bacteria—those which do not need oxygen—can survive)	Corrosive to ferrous metals.
Saline soils	Cause severe corrosion of aluminium and galvanised steel, and corrosion of lead if connected to copper.
Sulphate bearing soils	In anaerobic conditions—no oxygen—sulphate reducing bacteria convert cast iron into ferrous sulphide.

Trees and their effects on buildings

TREES AND THEIR EFFECT ON BUILDINGS

The dry weather of 1976 produced exceptional circumstances which resulted in a high number of claims being made to insurance companies or other parties for damage to buildings as a result of ground movement. In some cases this movement has been attributed to trees which have caused drying shrinkage in the ground by their root systems having reduced the moisture content. Damage to property or disturbance to occupants may have resulted from a number of actions of trees and shrubs.

(a) Root action causing ground shrinkage.
(b) Root action causing pressure in retaining walls or foundations.
(c) Root action causing failure in drains.
(d) Damage from falling leaves blocking gutters, drains or water gullies.
(e) Trees causing structural damage when they fall.
(f) Trees seeping sticky substances which stain paths, cars, driveways, etc.
(g) Trees which harbour insects, bees and wasps which can be unpleasant in a residential environment.
(h) Expenditure incurred in clearing up leaves each year.
(i) Creeper action causing mortar failure in brickwork and masonry.
(j) Creeper action causing failure to the face of weak block or brickwork.
(k) Creepers blocking outlets and openings such as gas ventilators, air bricks, flues, etc.

Root action

The majority of the damage to buildings which has been caused by trees and their root systems occur in the shrinkable clay soils commonly found in the South East of England.

The buildings which have been damaged by ground movement are those which have been constructed with limited foundation systems or where the depth of foundation is less than 1 metre below the ground level. All the information on tree root damage has to be related to the type of property that one is considering and the assumed foundation construction, type and depth. Where the property has a basement, the foundation level is usually at a greater depth than 1½ metres. At this depth the susceptibility of the foundations to fluctuations in the clay soil is small.

Safety

The safety factor, which is the distance that a specific type of tree should be from the property, refers to properties with a foundation depth of not less than 0.75m (2'6"). This safety factor gives an indication of the zone of influence of the tree roots over the range of the tree's development, from the minimum growth in clay soils in urban areas to the maximum growth in more ideal circumstances. The safety factor is the reasonable distance that a tree root will extend and be likely to cause damage to a building. It is not a guarantee that a building cannot be damaged at this distance, but an indication of the distance which property with the type of foundations referred to, should be to remain unaffected by root action in 99 cases out of 100. The information that has been gained by the research into the extent of damage caused by tree roots has validated the original contention that if a property was at a greater distance from a tree than the mature full grown height of the tree, there was very little risk of damage occurring to the property. The exceptions to this rule are Oak, Willow, Alder, Hawthorn, Poplar, Horse Chestnut and Elm. They are all able to damage a property that is further away from the tree than its mature height.

Estimating the age of a tree

When assessing the risk to a property caused by a specific tree type, one also has to estimate the extent of growth which has already occurred to the specific tree. By comparing the actual tree height with its possible mature height, one can assess the future extension of root activity that may occur. A simple method for calculating the height of the tree is set out. The likely maximum height for various trees in different areas is shown in the information sheets on trees.

It has now been suggested that a tree goes through specific cycles in its life and that in each of these cycles, the development of the tree root system varies.

For the first third of the age of the tree it is in the developing and growth period, during which it is pushing out its roots at a substantial rate and developing from a sapling to a full height tree.

For the middle third of its life, the tree consolidates its position and the extent of tree root growth is usually not very great. Should ground conditions vary, such as would occur during a drought or changes in the water table, then the tree root system will expand, but in stable conditions little movement should occur.

In the last third of its life the tree root system contracts gradually until the tree dies.

In order to be able to identify the position that the tree may be within the three stages of its life cycle, the estimated life span of each tree has been listed against each species. By checking the height and girth of the tree against its likely maximum in a particular environment, one may estimate if the tree is fully grown or still in its initial growth period. The likely damage that the tree may cause in the future will depend on whether the tree has reached full maturity.

Heave

Should it be necessary to recommend the removal of a tree because it will cause damage to the property and it is known that the foundations of the property are vulnerable because they are not of sufficient depth, serious consideration must be given to the risk of 'heave' occurring after a tree has been removed. 'Heave' is the term applied to the effect on ground areas surrounding tree roots after a tree has been cut down. The ground and roots become swollen with the moisture that had hitherto been transpired by the tree. On the assumption that the tree was young and healthy and not in its declining years, it may be necessary to recommend the reduction of the tree over a number of years.

Prior to the removal of any trees, it is important that the foundation systems of a property should be stabilised and where solid floors have been inserted which are in contact with the ground, consideration should be given to the risk of failure following tree removal. The swelling of ground beneath a solid floor may cause bulging of the material and set up stress in the building floor construction.

Similar care should be taken when new properties are being constructed on a site which was formerly heavily wooded. A surveyor inspecting new property should research the original ground conditions so that he may comment upon any risks that may be incurred in this way.

Dangerous trees

If one lists the most commonly encountered trees in the order in which they can cause damage by root action, those trees at the top of the list are those which have to be the greatest distance from a property for reasonable safety as they are regarded as being the greater risk. Those trees which are allowed to be nearer to a property because they are less likely to cause damage are placed lower down the order.

The list of trees in descending degree of risk is as follows:

1. Willow
2. Poplar
3. Oak
4. Elm
5. Horse Chestnut
6. Sycamore
7. Lime
8. Beech
9. Plane
10. Birch

It should be noted that in the case of the first three trees listed, they should be located more than 25 metres from a building which has vulnerable foundations. These trees will also cause the greatest amount of heave if they are felled and a property is within the various zones of influence. Where such trees occur, it is only common sense to arrange for their removal to be carried out over a period of years, by removing half the height in the first year and bringing the tree to ground level in the second or third year.

Drain damage by roots

In some cases trees cause damage to drains and drainage systems. The small roots are able to make their way into old pipe systems as the pipe joints were not very strong. The roots either enter through defects in the pipes laid over a hundred years ago, or through joint or pipe failures.

Failures may have occurred due to the roots exerting pressure on the pipes or by vibration set up by road traffic, building operations, railways, etc.

Once the minor root system has

TREES AND THEIR EFFECT ON BUILDINGS

TREES AND THEIR EFFECT ON BUILDINGS

penetrated a drain, the failure may be enlarged so that the larger roots can enter the pipes. This leads to a greater failure and a larger loss of water from the drainage system. The root system can also develop within the pipe, forming a net of narrow tendrils which eventually restrict the flow through the pipe system and cause blockages.

Tree damage caused when they fall

If one has to inspect the damage caused to a property by a fallen tree which was blown down on a stormy night, one cannot but be impressed by the size and power of that horizontal form. The weight of a mature plane tree lining a London street is over 14 tons in winter and in excess of that figure during the summer due to the extra weight of its foliage. Part of the tree will be doing over 40 miles an hour as it strikes the ground, which is similar to a double decker bus being driven into a building.

Trees collapse because they are no longer able to withstand the strain exerted by wind or other forces at a particular time.

The strain may be due to an exceptional storm, a lorry or similar large object colliding with it, or ground movement such as landslip or earthquake. Many of these items are beyond a surveyor's ability to predict, but the declining ability of a tree to withstand the normal strain of wind and weather can be assessed.

That declining ability may be due to age. If the tree is in the last few years of its life, it will be prone to failure as it is less able to withstand the elements.

Dutch Elm Disease

The tree may have been weakened by disease or the attack of Scolytid Beetles. The recent Dutch Elm Disease epidemic (the use of initials for all things resulted in the unfortunate label being placed on certain trees "Killed by DED") which altered the landscape in many parts of southern England, was carried by a beetle which burrowed through the bark and lived on the outer surface of the tree's sapwood. The disease causes a constriction in the sap supply tubes in the trunk, which leads to the suffocation or strangulation of the tree. The beetle is able to breed twice a year if the climate is warm.

Sanitation felling is having some success. The infected trees are immediately felled and the bark removed. The removal of the infestation in its infancy has saved many neighbouring trees. The greatest hope for saving the elm forests in the North and Scotland is the "trap tree" method which is achieving some success. A stricken elm is killed with an arboricide (cacodylic acid) which makes it especially attractive to the Bark Beetle. The beetles attack this tree, ignoring the surrounding sound timber and can then be killed in great numbers.

The elms which survive the disease, or regenerate after the disease, may be stunted and may not exceed 9 metres in height.

Disadvantages

The Lime tree is a popular tree for bees, perhaps because it flowers quite early.

The Common Lime also often drops a sticky liquid onto the ground below, a mould fungus turns these droppings black.

In spring, the Balsam Poplar exudes a sticky balsam which gives off a heavy sweet odour.

Falling branches

Even healthy trees can be a source of problems. For instance, all elms have a reputation for shedding their branches without warning.

The Victorians loved to plant Monkey Puzzle (Chile Pine) trees. These trees have a life of about 100 years and many failures will be occurring around this time as the trees reach the end of their life.

Fungus on trees

Razor Strop and Bracket Fungus live on dead birch trees, Fly Agaric on the still living roots of a dying tree.

The Razor Strop is a bulbous beige growth often on the side of the trunk, whilst Bracket is a rust-red half mushroom bracket projecting in clusters from the trunk.

Fly Agaric is a red topped mushroom with white patches growing close to the tree base.

The Weymouth Pine is vulnerable to a fungal disease called Blister Rust which lives on blackcurrant bushes.

TREES AND THEIR EFFECT ON BUILDINGS

Other arboreal oddities

Poplar – quick growing. The Balsam Poplar can grow up to 2 metres a year.
Italian Alder – rapid growth to form a compact tree.
Roble Beech – can grow 1.5-1.8 metres a year.
Sycamore – matures early within 50-60 years.

Details are set out for a number of common trees so that the surveyor may identify the trees around a property and assess the risk that the tree poses to the building, now and in the future. Each page is set out in the same format for ease of reference.

The notes and explanation of each category are set out in the following order:

Name Common name and Latin name.
Age range for this tree based on available information.
Various types of ground condition will affect the likely age achieved by each tree.
Height that a mature tree will attain in reasonable conditions:
(a) In shrinkable clay soils in an urban environment
(b) In ideal circumstances.
Girth. Likely girth of a mature tree.
Features. General items to aid recognition:
Flower: The flower, fruit, nut or other characteristic details of seed dissemination.
Trunk: General characteristics of the tree trunk and branch layout.
Leaf: Leaf location and arrangement on the branch, its shape and size.

Damage.
Danger zones for the tree in its current state of development.
Zone A. If the tree top is at a lower angle to the observer than this figure, the observer is in a position where the influence of the tree root **is extremely unlikely** to have any affect (see pages 192-193 which show how to estimate the angle required).
Zone B. If the top of the tree is about this angle above the observer's eye level, then the observer is at a point where the tree roots **will possibly cause damage** to property with foundations of moderate spread and to a depth not exceeding 2'6" (0.75m).
Zone C. When the top of the tree is at this angle above the observer's eye level, the observer is situated at a point where the tree root system **is highly likely** to cause damage to a property with foundations at a depth not exceeding 2'6".

All these zones relate to the tree at its existing state of development.

Safety factor. This is the distance that a specific tree should be from a property with no special foundations to be reasonably certain that no damage by the root spread of the mature tree will occur:
(a) Where the tree is in shrinkable clay soils in an urban environment.
(b) Where the tree is in good, well drained soil conditions in individual positions likely to result in maximum growth.

The pictures on these pages graphically illustrate the disastrous effect of a tree collapsing in a busy London square.

TREES AND THEIR EFFECT ON BUILDINGS

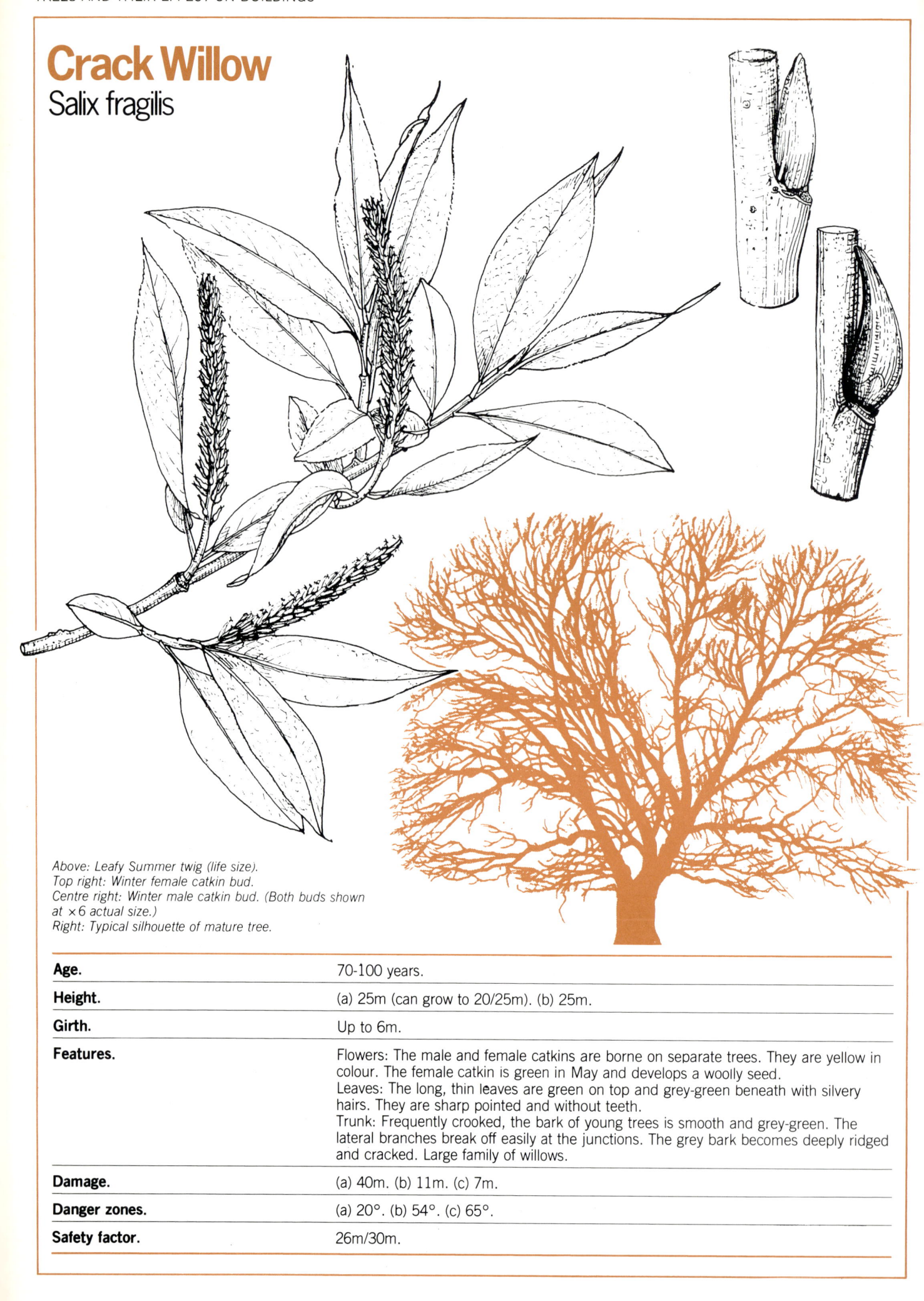

Crack Willow
Salix fragilis

Above: Leafy Summer twig (life size).
Top right: Winter female catkin bud.
Centre right: Winter male catkin bud. (Both buds shown at ×6 actual size.)
Right: Typical silhouette of mature tree.

Age.	70-100 years.
Height.	(a) 25m (can grow to 20/25m). (b) 25m.
Girth.	Up to 6m.
Features.	Flowers: The male and female catkins are borne on separate trees. They are yellow in colour. The female catkin is green in May and develops a woolly seed. Leaves: The long, thin leaves are green on top and grey-green beneath with silvery hairs. They are sharp pointed and without teeth. Trunk: Frequently crooked, the bark of young trees is smooth and grey-green. The lateral branches break off easily at the junctions. The grey bark becomes deeply ridged and cracked. Large family of willows.
Damage.	(a) 40m. (b) 11m. (c) 7m.
Danger zones.	(a) 20°. (b) 54°. (c) 65°.
Safety factor.	26m/30m.

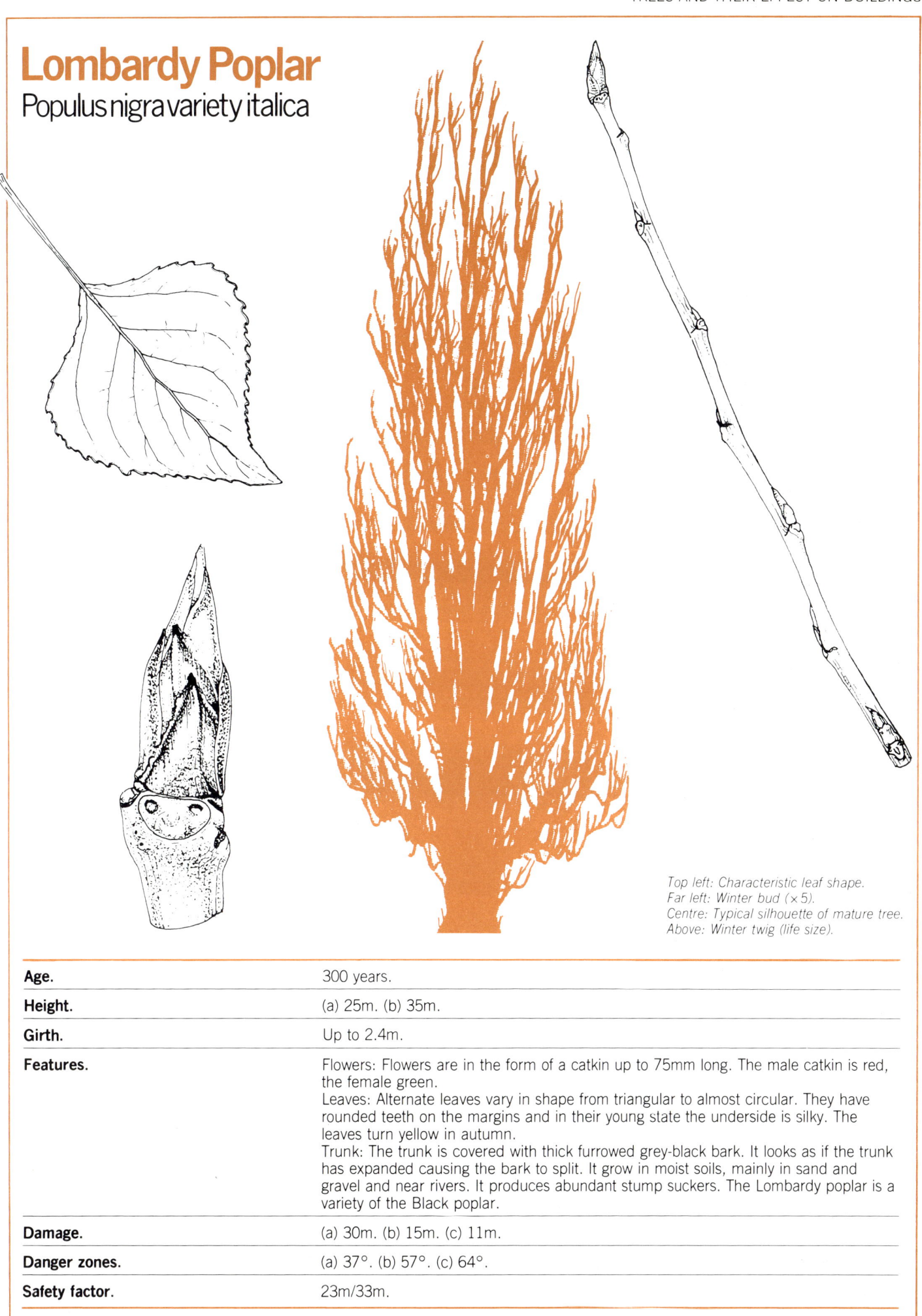

Lombardy Poplar
Populus nigra variety italica

Top left: Characteristic leaf shape.
Far left: Winter bud (×5).
Centre: Typical silhouette of mature tree.
Above: Winter twig (life size).

Age.	300 years.
Height.	(a) 25m. (b) 35m.
Girth.	Up to 2.4m.
Features.	Flowers: Flowers are in the form of a catkin up to 75mm long. The male catkin is red, the female green. Leaves: Alternate leaves vary in shape from triangular to almost circular. They have rounded teeth on the margins and in their young state the underside is silky. The leaves turn yellow in autumn. Trunk: The trunk is covered with thick furrowed grey-black bark. It looks as if the trunk has expanded causing the bark to split. It grow in moist soils, mainly in sand and gravel and near rivers. It produces abundant stump suckers. The Lombardy poplar is a variety of the Black poplar.
Damage.	(a) 30m. (b) 15m. (c) 11m.
Danger zones.	(a) 37°. (b) 57°. (c) 64°.
Safety factor.	23m/33m.

TREES AND THEIR EFFECT ON BUILDINGS

TREES AND THEIR EFFECT ON BUILDINGS

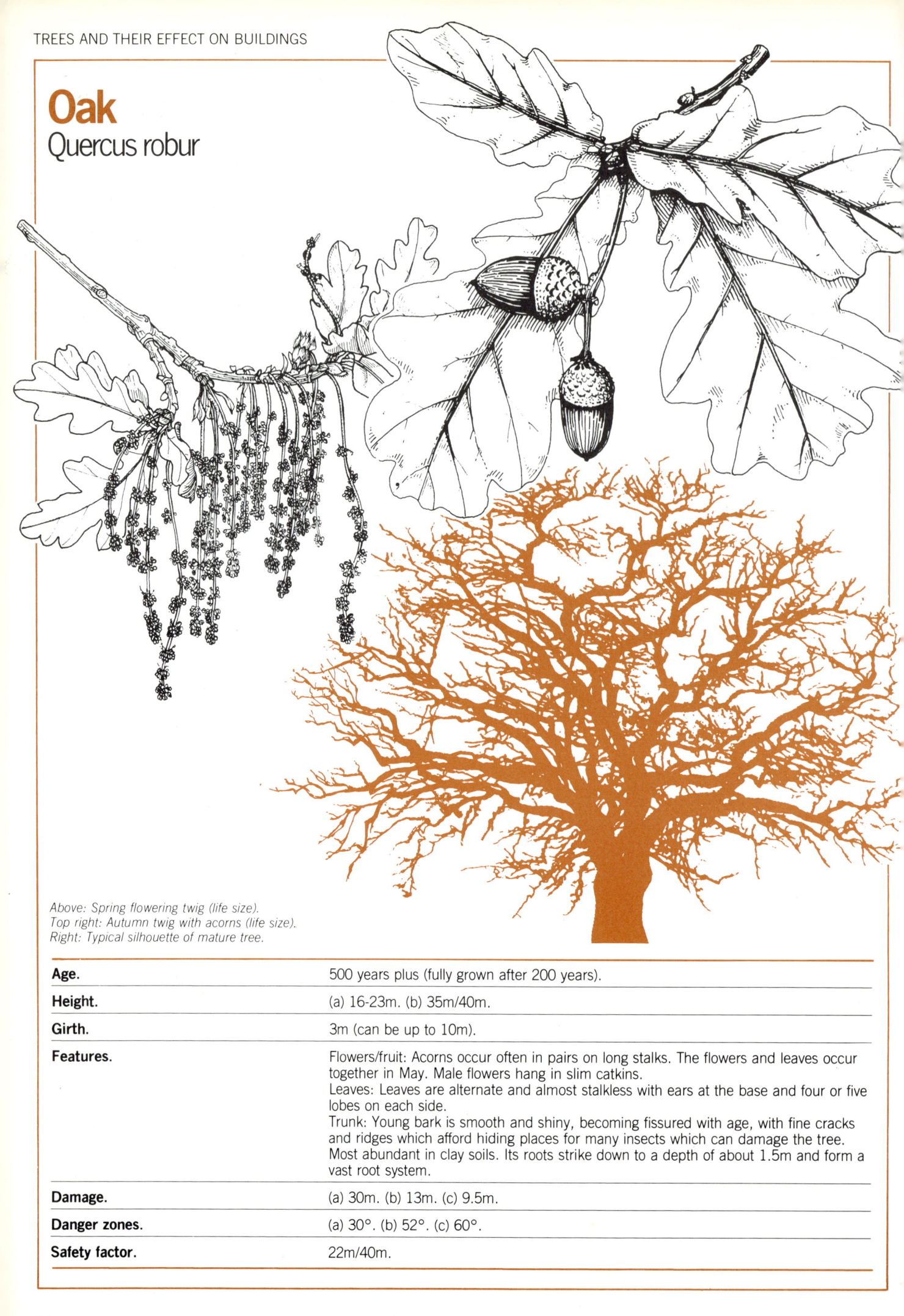

Oak
Quercus robur

Above: Spring flowering twig (life size).
Top right: Autumn twig with acorns (life size).
Right: Typical silhouette of mature tree.

Age.	500 years plus (fully grown after 200 years).
Height.	(a) 16-23m. (b) 35m/40m.
Girth.	3m (can be up to 10m).
Features.	Flowers/fruit: Acorns occur often in pairs on long stalks. The flowers and leaves occur together in May. Male flowers hang in slim catkins. Leaves: Leaves are alternate and almost stalkless with ears at the base and four or five lobes on each side. Trunk: Young bark is smooth and shiny, becoming fissured with age, with fine cracks and ridges which afford hiding places for many insects which can damage the tree. Most abundant in clay soils. Its roots strike down to a depth of about 1.5m and form a vast root system.
Damage.	(a) 30m. (b) 13m. (c) 9.5m.
Danger zones.	(a) 30°. (b) 52°. (c) 60°.
Safety factor.	22m/40m.

TREES AND THEIR EFFECT ON BUILDINGS

English Elm
Ulmus procera

Far right: Leafy twig (life size).
Right: Winter twig (life size).
Below: Spring flowering twig (life size).
Bottom right: Typical silhouette of mature tree.

Age.	Has been known to exceed 500 years.
Height.	(a) 25m. (b) 30m.
Girth.	Seldom exceeds 6.5m.
Features.	Dull dark green leaf, in autumn turns orange, then to pale yellow. Leaves have hairs covering the midribs which possess a minor irritant – similar to the nettle to which it is related. Dutch Elm disease is caused by scolytid beetle which burrows under bark and introduces a fungus which blocks tree's sap.
Damage.	(a) 25m. (b) 12m. (c) 8m.
Danger zones.	(a) 42°. (b) 62°. (c) 70°.
Safety factor.	19m/23m.

TREES AND THEIR EFFECT ON BUILDINGS

Horse Chestnut
Aesculus hippocastanum

Right: Characteristic 'palmate' leaf with Autumn fruits or 'conkers' (× 2/3).
Below: Winter twig with 'sticky buds' and distinctive horse-shoe-shaped leaf scar (× 2/3).
Bottom right: Spectacular Spring flower (× 2/3).
Bottom left: Typical silhouette of mature tree.

Age.	200 years.
Height.	(a) 16-25m (usually about 20m). (b) 27m.
Girth	Up to 2.8m.
Features.	The trunk of old trees is often twisted, always in a righthand direction and the bark peels in thin plates. The tree thrives best in rich moist soils but it is also tolerant of poor light and pollution. The leaf has five to seven large, thick stalkless leaflets with pronounced veins and a long tapering base. The spiney fruit (conker) is enclosed in a pale green spiked shell which falls, if it is allowed by the local children, about October.
Damage.	(a) 23m. (b) 10m. (c) 7.5m.
Danger zones.	(a) 38°. (b) 60°. (c) 67°.
Safety factor.	17m/23m.

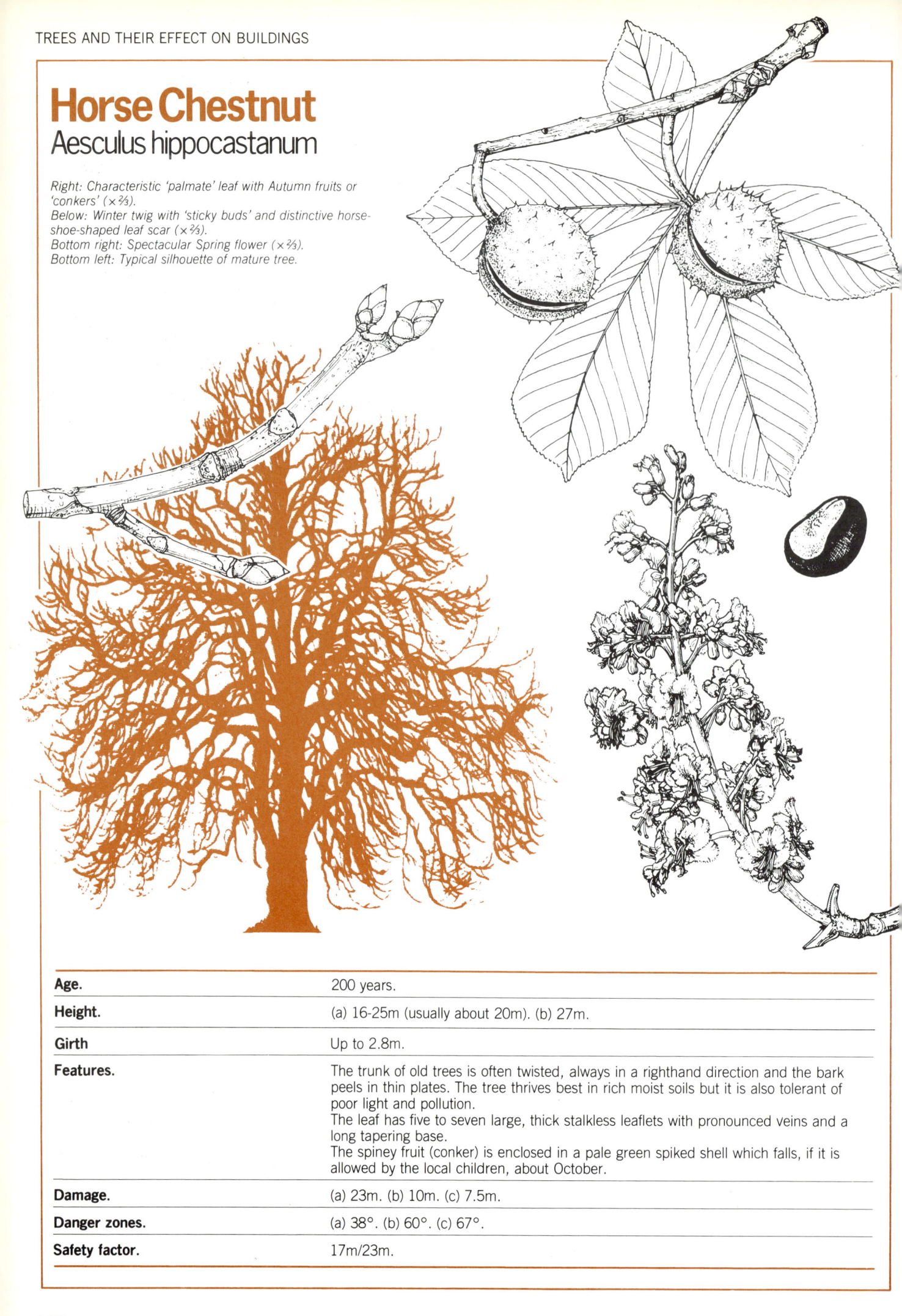

TREES AND THEIR EFFECT ON BUILDINGS

Sycamore
Acer pseudoplatanus

Below: Spring flowering twig (× ⅔).
Right: Characteristic leaf with Autumn fruits (life size).
Below: Typical silhouette of mature tree.
Bottom right: Winter twig (life size).

Age.	150-200 years (fully grown 50-60 years).
Height.	(a) 20-24m. (b) 30-35m.
Girth.	1.5m.
Features.	Fruit: Scimitar shaped about 25mm long. Red-brown in colour produced after tree is 20 years old. Leaves: Leaves are opposite and five fingered, approx 200mm across; the lower lobes are not fully separated. Upper sides are dark green. The edges are unequally toothed. Frequent black patch on leaf due to fungus called 'Tar spots'. Trunk: Grey fissured bark ages to pinkish-brown. It peels off in small flat plates. Prefers a cool climate and well drained soil. Has a large spreading root system.
Damage.	(a) 20m. (b) 9m. (c) 6m.
Danger zones.	(a) 45°. (b) 66°. (c) 73°.
Safety factor.	15m/22.5m.

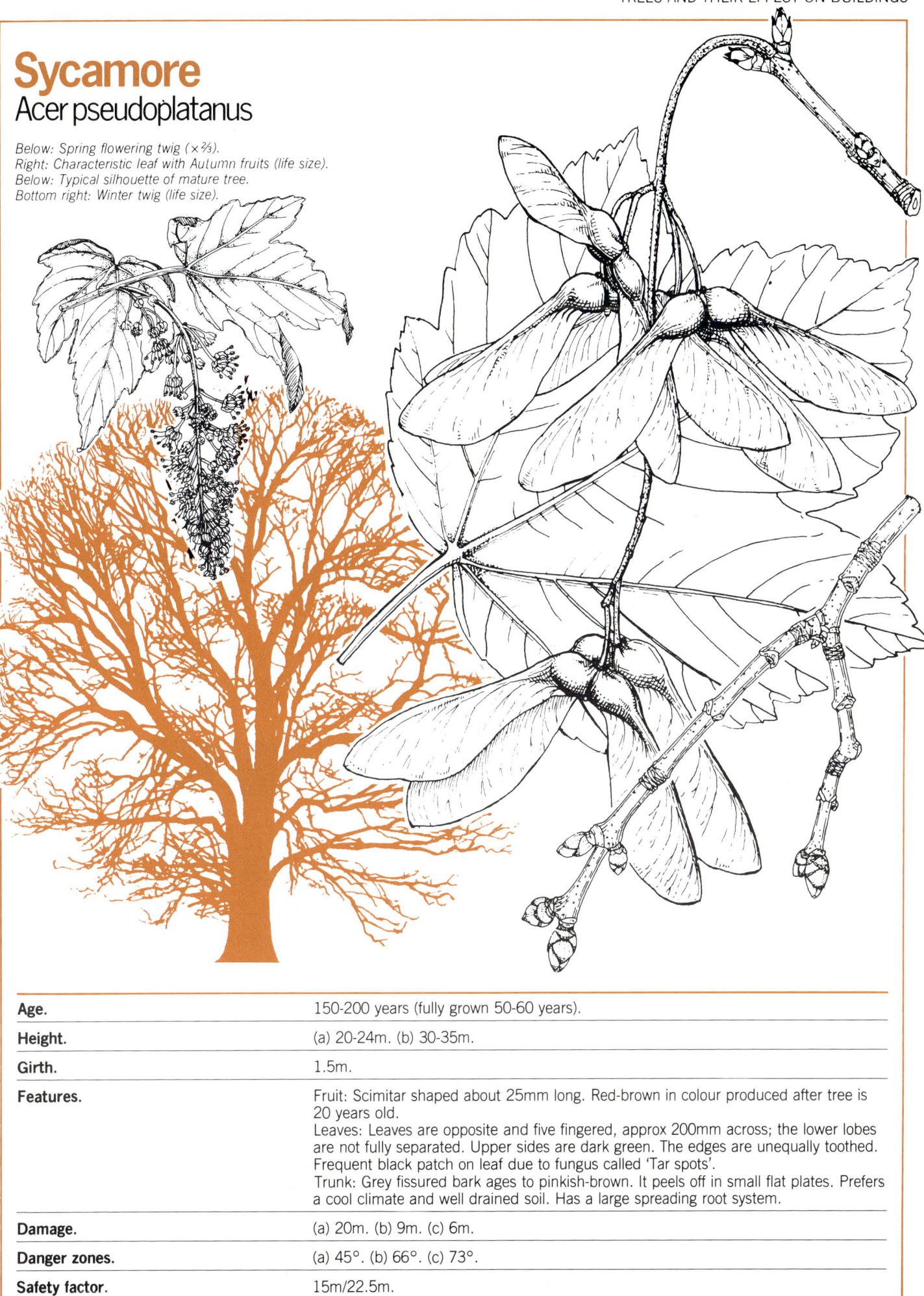

TREES AND THEIR EFFECT ON BUILDINGS

Lime
Tilia europaea

Left: Leaves and Autumn fruits (life size).
Centre: Winter twig (life size).
Below: Leaves with Summer flowers (life size).
Bottom left: Typical silhouette of mature tree.

Age.	500 (1000) years.
Height.	(a) 21-24m. (b) 30m.
Girth.	4.5m up to 7.5m.
Features.	Flower: yellowish-white flower has an abundance of nectars and a strong sweet smell. These are succeeded by globose little fruits about 8mm across, yellow in colour and covered in down. Leaf: Leaves are broad with sharply-toothed edges and are about 50-100mm long with almost straight bases: The underside has white or buff coloured hairs in the vein junctions. Aphids feed on leaves causing sticky leaf sap to drip; may leave red nail galls on leaf. Trunk: The young bark is smooth and grey becoming fissured as it gets older. Deeply rooted tree, with straight tall stems. They lose their leaves early in autumn.
Damage.	(a) 20m. (b) 8m. (c) 6m.
Danger zones.	(a) 48°. (b) 57°. (c) 75°.
Safety factor.	14m/17.5m.

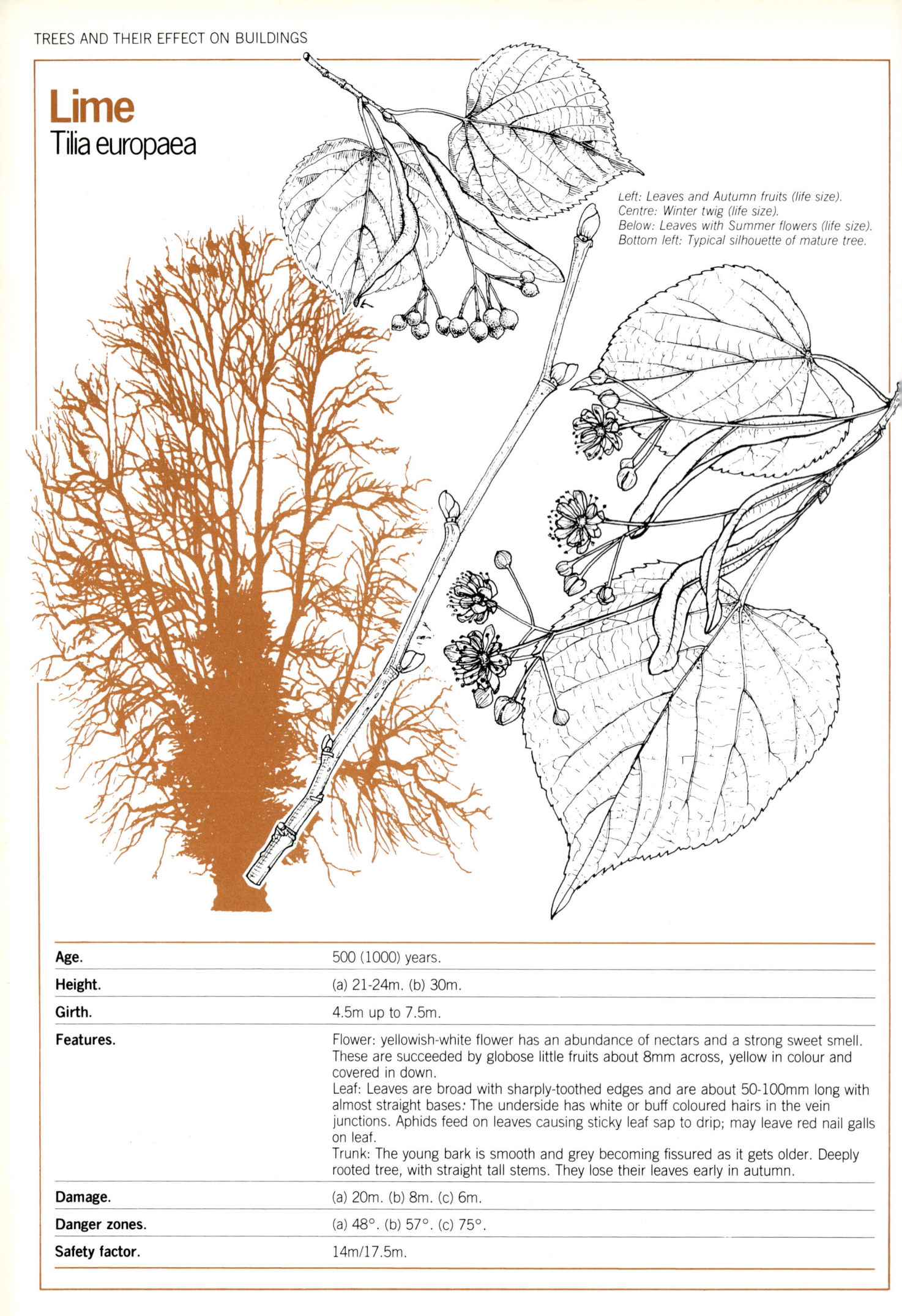

TREES AND THEIR EFFECT ON BUILDINGS

Beech
Fagus sylvatica

Left: Autumn fruiting twig (life size).
Bottom left: Winter twig (life size).
Above: Spring flowering twig (life size).
Below: Typical silhouette of mature tree.

Age.	In excess of 250 years.
Height.	(a) 20m. (b) 30m.
Girth.	Up to 6.2m.
Features.	Leaves are alternate, shiny green on both sides with a wavy margin and six to seven parallel veins. The leaves turn orange or red-brown in autumn. Grows as a preference in areas with a lime soil. The trunk is smooth silver-grey.
Damage.	(a) 15m. (b) 9m. (c) 6m.
Danger zones.	(a) 50°. (b) 63°. (c) 72°.
Safety factor.	12m/18m.

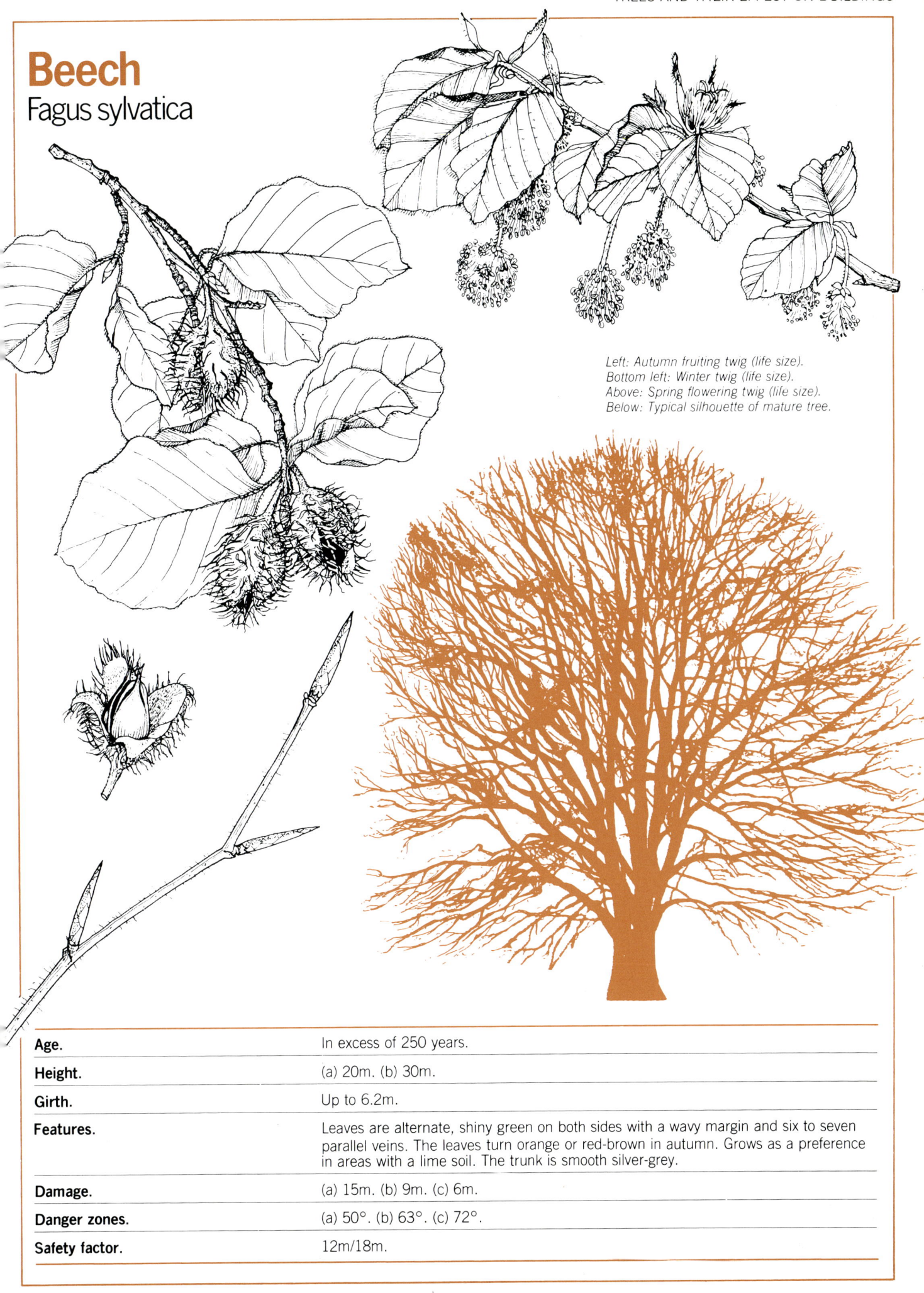

TREES AND THEIR EFFECT ON BUILDINGS

Plane
Platanus acerifolia

Below: Winter twig (life size).
Right: Characteristic leaf shape and Autumn fruits (life size).
Bottom: Typical silhouette of mature tree.

Age.	Over 200 years.
Height.	(a) 25-30m. (b) 30m.
Girth.	3.6m (7.5m has been recorded).
Features.	Fruit: The catkins of the plane take the forms of balls, and the fruits are rough balls (like old buttons). The fruit remains on the tree through the winter. Leaves: The leaves are alternate with five lobes and measure about 150mm across. No buds can be seen in summer as they are enclosed in the swollen bases of the leaf-stalks. Trunk: The rigid even layer of the bark, usually light grey (usually darker in a city due to grime), is discarded in large and small flakes which leave a smooth yellow patch behind.
Damage.	(a) 15m. (b) 7.5m. (c) 5.5m.
Danger zones	(a) 59°. (b) 73°. (c) 77°.
Safety factor.	12.5m/15.0m.

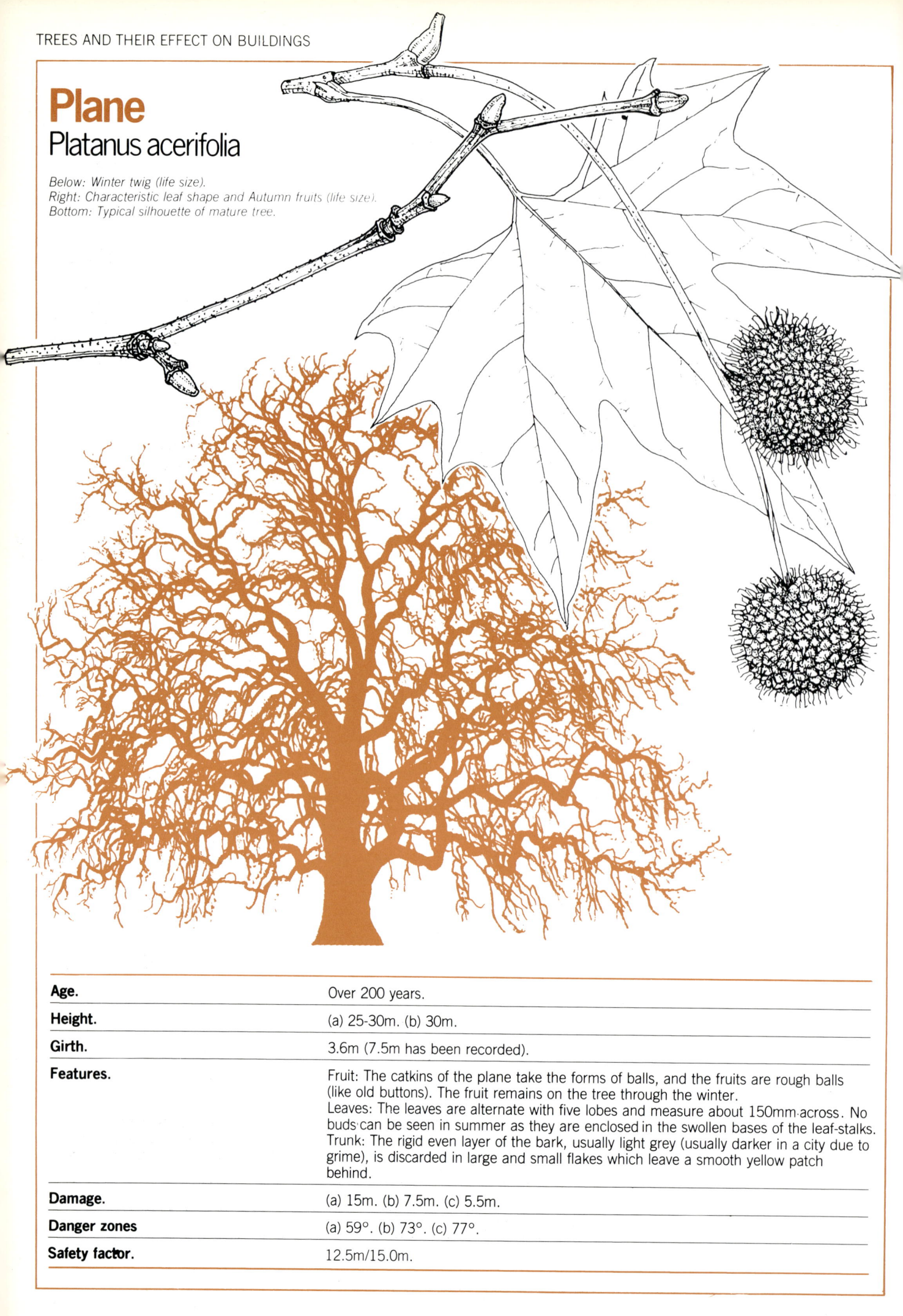

TREES AND THEIR EFFECT ON BUILDINGS

Hairy Birch
Betula pubescens

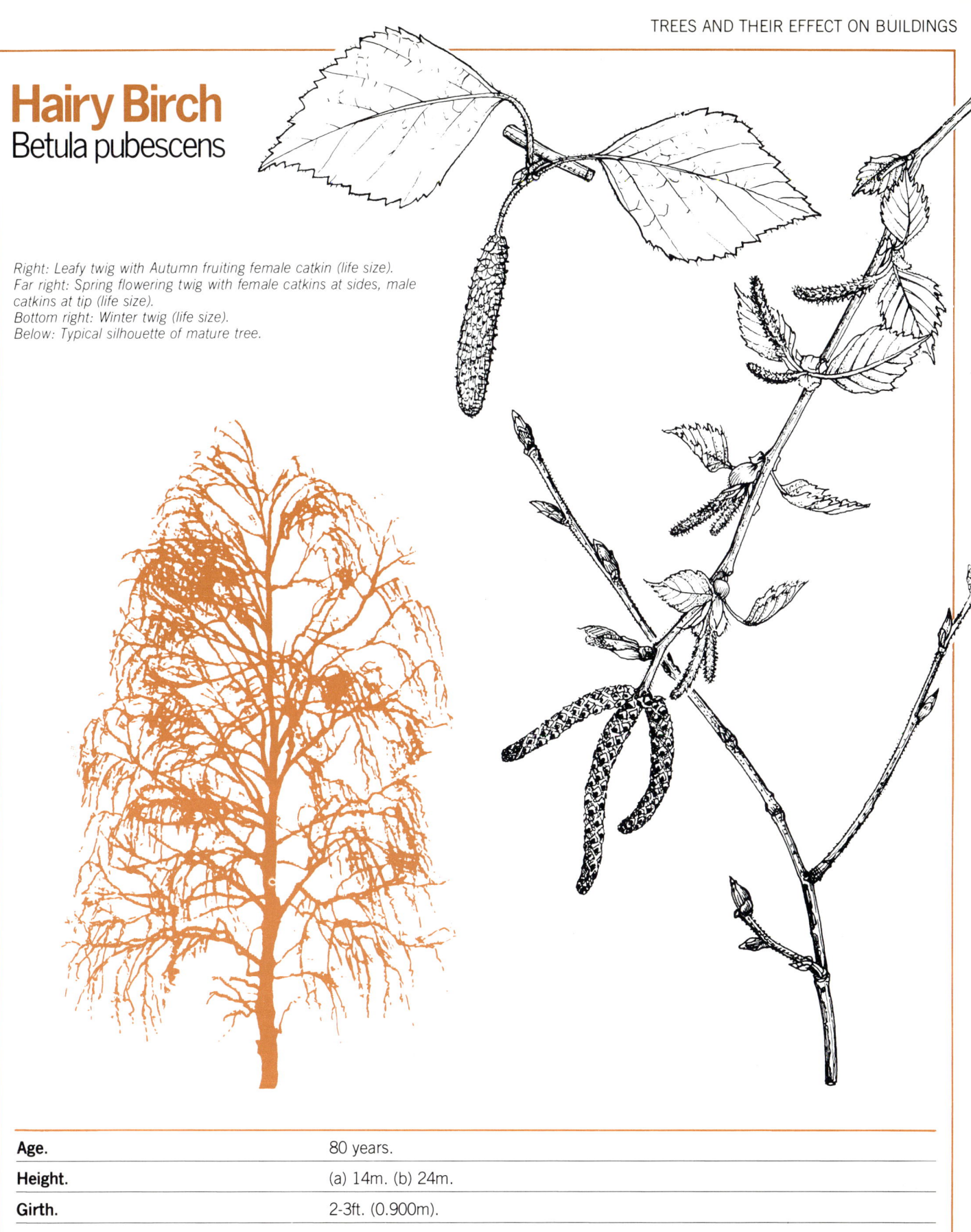

Right: Leafy twig with Autumn fruiting female catkin (life size).
Far right: Spring flowering twig with female catkins at sides, male catkins at tip (life size).
Bottom right: Winter twig (life size).
Below: Typical silhouette of mature tree.

Age.	80 years.
Height.	(a) 14m. (b) 24m.
Girth.	2-3ft. (0.900m).
Features.	Birch is shallow rooted on clay soils. The bark is more enduring that its timbers, because it is regularly shed. Bark is mainly silver-white giving a slender appearance to the tree. The glossy triangular leaves vary from a triangular form to a pointed oval, the edges are deeply toothed.
Damage.	(a) maximum root spread 10m. (b) 7m. (c) 4m.
Danger zones.	(a) 50°. (b) 60°. (c) 72°.
Safety factor.	8.5m/14.5m.

191

20°
25°
30°
35°
40°
45°
50°
55°
60°
65°
70°

Align this end of line with eye level mark on tree

Align this end of line with your eye

To calculate tree height align this line with top of tree

TREES AND THEIR EFFECT ON BUILDINGS

How to estimate the height of a tree

The height of a tree may be estimated by recording the angle that the top of the tree makes with a horizontal line taken from the base of the tree.

In order to make the estimation of the tree height as simple and painless as possible the calculation of the tree height is based on the measurement of a horizontal length. If you are an estate agent the accurate measurement of the horizontal distance may mean pacing out the length, but a tape, staff or sonic tape (as described in the Equipment section) will give you a dimension you can be more confident in.

All dimensions are based upon the tree's height above your eye level so don't forget to add your eye height onto your horizontal length.

In order to obtain a horizontal line place a mark on the trunk at eye level. Carefully retreat, having duly examined the ground behind you for hazards, small children, etc. When you have the top of the tree at an angle of 45° to your horizontal eye level line, the distance you are from the tree plus your eye height equals the tree height.

Tree height = distance + eye height

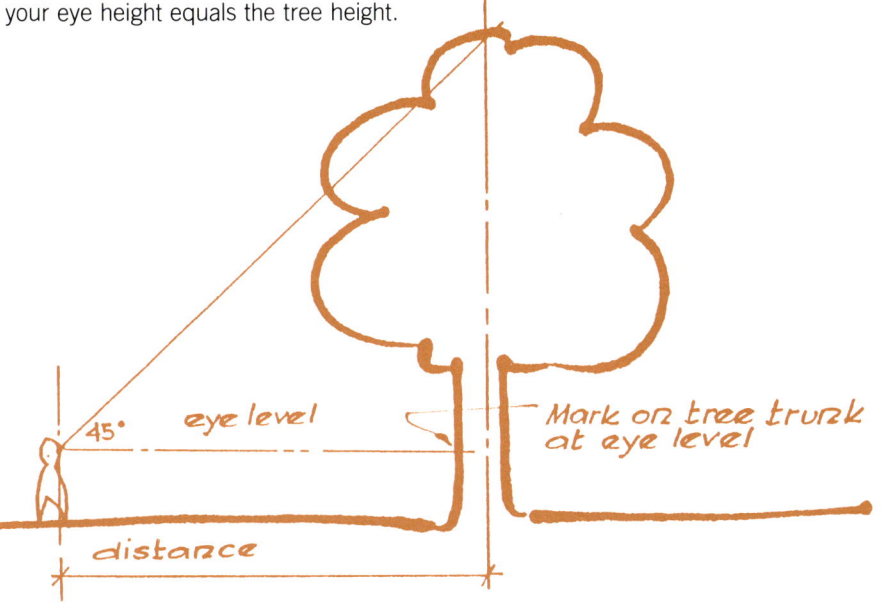

How to calculate the danger zones

In order to find out the extent of the danger zones find out the angle that is given for a particular zone for the tree you have identified. Using the page on the left, retreat until the angle to your eye level horizontal line made by the top of the tree is that given in the table. You are then standing at the edge of the particular zone.

An allowance for your eye height has been made in setting up these angles for each tree so that there is no need to make any adjustments.

	Zone A	Zone B	Zone C	Minimum safe distance from building
	Likelihood of damage			
	Remote	Possible	Probable	
Crack willow	20°	54°	65°	26m
Lombardy poplar	37°	57°	64°	23m
English elm	42°	62°	70°	19m
Oak	30°	52°	60°	22m
Horse chestnut	38°	60°	67°	17m
Sycamore	45°	66°	73°	15m
Lime	48°	57°	75°	14m
Beech	50°	63°	72°	12m
Plane	59°	73°	77°	11m
Birch	50°	60°	72°	8·5m

The importance of accurate observation

Some people who are involved in the examination of buildings may query the relevance of being able to identify the various agents of timber failure.

There has been a trend for members of the surveying profession to abdicate their responsibilities to specialists when they have discovered timber failures.

The surveyor should be able to recognise the various species of trees. The ability to identify at least the more commonly occurring trees is vital in enabling the surveyor to make a considered observation on the likely effect the proximity of a particular tree and its roots may have on the building he is inspecting.

Each fungus will produce certain reactions in timber. Certain of the timber 'infections' will not damage the timber, some will cause early strength loss, some are capable of living in drier timber, and some only in wet wood.

The cost of repairing a building will vary considerably depending upon the type of fungal attack that may be present. Each fungus requires moisture to be present. No fungus will propogate in the first instance if the moisture content is less than 20% (moisture content is the weight of water divided by the weight of the wood).

The photograph on this page shows two pieces of wood which have suffered fungal decay. In both cases the timber is damaged, and the appearance of the failure is very similar. The moisture content of each piece of timber would enable a diagnosis to be made between *Serpula lacrymans* (dry rot) and *Coniophora puteana* (wet rot) if the fungus was still active. Each timber shows the characteristic failure which would be attributed to *Serpula lacrymans*. If the failure was dry rot induced, one would have to allow for the extra cost of cutting out the failure and the sound timber beyond the infestation. If the failure was wet rot, then the eradication of the dampness would result in the demise of the fungal attack.

Dry rot does not require a lot of moisture in the timber for it to propagate. Once established, it is able to spread to timber having a moisture content of only 14-15%. It is this risk of spreading that makes it so dangerous. Wet rot can only propagate and survive in wet wood. It requires a minimum moisture content in excess of 40%. Dry rot cannot propagate or survive in wood that is wet. For this reason the moisture content of these two pieces of wood would help to identify the top piece of wood as being damaged by dry rot and the lower by wet rot.

The various beetles will also cause different types of timber damage. The correct identification of the beetle will enable an accurate assessment to be made of the likely timber damage. It may mean that a more detailed inspection should be carried out. For instance, Death Watch beetle could be an indicator of the presence of a more serious fungal attack as it is often linked with dry rot.

The recognition of the specific failure is part of the surveyor's duty, because without that specific identification he is unable to comment accurately upon the consequences of the infestation.

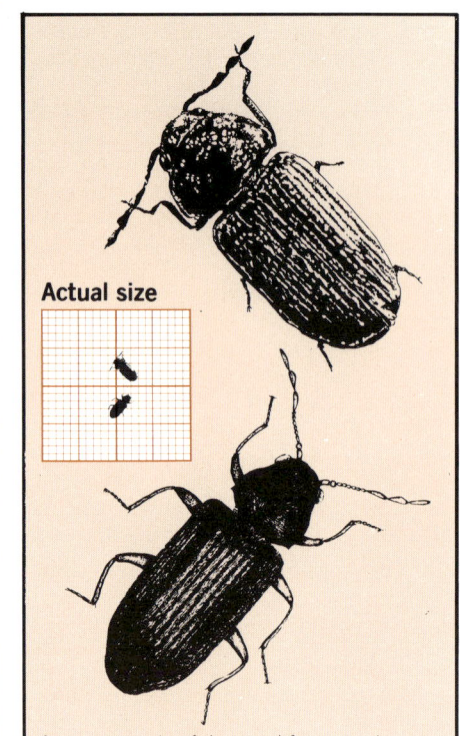

As an example of the need for correct identification, two beetles are shown here which would appear very similar if encountered during a survey. Closer examination would reveal, however, that the top one is the harmless biscuit beetle, while the bottom one is the furniture beetle—extremely destructive in timber structures!

Specialist assistance is available to the surveying profession in connection with timber failures. In many cases these 'specialists' are unqualified employees of companies who sell the service of timber treatment and eradication of timber diseases. In too many cases the advice given is not of a high quality. The surveyor, who may place too much reliance upon this 'specialist' advice in commenting upon the likely extent of the damage, is himself responsible for any inaccuracies in this advice.

The requirement for specialist comment, often requested by members of the legal profession, is a condemnation of the quality of the advice of the surveyor or a continuation of a misconception about the status of the 'specialist'.

The following pages set out the methods of recognition of various beetle and fungal attack.

Timber decay caused by insect attack

An insect like many other living creatures passes through a definite life cycle. The egg is laid in a suitable place by the female and hatches in the form of a small larva which then feeds for some time until it is ready for the next stage in the process of evolution.

This is the period of rest or pupation during which a reorganisation takes place. At the end of this period the insect breaks through the case in which it was enclosed during the resting period and emerges as the full adult, whether it be a beetle, moth or butterfly.

The destruction of wood is carried out sometimes by the juvenile form, the grub, and sometimes by the adult form. The manner of attack spread varies with different insects. Some will attack only the living tree, and infested timbers, once felled, will not suffer further reinfestation. Some will only eat decayed wood. Others will only proliferate in the warmer climates and are unable to survive being exported to colder climates.

With a temperate climate, some insects can multiply rapidly. It has been estimated that one pair of bark beetles in this country can produce more than two and a half thousand descendants in one year. In the tropics, the multiplication of the so called white ant, far exceeds the reproduction of the bark beetle.

Treatment

The treatment of insect attack may be achieved in three ways.
1. The use of high temperatures which often occur during the seasoning process of timbers when they are kiln dried. This does not necessarily mean that the timber is immune from any further attack.
2. The insects can be suffocated.
3. They can be poisoned, either by using vapour or through the stomach of the insect.

Suffocation

This depends upon the blocking of the pores in the outer covering of the insect. These are the small openings which are found all over the insects' bodies. Water will not block these openings. Insects do not breathe through one single opening as the human being does, but depend on a multitude of tiny holes in the outer layer. Some liquids wet the fine hairs that surround the insect and usually protect it. In this way the wet hairs block the openings around the body and suffocate the insect. Soapy water kills the common household greenfly in this way.

Poison

Poison vapours are becoming more and more prevalent. Many of them in larger quantities, also have an effect on human beings and may damage parts of a building. For instance, sulphur dioxide, which is obtained by burning sulphur, corrodes iron, whilst ammonia gas will affect the colour of timber. In the case of oak, the ammonia causes the wood to darken.

Stomach poisoning

This is used in the tropics against termites. Timber is impregnated with tasteless preservatives containing arsenic or lead salts, which enter the stomach of the insect when they are eaten.

TIMBER DECAY CAUSED BY INSECT ATTACK

Types of preservative relating to dealing with woodworm

There are two points to be borne in mind in selecting an effective treatment for an infestation of woodworm. The first of these is the toxic effect on human beings and the second is the effective dose which will be effective in 100% of cases. The aim of any application is the elimination of *Anobium puncatum* from wood with virtual certainty and that the wood should be impervious to future attacks for many years. The majority of fluids are oil based but the recent high cost of petroleum products has led to research into water based fluids but none are available for retail sale at the present time. Oil based fluids usually contain the insecticide Lindane or, more rarely, Dieldrin. Both of the insecticides are extremely long-lived and if properly formulated and applied will last for over 30 years.

Smoke Generators

In the past Lindane smoke generators have been used in an attempt to kill woodworm. This method was not particularly effective since the small particles of insecticide settled only on horizontal or shallow sloping surfaces leaving the remaining timbers unprotected. The insecticide did not penetrate the timber to kill the larvae.

Dichlorvos Strips

Dichlorvos impregnated resin strips have recently been advocated. In unlined roof spaces the extra ventilation reduces the amount of inesticide in the air and few beetles are killed. Conditions in modern lined roof spaces are such that most adult beetles are killed before they can complete their exit hole. Eggs laid on the surface will be killed before hatching but those laid deep in old flight holes will produce viable larvae. The strips used every year, June to August, may gradually eliminate an infestation but are not an adequate replacement of a liquid insecticide.

Fumigation

It is rarely necessary to fumigate a property to control wood boring beetles. The method is extremely expensive since the building must first be made gas tight. The gas used is methyl bromide and phosphine. These are extremely poisonous and can only be applied by specialists. This treatment will only kill the insects within the building timbers at the time of treatment, the timbers are open to reinfestation as soon as the gas has dispersed.

Treatment of Other Insect Attack

After the drying out of the excess moisture content which would have occurred in the timber. Timbers which are to remain would have to be pressure impregnated with fluid. The best way of achieving this is to insert plastic injectors containing a non-return valve which remain permanently in the wood. The fluid is then injected at high pressure.

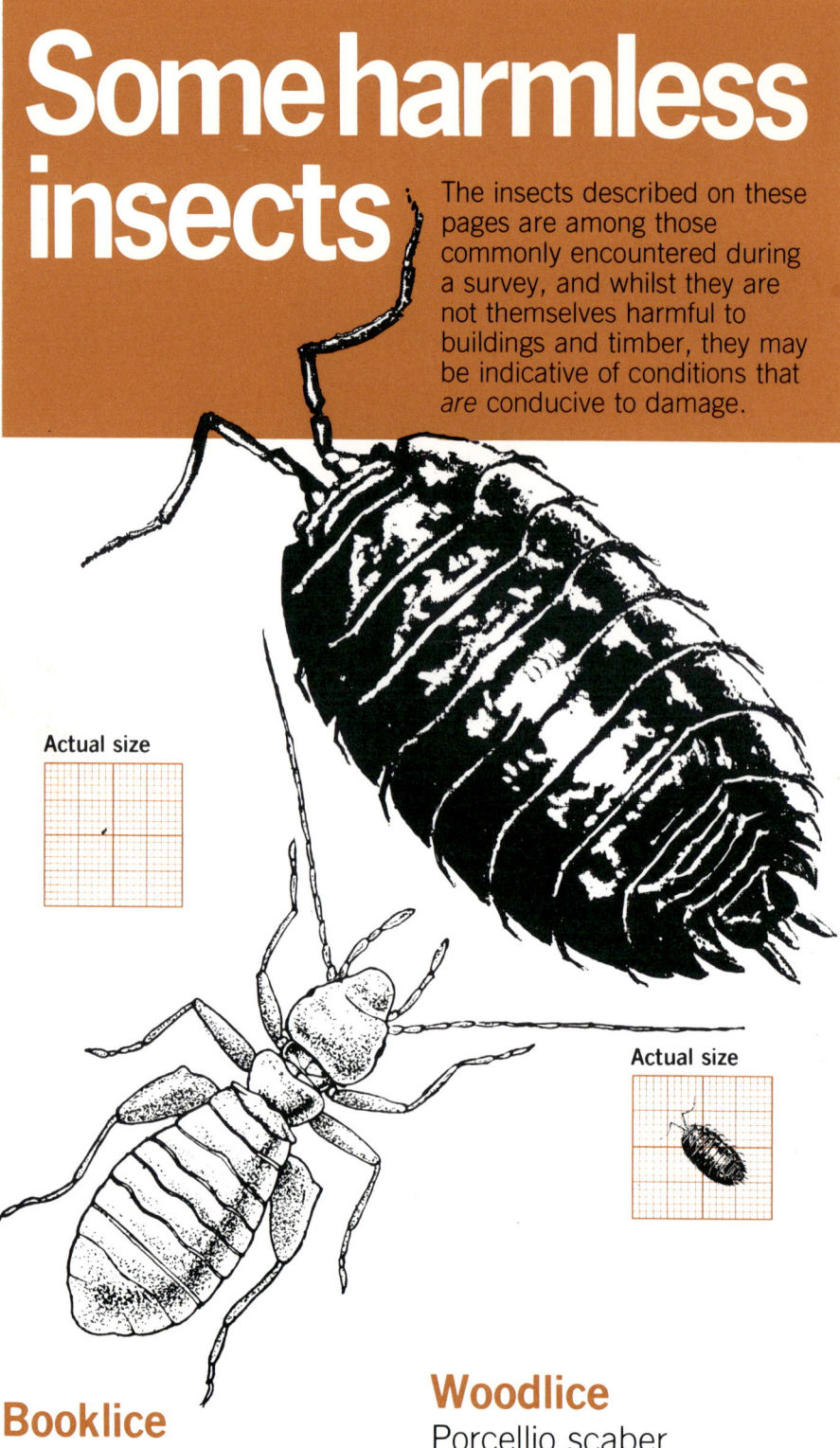

Some harmless insects

The insects described on these pages are among those commonly encountered during a survey, and whilst they are not themselves harmful to buildings and timber, they may be indicative of conditions that *are* conducive to damage.

Actual size

Actual size

Booklice
Psocids

Booklice or psocids are approximately 1mm long and feed on minute fungi in damp surfaces. The booklice eggs are laid by the famales amongst the dust in crevices in woodwork. After a short time the young psocids hatch. They do not undergo a larval or grub stage and take only a few weeks to reach the adult stage. To eradicate the booklouse, avoid the warm humid conditions they require for survival and provide plenty of ventilation. Newly built houses with a higher moisture content provide ideal breeding ground for this type of insect which has a creamy white appearance.

Woodlice
Porcellio scaber
Armadillidium vulgare
Omniscus asellus

The woodlouse is a crustation. It is characterised by its ten tile-like segments which articulate with each other. It has seven pairs of walking legs. It feeds on vegetable rubbish and frequents damp and dark places. It is often found beneath garden paving or rotting wood. Very humid conditions are essential for its existence. It causes no damage to sound wood but is often found where wood is in the last stages of fungal decay. It is a nuisance but completely harmless in the home.

TIMBER DECAY CAUSED BY INSECT ATTACK

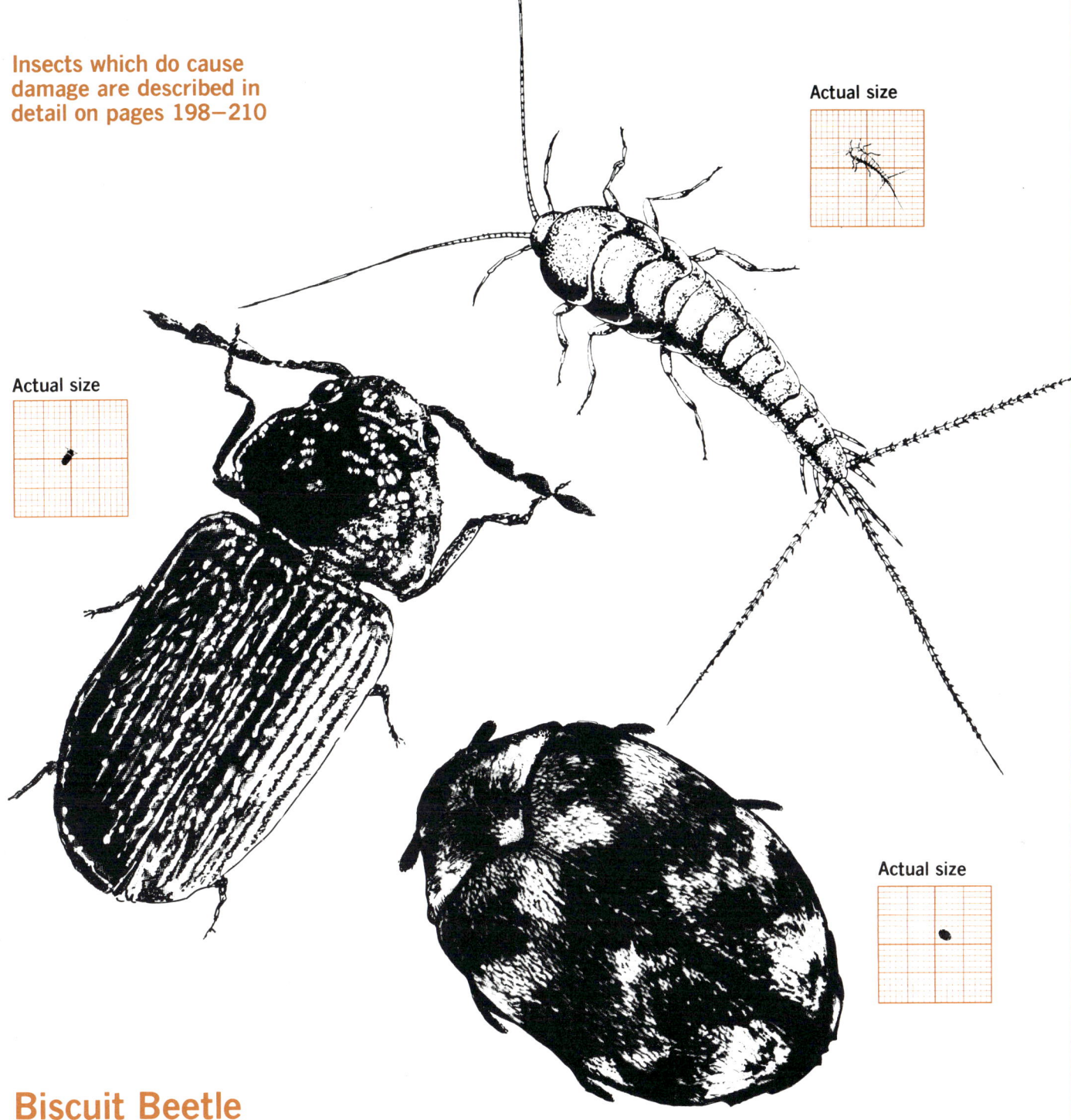

Insects which do cause damage are described in detail on pages 198–210

Actual size

Actual size

Actual size

Biscuit Beetle
Stegobium paniceum

This is a close relative of the common furniture beetle and death watch beetle. It is not a wood boring insect but is a common pest in many stored foods. It has a slightly more round appearance than the common furniture beetle and is a little more reddish in colour. The punctures in the shell are not as deep as occur in *Anobium punctatum*. It is covered with short fair hairs when it is freshly emerged. It is about 2.5mm in length which is just a little smaller than the common furniture beetle. To eradicate it one should destroy the infested food and spray the inside of the storage area with an odourless oil-soluble insecticide.
Woodworm killing fluid must not be used where food is stored.

Carpet Beetle
Anthrenus verbasci

The larvae of the carpet beetle attack woollen materials and furnishing. The grubs are usually long and slender approaching 10mm. They are greyish-brown or light brown in colour with tufts of long hairs from the tail region. It is often found under the edge of carpets. The adult beetle is about 5mm long and is black in colour. It has a number of greyish spots, two being on the wing case in the middle of the back. It is a tenacious little beetle and requires regular treatment. It is very difficult to eradicate.

Silverfish
Lepisema saccharina

The silverfish insect is about 12mm (½″) long and can be recognised by its rapid darting runs, usually in a bathroom or kitchen. There are three long bristles emerging from the tail end. It is usually found in warm damp places and upon being discovered it rapidly seeks concealment in dark corners. It does no damage to woodwork and can be treated by dusting with powder insecticide which needs to be distributed into all the crevices.

TIMBER DECAY CAUSED BY INSECT ATTACK

House Longhorn Beetle
Hylotrupes bajulus

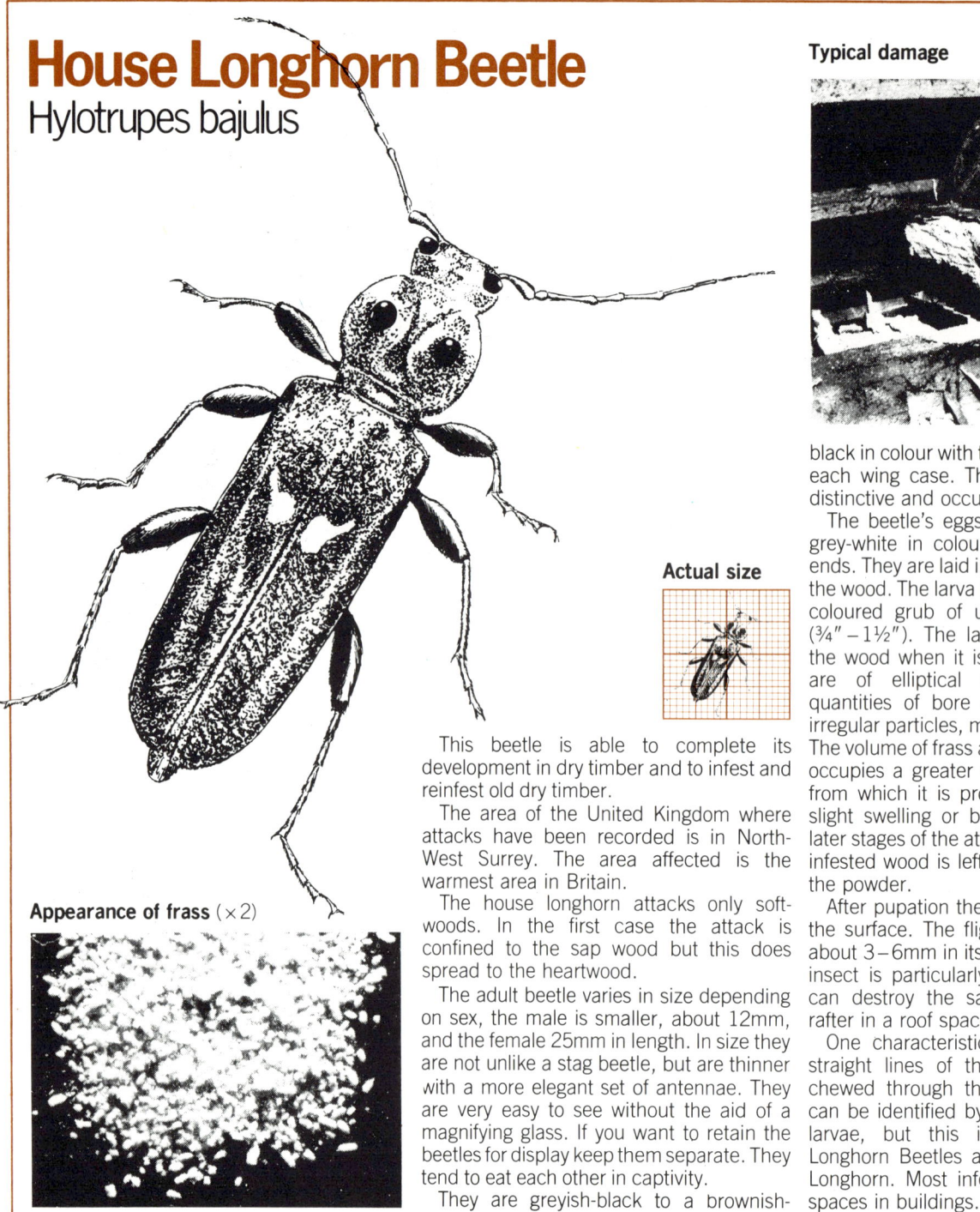

Typical damage

Actual size

Appearance of frass (×2)

This beetle is able to complete its development in dry timber and to infest and reinfest old dry timber.

The area of the United Kingdom where attacks have been recorded is in North-West Surrey. The area affected is the warmest area in Britain.

The house longhorn attacks only softwoods. In the first case the attack is confined to the sap wood but this does spread to the heartwood.

The adult beetle varies in size depending on sex, the male is smaller, about 12mm, and the female 25mm in length. In size they are not unlike a stag beetle, but are thinner with a more elegant set of antennae. They are very easy to see without the aid of a magnifying glass. If you want to retain the beetles for display keep them separate. They tend to eat each other in captivity.

They are greyish-black to a brownish-black in colour with two paler grey marks on each wing case. These eye-like spots are distinctive and occur at its mid-length.

The beetle's eggs are long, yellowish to grey-white in colour, with slightly pointed ends. They are laid in cracks and crevices in the wood. The larva is greyish white, a flesh-coloured grub of up to 20–37mm long (¾″ – 1½″). The larva bores straight into the wood when it is hatched. The tunnels are of elliptical cross section. Large quantities of bore dust are produced of irregular particles, mainly cylindrical in size. The volume of frass and the air mixed with it occupies a greater volume than the wood from which it is produced. This causes a slight swelling or blister to appear in the later stages of the attack. The exterior of the infested wood is left as a veneer enclosing the powder.

After pupation the beetle bites its way to the surface. The flight hole is oval, being about 3–6mm in its longer dimension. The insect is particularly voracious. Six larvae can destroy the sapwood in a common rafter in a roof space.

One characteristic of the beetle is the straight lines of the galleries or tunnels chewed through the wood. The galleries can be identified by the bite marks of the larvae, but this is common in many Longhorn Beetles and not just the House Longhorn. Most infestation occurs in roof spaces in buildings.

Adult insect	10–25mm long beetles. Antennae grey-black-brown in colour. Female beetles usually larger than the male. Flight season – July to September.
Eggs	12mm long spindle-shaped, are laid in splits and cracks on softwood sapwood.
Larvae	Greyish-white well segmented larva, with small retracted head. Body tapers toward the 'tail'. May be 30mm long when fully grown.
Nature of boring	By larvae only in sapwood of softwoods. Tunnels are large and packed with dust containing cylindrical pellets. Exit hole 6–9mm diameter.
Wood species attacked	Sapwood of softwoods attacked, chiefly structural roof timbers.
Remarks	Confined to certain areas of Surrey and Berkshire. Life cycle may vary from 3–11 years.
Frass as seen by naked eye	Fine dust mixed with barrel shaped pellets.
Feel of frass	The pellets stand out and feel coarse.
Appearance of frass viewed through microscope	Large cylindrical pellets with well squared ends (barrel-shaped), about 1mm long.

TIMBER DECAY CAUSED BY INSECT ATTACK

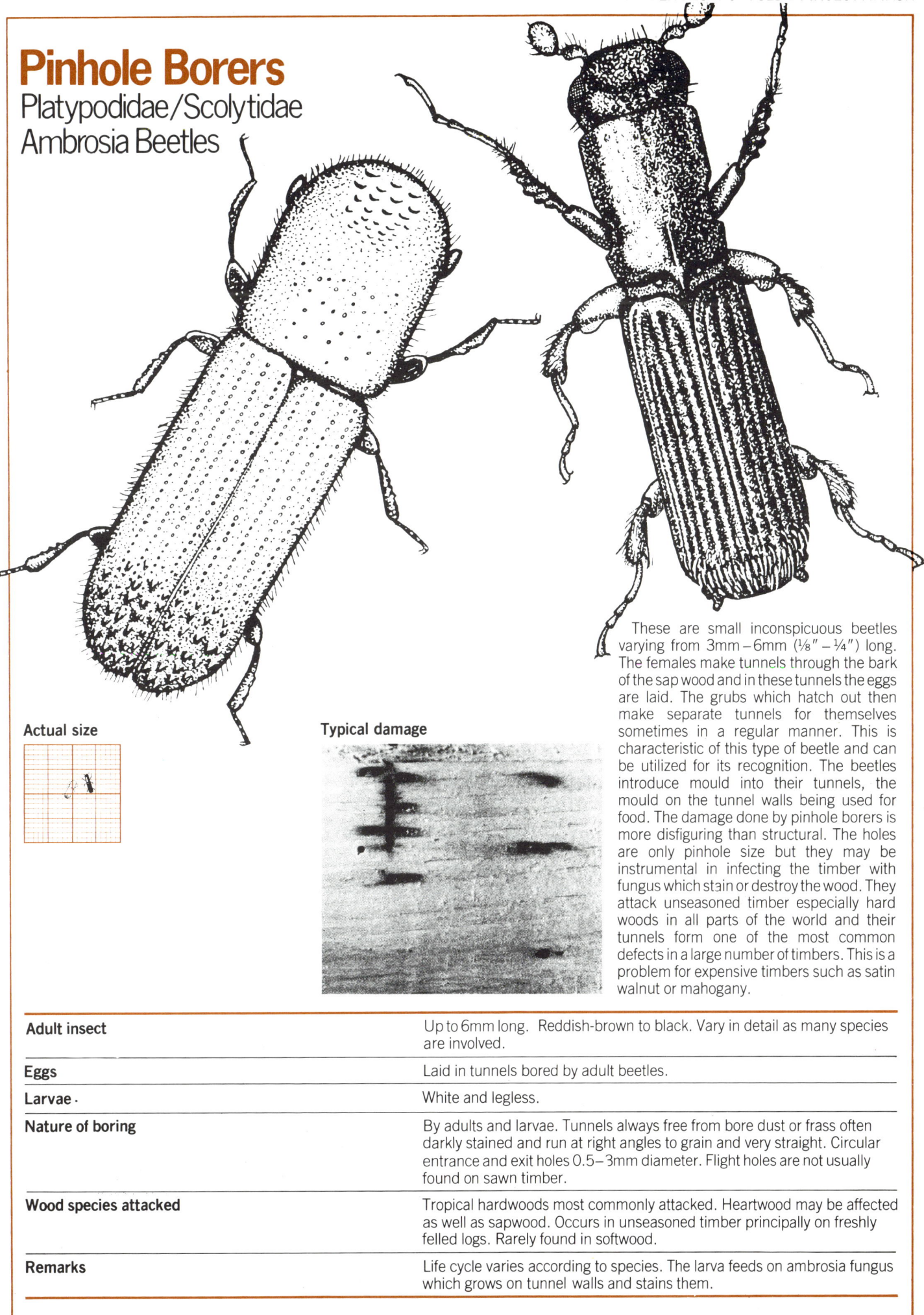

Pinhole Borers
Platypodidae/Scolytidae
Ambrosia Beetles

Actual size

Typical damage

These are small inconspicuous beetles varying from 3mm–6mm (⅛"–¼") long. The females make tunnels through the bark of the sap wood and in these tunnels the eggs are laid. The grubs which hatch out then make separate tunnels for themselves sometimes in a regular manner. This is characteristic of this type of beetle and can be utilized for its recognition. The beetles introduce mould into their tunnels, the mould on the tunnel walls being used for food. The damage done by pinhole borers is more disfiguring than structural. The holes are only pinhole size but they may be instrumental in infecting the timber with fungus which stain or destroy the wood. They attack unseasoned timber especially hard woods in all parts of the world and their tunnels form one of the most common defects in a large number of timbers. This is a problem for expensive timbers such as satin walnut or mahogany.

Adult insect	Up to 6mm long. Reddish-brown to black. Vary in detail as many species are involved.
Eggs	Laid in tunnels bored by adult beetles.
Larvae	White and legless.
Nature of boring	By adults and larvae. Tunnels always free from bore dust or frass often darkly stained and run at right angles to grain and very straight. Circular entrance and exit holes 0.5–3mm diameter. Flight holes are not usually found on sawn timber.
Wood species attacked	Tropical hardwoods most commonly attacked. Heartwood may be affected as well as sapwood. Occurs in unseasoned timber principally on freshly felled logs. Rarely found in softwood.
Remarks	Life cycle varies according to species. The larva feeds on ambrosia fungus which grows on tunnel walls and stains them.

TIMBER DECAY CAUSED BY INSECT ATTACK

Powder Post Beetle
Lyctus brunneus

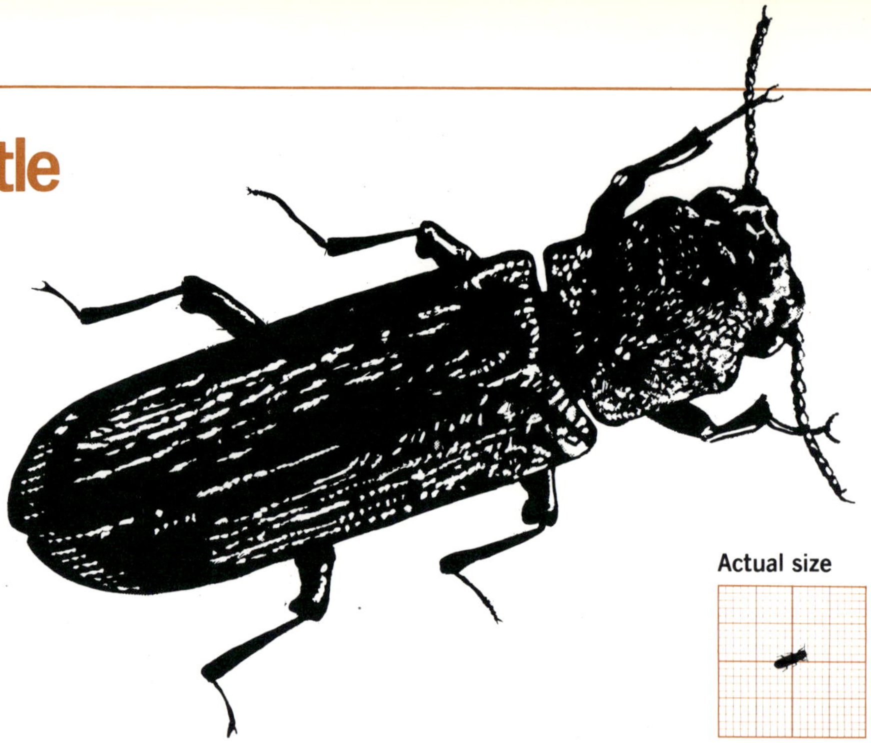

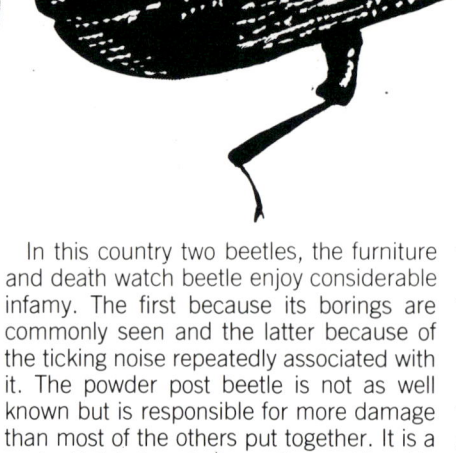

Typical damage

Appearance of frass (×4)

Actual size

In this country two beetles, the furniture and death watch beetle enjoy considerable infamy. The first because its borings are commonly seen and the latter because of the ticking noise repeatedly associated with it. The powder post beetle is not as well known but is responsible for more damage than most of the others put together. It is a pest which is found all over the world and is as destructive in Australia as it is here. It is a small inoffensive looking beetle with a narrow body about 5mm (3/16″) long.

There are six species of the *Lyctus* found in the United Kingdom. Only one of these species (*Linearis*) is truly indigenous, the others having been imported. Four species have been introduced since the 1950's and are much less common.

The female lays her eggs in the late summer and selects a timber which has pores large enough for her egg laying tube to enter and which in addition has plenty of starch. In the United Kingdom she would choose the sap wood of a hardwood such as ash, oak, or walnut. Softwoods are not attacked. After she has tested the wood for the presence of starch she lays the egg in the pores. She will lay up to 50 eggs.

When the egg hatches the grub feeds on the egg yoke until it is large enough to fill the hole into which it had been placed. It then moves by gripping the sides of the opening until it can pierce and enter the wood. They then move inwards eating the wood as they go. They digest the starch, excreting the rest unchanged as a fine dust. The result is that in a very short time the interior of the wood is reduced to a fine powder enclosed in the thin shell of apparently sound timber.

When full grown the grub comes to rest in a small chamber which it makes for itself under the surface of the wood and there it remains until it changes its form and emerges as an adult the following summer. The beetle only attacks the starchy seasoned sap woods with large pores and while the heat of kiln drying and seasoning kills any beetle in the wood it does not remove the starch and therefore does not render it immune from future attack.

The insect was given the name Powder Post because damaged timbers were left with a thin veneer of sound timber which would disintegrate at a touch.

Adult insect	About 5mm long. Reddish-brown to black body. Slender body. No hooded thorax. Antennae short ending in a two-jointed club. Emergence chiefly in June to August.
Eggs	Laid in vessels of sapwood of larger pored hardwoods.
Larvae	Small three-jointed legs. One brown oval spot on each side of the 8th segment. When fully grown, up to 6mm long.
Nature of boring	By larvae only. Tunnels filled with fine flour-like dust unstained. Sapwood may be completely powdered. Exit holes 1.5mm diameter. No blackening of tunnels. Tunnels usually follow grain.
Wood species attacked	Sapwood of larger pored hardwoods. Mainly a pest of timber during drying or when recently seasoned.
Remarks	Life cycle can be as short as one year. Usually 1–2 years. Not an insect commonly causing damage inside a building.
Frass as seen by naked eye	Very fine dust.
Feel of frass	Like rubbing talcum powder.
Appearance of frass viewed through microscope	Very fine powder. No characteristic shapes.

Common Furniture Beetle
Anobium punctatum

Actual size

Appearance of frass

Typical damage

This is a little brown beetle slightly less that 5mm (¼") in length. It comes out of the infested wood in the summer time around about June or July. Like most beetles it is able to fly. After the insects mate the female lays her eggs in small cracks and crevices in the surface of the wood. She will not lay them on smooth varnish or waxed surfaces and so will generally attack furniture from the underside where the wood is rough and unpolished.

When the grubs hatch out they move into the wood. When they get well into the timber they tend to move along the grain. As they chew the wood it passes through the body so that the tunnel they leave behind is filled with frass, the excretion of the beetle. This is mixed with small splinters of wood which were dropped from their jaws during their chewing the tunnel in the wood. This frass is composed of small cylindrical granuals which are quite characteristic and can easily be seen with the naked eye or with a hand lense. The frass of each insect varies and is a useful item in the identification of individual beetles in infected timber.

The grub remains in the larva stage for not less than two years; the average length of the larval stage being from 3–4 years.

The larva is greyish-white and is covered with fine hairs. The head is browner and the jaw section darker. Most movement is along the grain but link chambers enable the insect to move into a different growth ring.

In the early spring in the year in which the larva matures, it bores towards the outside of the wood, and builds a chamber below the surface. It then changes into a pupa or chrysalis. After 6–8 weeks it emerges as an adult beetle. The beetle then rests whilst the exterior hardens. The beetle then bites its way out into the open, leaving a flight hole as it emerges for the first time from the timber.

The identification of current attack may be made by the examination of these flight holes to see if there is a light powdering of dust immediately inside and just around the opening. Infestation of woodworm have often been treated. A close examination of the flight holes reveals any recent activity of the insect in the wood.

The adult beetle stage lasts for two or three weeks. During that time mating takes place. (It's a bit now or never.) The female will lay her eggs to start the cycle all over again. The female does not often leave the wood, but allows the more interesting part of her anatomy to protrude from the flight hole, awaiting the ministrations of any male beetle in the area.

Adult insect	3–5mm long. Thorax hood over head. Squat body of reddish brown colour. Rows of small pits on wing cases. Last three joints of the antennae larger than the others. Beetles fly May to August.
Eggs	Laid in cracks, joints, crevices, unprotected end grain and old exit holes.
Larvae	Three pairs of five-jointed legs. Body up to 6mm long
Nature of boring	By larvae only. Borings packed with frass which contains ellipsoidal pellets having a 'gritty' feel. Exit holes about 1.5mm diameter. No blackening or staining of tunnels which usually run with grain. Wood honeycombed.
Wood species attacked	Softwood and hardwoods. Mostly in sapwood. Attacks structural timber, joists, rafters, plywood etc, as well as furniture. Often found in old seasoned timbers.
Remarks	Life cycle usually at least three years.
Frass as seen by naked eye	Granular appearance. Individual pellets just visible. Ellipsoidal shape.
Feel of frass	Slightly gritty; like very fine sand.
Appearance of frass viewed through microscope	Cylindrical pellets pointed at one or both ends. Frass composed entirely of pellets.

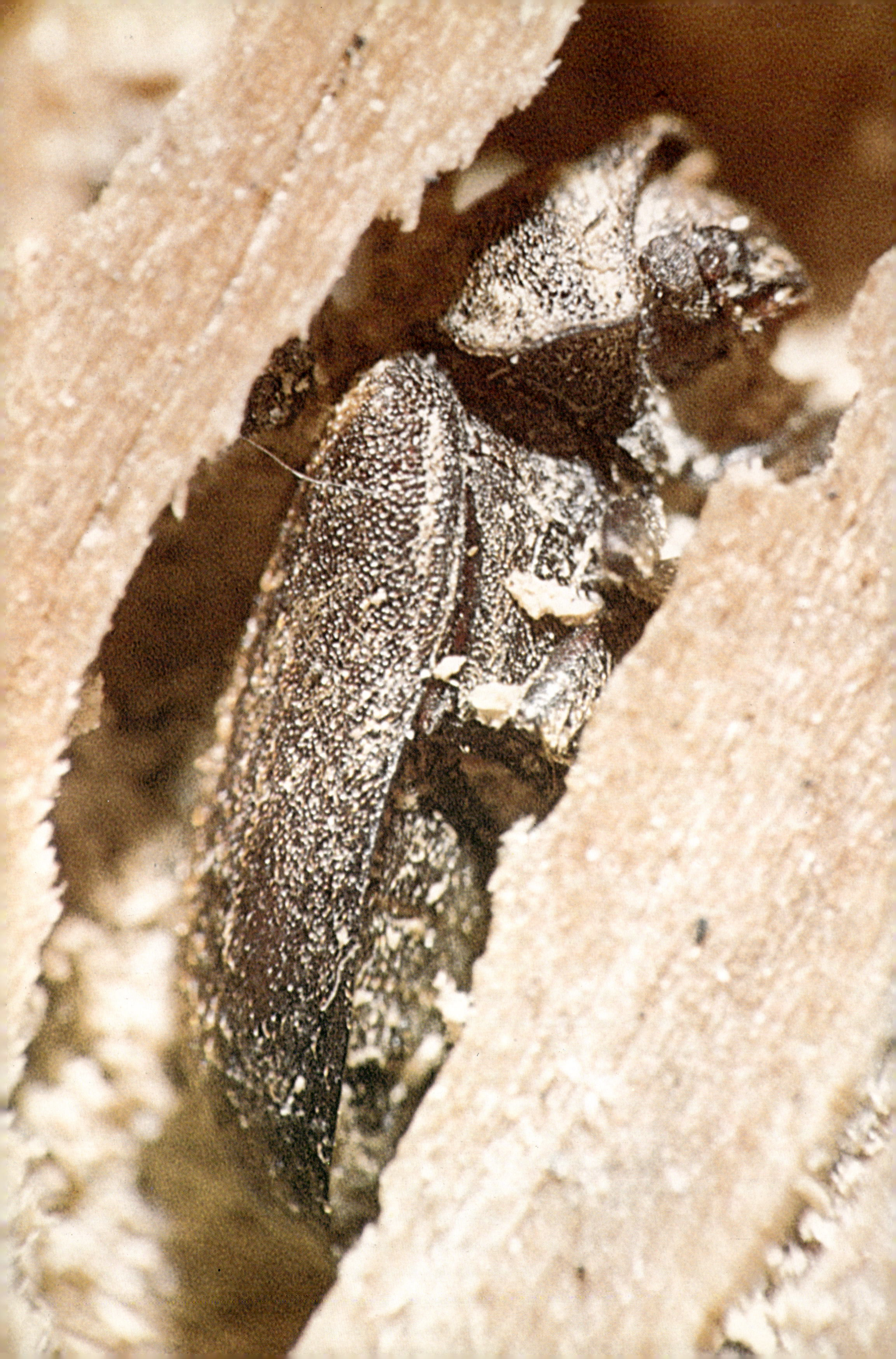

Death Watch Beetle
Xestobium rufovillosum

Actual size

Typical damage

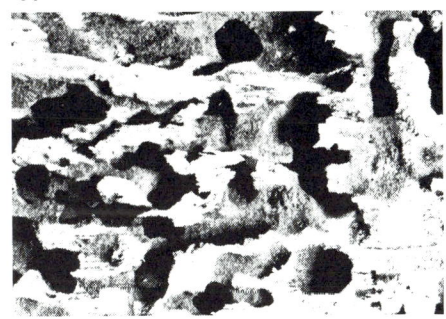

Appearance of frass (×4)

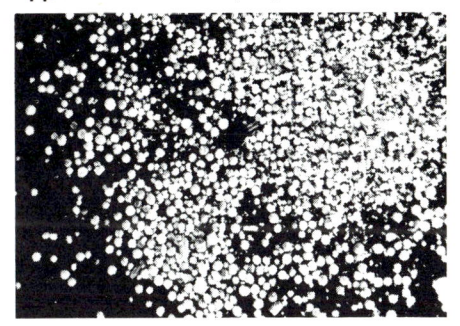

This is an indigenous British insect. It was named *Xestobium rufovillosum* in 1774.

Outside it attacks the dead wood in trees or branches of several hardwoods where fungal decay is present.

Inside buildings it will only attack mature hardwoods where there is some presence of fungal decay.

Although it has a reputation for the damage it causes it is not a major agent of timber destruction in the United Kingdom.

The adult beetle is between 5 and 7mm long (¼"). The female is slightly larger. It resembles in size a small ladybird. It is larger than the common furniture beetle. It is chocolate-brown in colour with mottled markings on its back. These can be seen with a magnifying glass.

The hood over the insect's head is more widely flanged but follows the side of the head like a maid's cap. The furniture beetle has a wider flange covering to its head which is wider than the death watch beetle.

The tapping of the adult is believed to be involved with mating. Both sexes tap by striking their heads against the wood about 12 times. This staccato tapping takes about 1¼ seconds and the beetle can be stimulated by imitating the sound with a pencil point tapping wood.

The eggs are laid in small clusters on the surface of rough woods. They are sticky when first laid and adhere to each other. The eggs are twice as large as the common furniture beetle but similar in appearance. About 50 eggs are laid and this stage lasts 2–5 weeks.

When the larva hatches it moves rapidly about over the surface of the wood before finding an old flight hole or crack by which it can enter the wood. The length of the larval stage varies. It is longer outside than inside buildings and can vary from two to ten years.

The frass of the death watch is very easy to identify with a magnifying glass. Lying within the pale brown dust are larger oval pellets which are frequently described as bun-shaped. They are about the size and feel of grains of sand.

During the summer when the larva has reached maturity it burrows towards the surface and makes a chamber, sealing the open end with frass. The change to adult beetle takes about four weeks, but the beetle does not emerge until the following spring. The flight holes are circular and about 2.5mm (¹⁄₁₀") in diameter.

The beetle is a sluggish individual who rarely flies. It prefers to crawl on the surface of the wood. If it falls it drops to the floor.

The identification and location of death watch beetle within buildings can be aided by beetle hunts. These are a useful method of determining the location of the attack. Volunteers record the location of the beetles that have fallen from the roof timbers and accurately pinpoint them by coloured stickers indicating when and where found. This information is recorded on a plan of the building. From this it is possible to locate the infestation within the wood immediately above and also estimate the extent of the infestation.

It is now known that these beetles do not infest sound timber. They attack timbers infected by fungus. Traces of one fungus or another are always found in timber infected with death watch beetle.

Although they are capable of flying they do so rarely and this limits the spread of infestation. The flight holes are larger than the majority of insect flight holes. Fungal decay is a necessary prerequisite for attack by death watch beetle. If oak timber is attacked it does not necessarily mean that the timber is decayed to a point of substantial loss of strength. Careful examination of infested timber will reveal if timber replacement is necessary.

Adult insect	6–8mm long. Chocolate brown, squat bodied, mottled, with patches of short yellowish hairs on thorax and wing cases which have no pits. Thorax hooded over head, broad and flanged. Last three joints of antennae enlarged. Beetles fly March to June.
Eggs	Laid in cracks and crevices in decayed wood, also in old exit holes.
Larvae	Up to 8mm long. Hairy appearance.
Nature of boring	Borings filled with frass containing coarse bun-shaped pellets. Tunnels unstained, exit holes circular about 2.5mm diameter, and line of tunnels usually along grain leaving a honeycomb of damaged timber.
Wood species attacked	Almost entirely confined to sap and heart of hardwoods especially oak. The wood must be partially decayed by a fungus. Confined to old timbers usually located in old buildings.
Remarks	Life cycle usually 4–5 years but may extend to 10.
Frass as seen by naked eye	Bun-shaped pellets. Easily seen as individual grains, sand-coloured and grain size in dust or powder.
Feel of frass	Gritty; like coarse sand.
Appearance of frass viewed through microscope	Bun-shaped pellets less than 1mm in diameter.

TIMBER DECAY CAUSED BY INSECT ATTACK

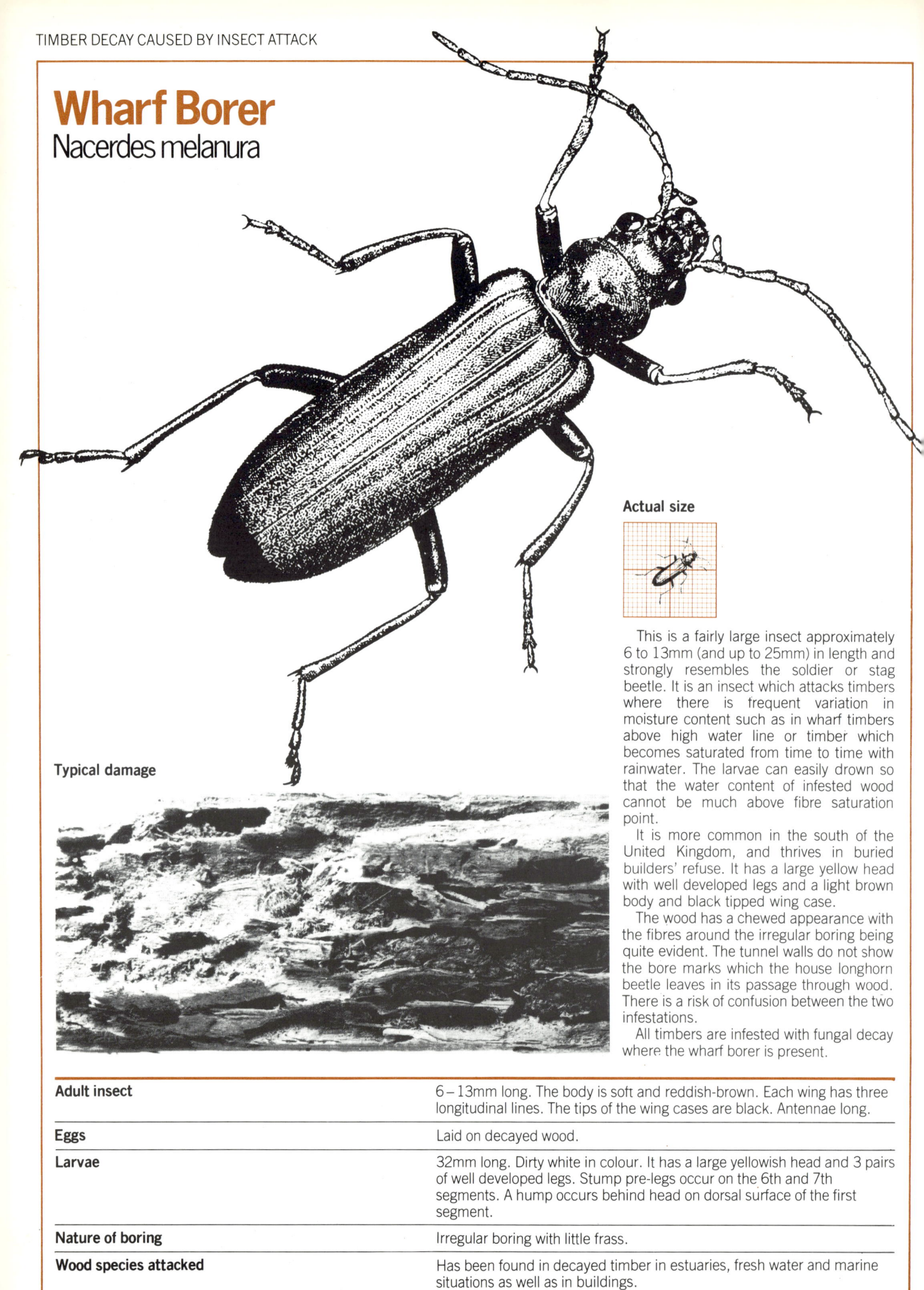

Wharf Borer
Nacerdes melanura

Actual size

Typical damage

This is a fairly large insect approximately 6 to 13mm (and up to 25mm) in length and strongly resembles the soldier or stag beetle. It is an insect which attacks timbers where there is frequent variation in moisture content such as in wharf timbers above high water line or timber which becomes saturated from time to time with rainwater. The larvae can easily drown so that the water content of infested wood cannot be much above fibre saturation point.

It is more common in the south of the United Kingdom, and thrives in buried builders' refuse. It has a large yellow head with well developed legs and a light brown body and black tipped wing case.

The wood has a chewed appearance with the fibres around the irregular boring being quite evident. The tunnel walls do not show the bore marks which the house longhorn beetle leaves in its passage through wood. There is a risk of confusion between the two infestations.

All timbers are infested with fungal decay where the wharf borer is present.

Adult insect	6–13mm long. The body is soft and reddish-brown. Each wing has three longitudinal lines. The tips of the wing cases are black. Antennae long.
Eggs	Laid on decayed wood.
Larvae	32mm long. Dirty white in colour. It has a large yellowish head and 3 pairs of well developed legs. Stump pre-legs occur on the 6th and 7th segments. A hump occurs behind head on dorsal surface of the first segment.
Nature of boring	Irregular boring with little frass.
Wood species attacked	Has been found in decayed timber in estuaries, fresh water and marine situations as well as in buildings.

TIMBER DECAY CAUSED BY INSECT ATTACK

Wood Wasp
Sirex noctilio

Actual size

Typical damage

Appearance of frass (×2)

There are a number of species which are found in this country. It is a very large group. They are fairly large insects having four wings, the smaller set located to the rear are hooked onto the larger front wings.

Certain wasps and bees will bore into the soft mortar of masonry walls and can damage the stability of old buildings. This tends to occur on south facing walls in areas where there is not an abundance of trees.

The most common types are Sirex species (*S. juvencus*, *S. noctilio* and *S. cyaneus*. The female is steely blue in colour with a prominant egg laying tube extending from the end. The male has a brownish underside with black markings at each end.

The female lays her eggs by boring a hole with the protruding tube into the bark of softwood trees. She lays a few eggs in each hole but also secretes a fluid which causes a white-pocket rot in many coniferous trees. the wood wasp cannot survive without this rot.

The larvae are whitish cylindrical grubs with three blunt projections which serve as legs.

The life cycle varies between two and three years, and will be longer if the timber has been converted and used in buildings.

They do not re-infest dry wood inside buildings.

Adult insect	Up to 38mm long, male is yellow and black, female blue-black in colour. Two pairs of membraneous wings. Flight season – summer months.
Eggs	Eggs are laid in sapwood through bark which must be present. Logs and sickly growing trees affected.
Larvae	Up to 25mm when fully grown. There is a sharp spine on end of body.
Nature of boring	By larvae only, producing tunnels and circular exit holes 6–8mm diameter. Tunnels are not stained and are usually curved.
Wood species attacked	Sapwood or softwoods principally attacked such as pine, spruce, fir, larch. Heartwood may be affected. Attack commences by egg laying through the bark of sickly growing trees, or freshly felled logs.
Remarks	Life cycle usually takes 1–2 years and adults may emerge from wood during seasoning or even in use.
Frass as seen by naked eye	Frayed sawdust particles. Densely packed in the tunnels.
Feel of frass	Like very fine sawdust and feels coarse.
Appearance of frass viewed through microscope	Shredded particles of wood.

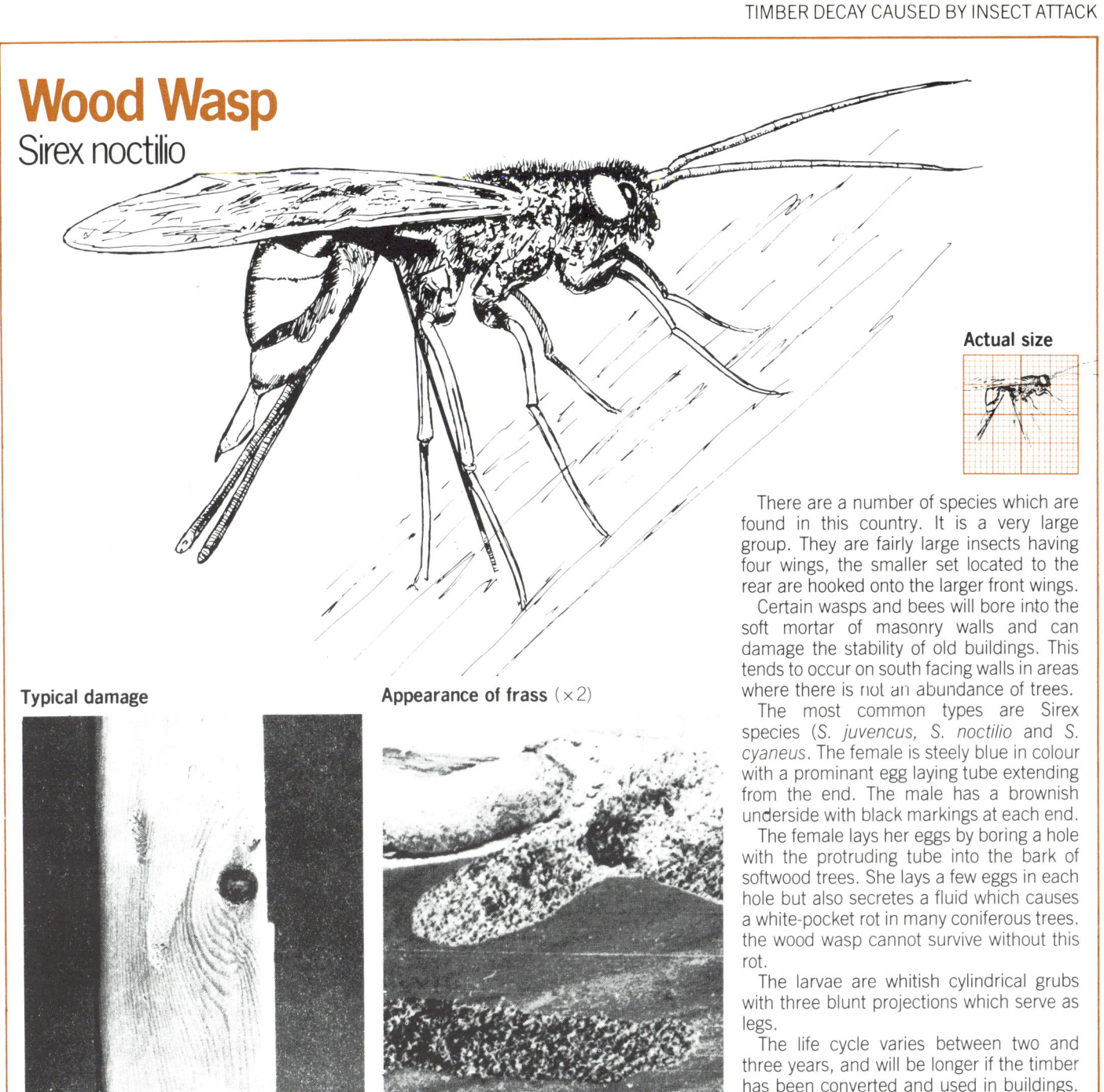

TIMBER DECAY CAUSED BY INSECT ATTACK

Wood boring Weevils
Euophryum confine

Actual size

Appearance of frass (×12)

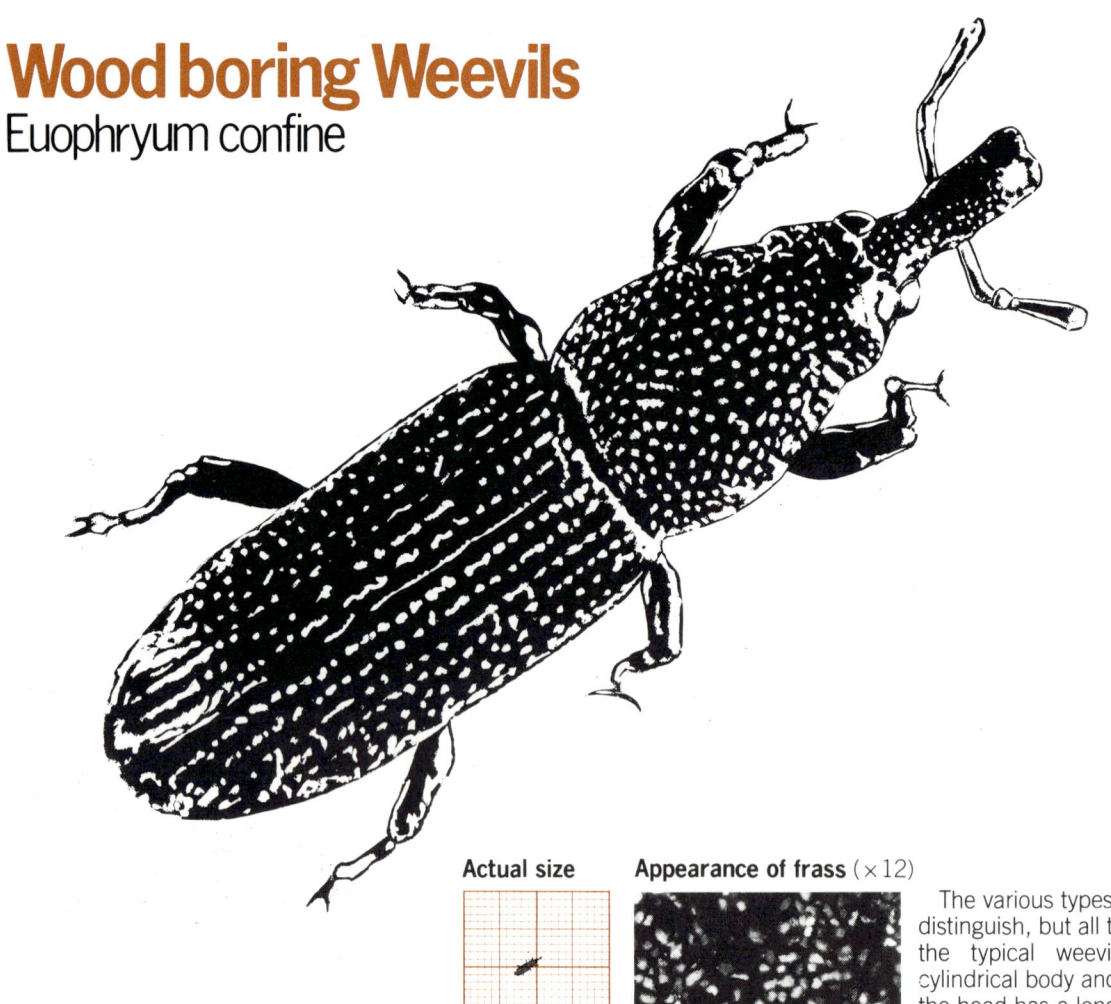

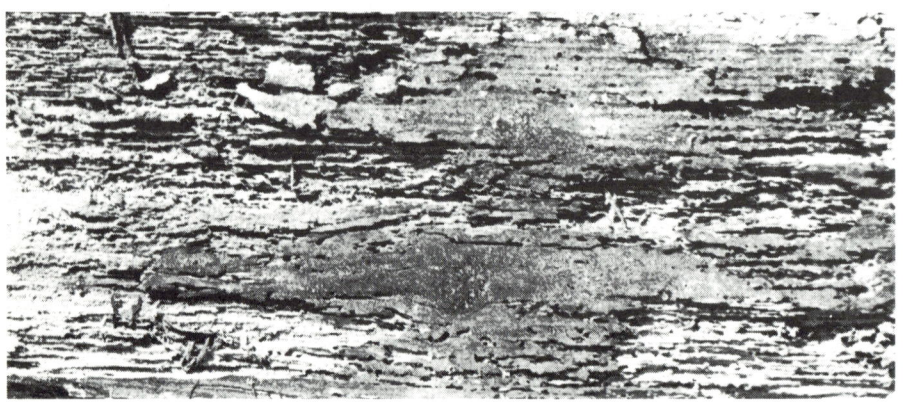

The various types of weevil are difficult to distinguish, but all the various species have the typical weevil appearance with a cylindrical body and short legs. The front of the head has a long snout. They tend to be between 2.5 and 4.5mm long and are more common in the London area. They do occur throughout Great Britain. Damp timber is essential as well as badly ventilated conditions and the weevil attacks both hard and soft wood.

Weevil infestation is not known to spread in timber which has not suffered from some stage of fungal decay so that the control of the infestation is normally to be achieved by the normal treatment of the fungal decay or eradication of the dampness in the timber.

The flight holes are about 1mm in diameter and are ragged in outline. The colours of the beetle varies between the various species. It can be black, or reddish-brown in colour.

Adult insect	2.5–4.5mm long. Red-brown to black bodies. Head is prolonged into a short snout. Elbowed antennae arise halfway along the snout. Wing cases cover the abdomen completely.
Eggs	Laid on wood usually decayed.
Larvae	White curved wrinkled and legless.
Nature of boring	By adults and larvae. Borings filled with frass containing pellets which are smaller than those of *Anobium*. Exit holes ragged in outline, up to 1.5mm diameter.
Wood species attacked	Softwoods and plywood may be affected.
Frass as seen by naked eye	A fine dust with a very fine granular appearance.
Feel of frass	Slightly more gritty than *Lyctus*, but still fine.
Appearance of frass viewed through microscope	Very small pellets. Just possible to make out shape as being round-cyclindrical. Frass composed entirely of pellets.

TIMBER DECAY CAUSED BY INSECT ATTACK

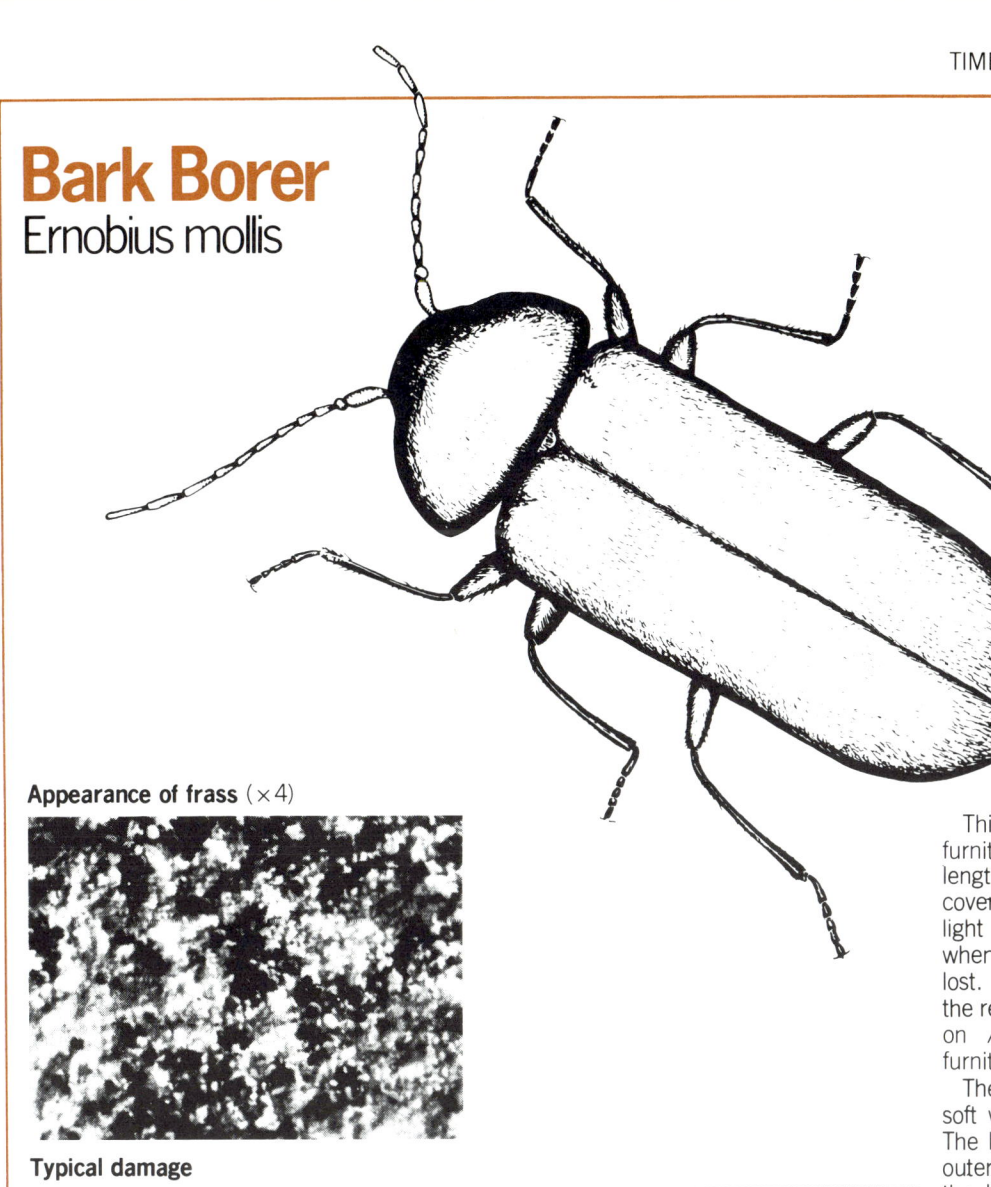

Bark Borer
Ernobius mollis

Actual size

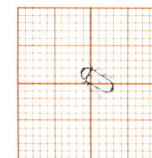

Appearance of frass (×4)

Typical damage

This species is rather like the common furniture beetle being about 6mm in length. When it is freshly emerged it is covered in short golden hairs which give it a light golden colour. This colour darkens when some of the light coloured hairs are lost. It does not have lines of small holes in the rear part of the body which can be seen on *Anobium punctatum*, the common furniture beetle.

The eggs are laid in summer in the bark of soft wood. It does not attack hard wood. The larvae then tunnels into the bark and outer ring of sap wood. This often causes the bark to fall off, exposing the tunnels which are packed with frass. The adult beetle bores out of the bark leaving a round flight hole of approximately 2.5mm in diameter. Softwood is only attacked when the bark is still attached. Little structural damage is caused by the insects and control is easily effected by stripping off the bark where it is infested. The infestation will die out following this action.

Outbuildings, such as barns and storage sheds seem more vulnerable to attack.

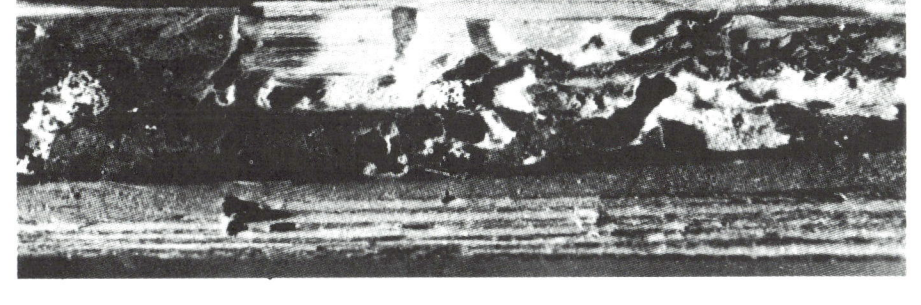

Adult insect	3–6mm long. Red chestnut-brown in colour. Beetles covered with silky hairs. Flight season – June to August.
Eggs	Laid in the bark of softwoods.
Larvae	White curved larva covered with fine golden hairs. Up to 6mm long when fully grown.
Nature of boring	By larvae only. Borings confined to bark and outermost zone of sapwood. Frass brown and white containing spherical pellets smaller than those of death watch beetle. Exit holes about 2.5mm diameter.
Wood species attacked	Confined to wood with bark. Attack will die out if bark is removed.
Remarks	Life cycle lasts about one year.
Frass as seen by naked eye	Bun-shaped pellets. Similar to that of death watch beetle, but smaller. Some are dark in colour being composed wholly or mainly of bark.
Feel of frass	Gritty, like sand.
Appearance of frass viewed through microscope	Bun-shaped pellets (0.5mm) both light and dark. Smaller than death watch beetle.

TIMBER DECAY CAUSED BY INSECT ATTACK

Bostrich Beetle
Bostrichidae

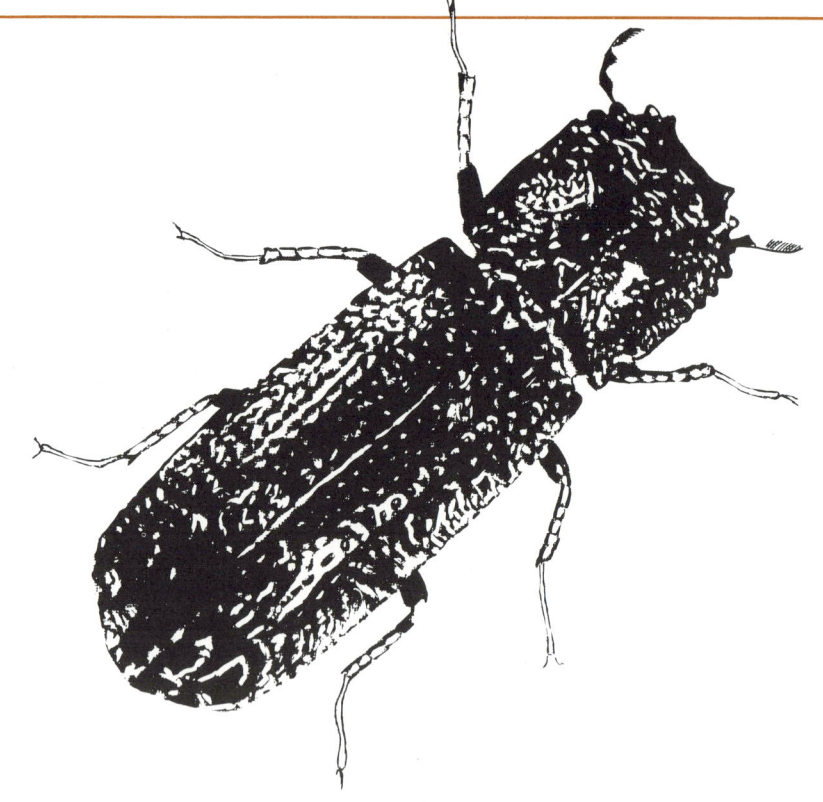

The *Bostrichidae* are joined with the *Lyctidae* to form the powder post beetle family. In the United Kingdom it does little damage but is a major problem in the tropics. The larvae have a curved body and remain 'C' shaped when removed from the wood. The damage caused to the timber is similar to that caused by *Lyctus*. The galleries are larger and more circular in cross section. They are generally only found in the sapwood of hardwoods rich in starch. Most species are larger than the *Lyctus*, being up to 25mm long. They are mainly a pest of the freshly felled tree, and will die out as the timber dries, and will not reinfest dry timber. A common insect infesting bamboo.

Actual size

Appearance of frass (×12)

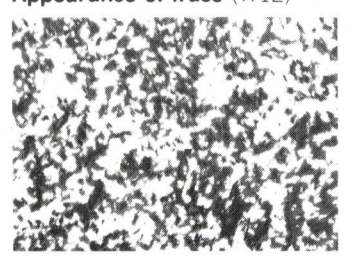

Typical damage

Adult insect	Between 6–25mm long. Thorax hooded over head and sculptured. Short antennae with three-jointed club, often serrated. Cylindrical body.
Eggs	Laid in wood by adult beetle boring special egg tunnel.
Larvae	Larger than *Lyctus*. No brown spot. Flight hole 1.4mm diameter.
Nature of boring	By beetles and larvae. Larval tunnels filled with flour-like dust. Egg tunnel free of frass. All tunnels unstained.
Wood species attacked	Sapwood of hardwoods or recently dried timber attacked.
Remarks	Mainly in tropical timber. Hardwoods mainly attacked.
Feel of frass	Flour-like powder.
Appearance of frass viewed through microscope	Very fine powder, no characteristic shape.

TIMBER DECAY CAUSED BY INSECT ATTACK

Gribble
Limnoria lignorum

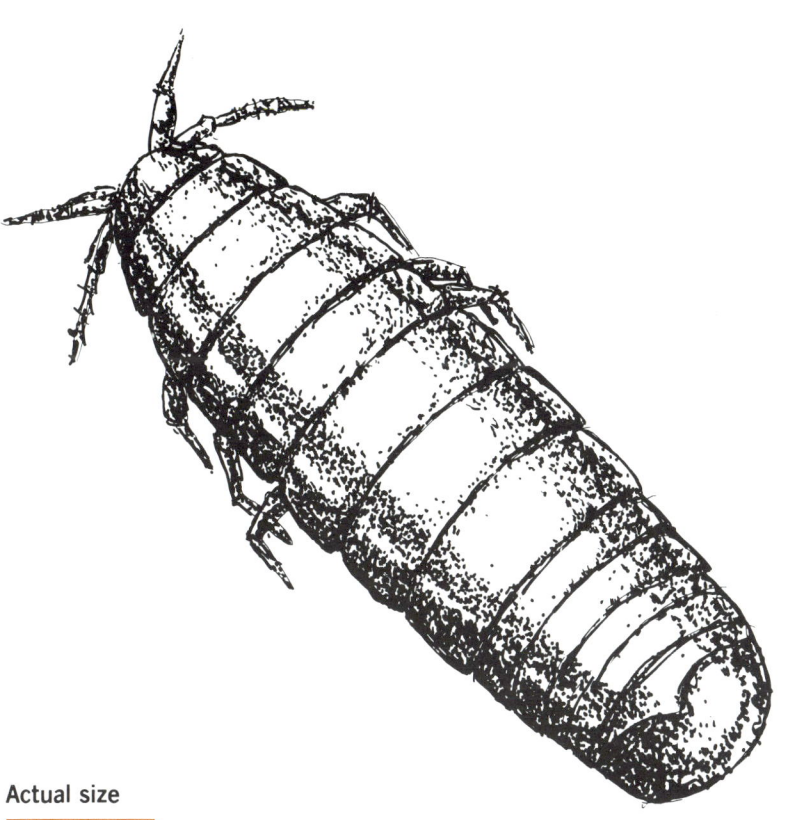

Actual size

Typical damage

In salt water the wooden bottoms of ships, wooden piles and any other woodwork in contact with the water may be attacked by animals that are not insects but are crustaceans or molluscs. The Gribble is a small crustacean which gradually destroys submerged woodwork usually in harbours. The Gribble is a tiny hairy shrimp-like animal about 5mm ($1/5$″) long. It doesn't burrow deeply into the timber but the constant surface excavation by thousands of the animals quickly errodes the timber even though the individual tunnels are not more than 50mm (2″) long. Since the burrowing is confined to the surface, floating logs attacked by Gribbles are not deeply affected. The interior of the log can sometimes be salvaged by cutting off the damaged outside potion. Damage is sometimes found on imported timber.

Each gallery houses a male and a female who does all the work of burrowing. Mating takes place in the tunnel. Usually no more than 12 lavae are produced at a time and they then start burrowing side tunnels from the parent. When fully grown they leave the tunnel and swim through the water to find fresh timber to attack.

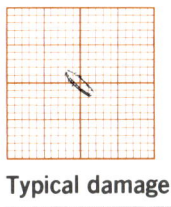

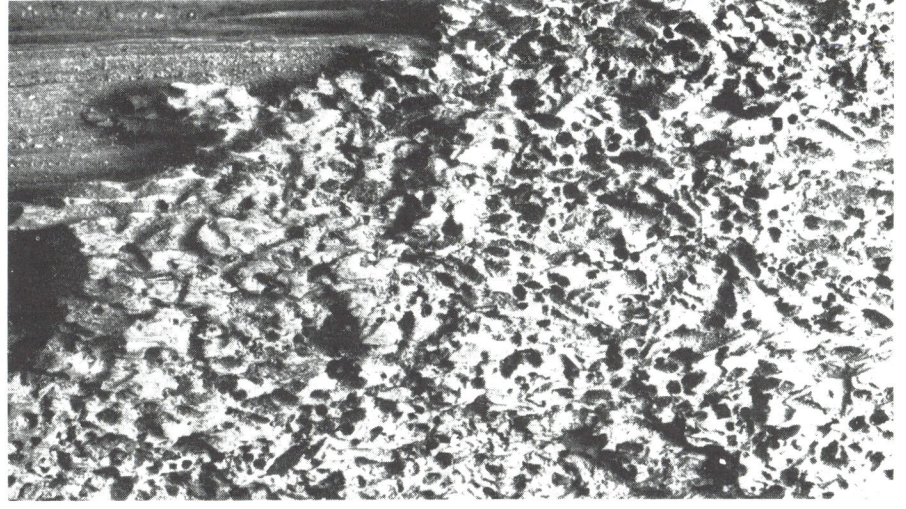

Adult insect	3–5mm long. Looks like common woodlouse. Grey-white in colour.
Eggs	No eggs produced, but are developed in the brood pouch of the female. They are hatched as immature adults.
Larvae	No larvae.
Nature of boring	50 mm long tunnels close to the surface of the wood, about 3mm in diameter.
Wood species attacked	Most timber except certain very resistant species.
Remarks	Requires a plentiful supply of oxygenated salty water. Damage to wood always on the surface.

209

TIMBER DECAY CAUSED BY INSECT ATTACK

Shipworm
Teredo navale

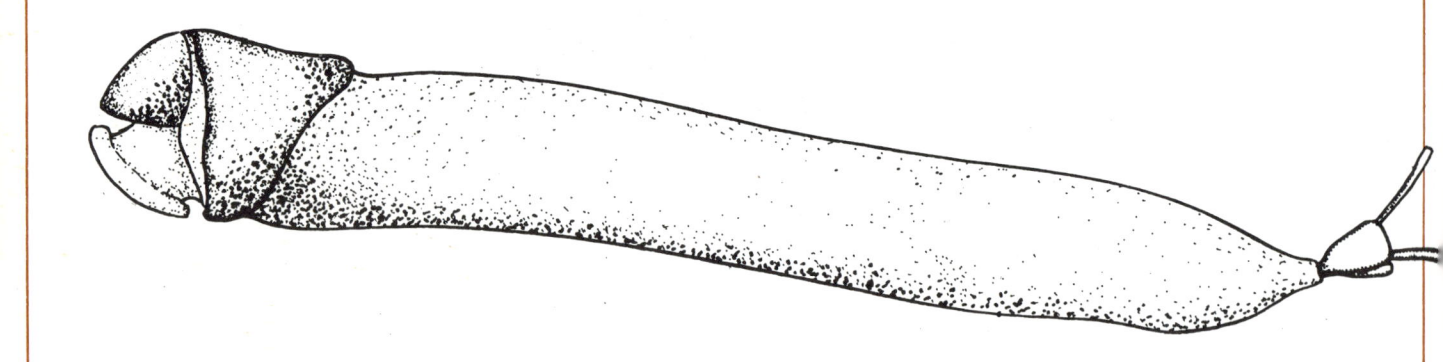

Typical damage

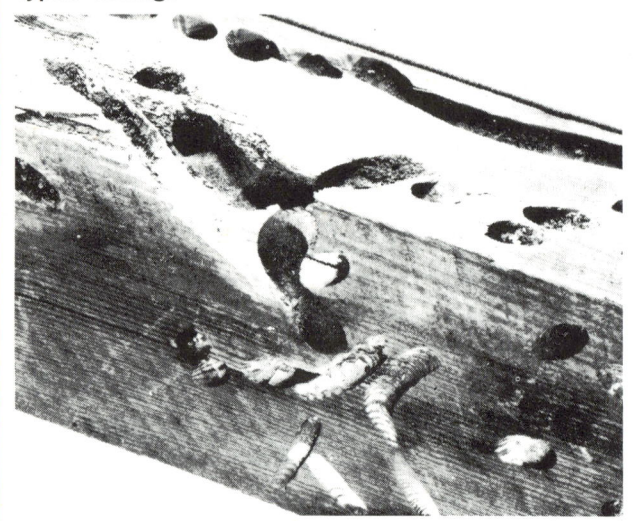

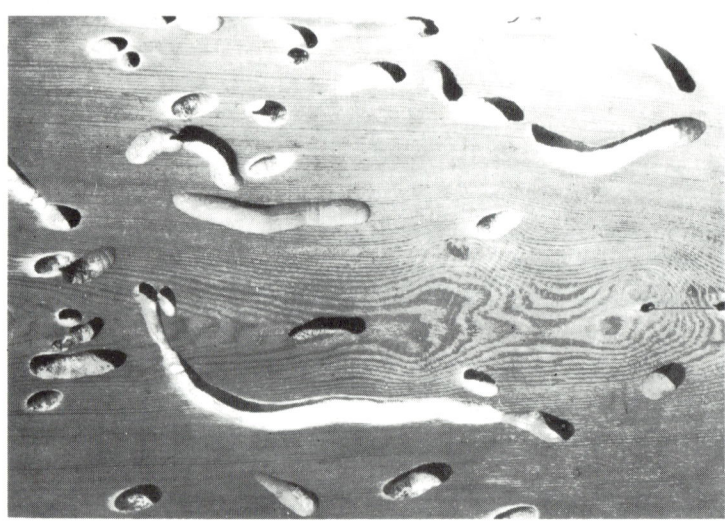

This is a bivalve mollusc like an oyster. Its eggs are released into water in very large numbers and develop into a larvae stage which is able to propel itself. It develops, still in the water, until it ultimately attaches itself to a wooden surface. The larva burrows into the wood by boring a hole using a rocking action of the hardened tips of the larva. As the burrow lengthens so the shipworm expands, because the worm is always the same length as the burrow. They can extend to lengths of up to 1 metre, but this will depend upon the number of *Teredo* in the wood.

The *Teredo* is able to survive quite short periods of drying out, up to a fortnight. It can only survive in salty water. It prefers the warmer water often found in harbours.

Adult insect	Adults can be up to 500mm long. Greyish-white soft body with two small shells at the head end.
Eggs	Retained by the adult until they hatch.
Larvae	Microscopic, consisting of two shells which enclose the body. They swim and crawl on the surface of timber to find a suitable site for entry.
Nature of boring	Tunnels generally about 6mm diameter. Lined with a white chalky deposit, except at the end. Entrance holes about 0·5mm diameter. Tunnels tend to run along the grain, are not filled with frass.
Wood species attacked	Most softwoods and hardwoods except certain resistant species. Both sapwood and heartwood attacked.
Remarks	Confined to timber in sea water and estuaries with near sea salinities. Larvae responsible for spread of infestation.

Timber decay caused by fungus

TIMBER DECAY CAUSED BY FUNGUS

There is a long tradition in the use of wood in building; in houses, boats and even aeroplanes. Even with the development in the use of other materials such as brick and concrete, wood remained as a major part of the construction process as well as in the final building.

Many housing estates are now being constructed by Wimpey, Barratts, and other developers with an internal wood frame, faced with brick or tile to the outside.

Laminated timber portal frames span many halls, swimming baths or sports complexes where the light weight/strength ratio of wood is of great value.

The development of the tree

During the growth of the tree there is a chemical reaction between the water which is brought to the leaves as sap from the roots and the carbon dioxide from the air which enters the leaves through minute pores. This chemical reaction known as photosynthesis creates glucose. The glucose is transported through the tissues of a tree to its growing area, where the molecules of the glucose are joined together to form long end-to-end chains, forming a new substance, cellulose.

The wood gains its lightness and strength from having been formed from numerous cells all of varying sizes and shapes and all linked together. The walls around these cells are built up from four layers. Each layer is composed of many cellulose chains bonded together when the tree is in growth. The sap is transported by the outer sections of the timber being that part nearest to the bark. As the wood tissues become older, being those parts located towards the middle of the tree, their ability to convey sap declines and they form a central core known as the heartwood. This is often more resistant to decay than the outer sections of sapwood.

It is the balance in the tree between this heartwood and the surrounding sapwood which is one of the main differences between a hardwood and a softwood. The two groups, hardwood and softwood, are defined by relation to the methods of seed distribution. Hardwoods are timbers converted from trees bearing flowers whereas softwood are the timbers converted from cones or needle shaped leaves.

The decay in timber results from the breaking down of the cellulose chain in the wood. The organisms which break down the wood are known as enzymes. Glucose is released when the cellulose chain breaks down and the enzymes feed upon it.

It is only when the cell wall contains enough water to allow the enzymes and glucose to defuse freely that a wood rotting fungi can survive. Timber which is below 20% moisture content is immune from attack by fungus.

Freshly cut timber has a moisture content of between 40 and 90% of its dry weight.

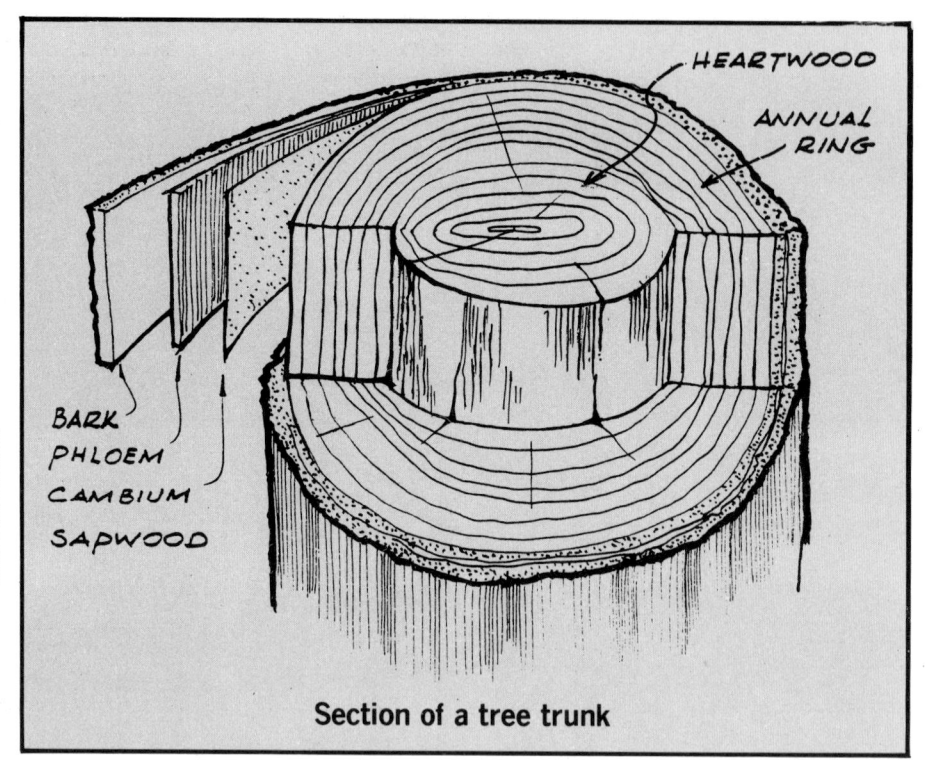

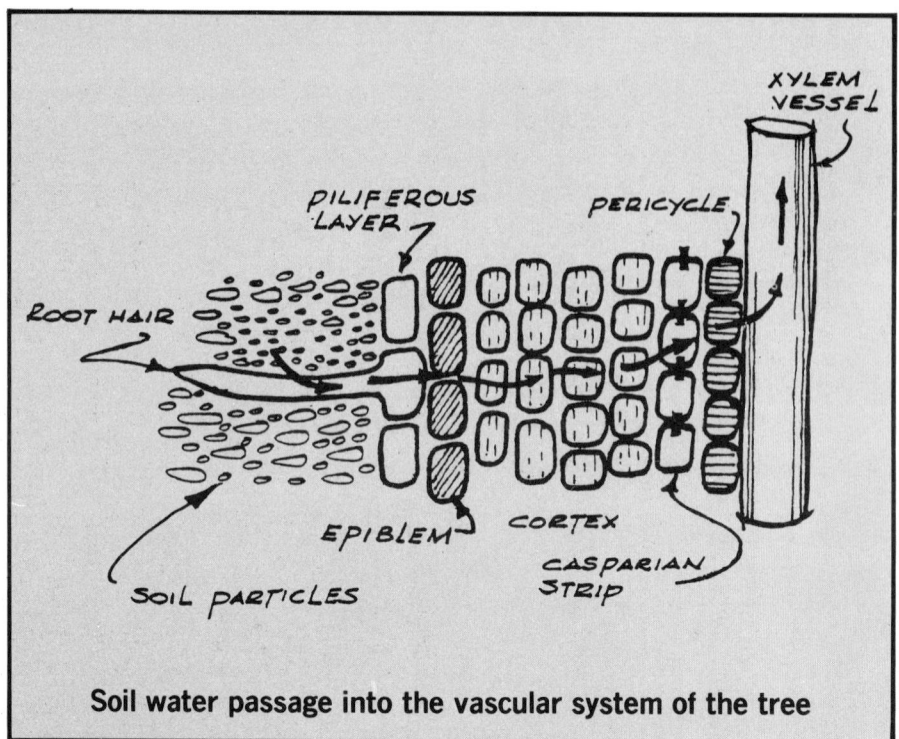

Section of a tree trunk

Soil water passage into the vascular system of the tree

Much of this moisture is lost soon after felling, because it is mainly the sap water held in the timber at the time that it was cut.

The remaining moisture is held within the fibres of the timber and this is somewhere between 25 and 30% of the dry weight of the timber. Further reductions are only obtained by seasoning the wood. In a natural state the timber would reach an equilibrium with the moisture content of the atmosphere. The moisture content can then be reduced to the region of 18%. Kiln drying means that most timbers are delivered to a building site with a moisture content in the region of 7 to 10%.

For moisture to be introduced to seasoned timbers to enable the propagation of a fungus there must be either a defect in the construction of the property or a failure in the components of the building.

A fungus differs from most plants in that it is not able to produce its own nutrients by photosynthesis and to survive must feed as a parasite on living plants or as saprophytes on the dead remains of trees. Fungi are normally composed of narrow tubes or filaments (hyphae) which grow by extending their tips and frequently branching. Sectional analysis of the hyphae of different fungi do not reveal sufficient variation to accurately identify each species. The identification of the type of disease has to be achieved by an examination of other characteristics of each fungus.

TIMBER DECAY CAUSED BY FUNGUS

Resistance of Home-grown Hardwoods to Decay

Timber	Common name	Average Loss in Dry Weight (as Percentage of Original Dry Weight) caused by Test Fungi after Four Months at 22°C			Class of Decay resistance
		Serpula lacrymans	Coniophora puteana	Fibroporia vaillantii	
Acer pseudoplatanus L.	Acer sycamore	26.6	20.8	—	Perishable
Aesculus hippocastanum L.	Horse chestnut	33.6	29.9	27.3	Perishable
Carpinus betulus L.	Hornbeam	21.4	11.8	9.4	Perishable
Castanea sativa Mill.	Sweet chestnut	Negligible	Nil	Negligible	Resistant
Fagus sylvatica L. (two tests)	Common beech	26.2 / 39.5	30.9 / 36.5	13.2	Not resistant to perishable
Ilex aquifolium L.	Holly	3.1	3.0	2.9	Not resistant
Juglans regia L.	Walnut	3.4	5.1	9.3	Moderately resistant
Pyrus torminalis Ehr	Pear	11.9	24.1	—	Perishable
Quercus robur L.	Oak	Negligible	4.0	—	Resistant
Ulmus hollandica Miller	Dutch elm	16.0	—	17.2	Not resistant
Ulmus procera Salisb.	English elm	15.5	21.6	1.9	Not resistant

As the hyphae grow they form a matt of material known as the mycelium. Most of the various forms of fungus have individual characteristics in the mycelium. The fungi reproduce by spores which are distributed from the reproductive structures known as fruiting bodies. These spores are readily dispersed by the air.

It is believed that the spores carry a small food reserve. When the spore settles it must find food quickly or perish.

Fungi fall into a number of classifications based upon the method of spore transportation. Most of the fungi which cause decay in timber are *Basidio mycetes*, because the spores are transported on small club-shaped structures known as *Basidia*.

The *Basidio mycetes* are further broken down into categories based on the shape and form of the *Hymenium*, the surface from which the seed is released.

1. The *Lephoraceae*
   Freely exposed on a flat skin-like surface.
2. *Hydnaceae*
   Spine-like outgrowths.
3. *Poly poraceae*
   The hymenium lining is inside pores or tubes.
4. *Agaricaceae*
   Plate-like gills under a cap-shaped pileus or mushroom.

If the fungus is able to remove the cellulose from the timber it darkens the wood and tends to cause the structure to break up into cubic shapes. This tends to be referred to as a brown rot.

If the fungus is able to attack both the cellulose and the walls of the cells then it tends to bleach the wood which breaks up into long strands or flakes and looks shredded. This is referred to as a white rot.

Most fungi propagate at temperatures between 0°C and 35°C. Most hyphae remain dormant below freezing point but are not killed; whilst there is a probability that they will die when temperatures are increased above 35°C. The temperature ranges for each fungus vary and where they are known they are set out in the detailed examination of each fungus.

The signs of fungus below the linoleum and the cuboid shape formed by the cracks across the grain suggest dry rot. You can check and test this on page 218.

213

TIMBER DECAY CAUSED BY FUNGUS

Viability of fungi after various period of exposure – time in days					
Fungus	Relative humidities – per cent				
	77.2%	62.3%	49%	33.4%	21.5%
Serpula lacrymans	dead after 343 days	dead after 212 days	dead after 212 days	dead after 152 days	feeble after 54 days dead after 81 days
Coniophora puteana	alive after 343 days	dead after 343 days	dead after 94 days	dead after 81 days	dead after 81 days
Fibroporis vaillantii	alive after 343 days	alive but feeble after 343 days	still viable after 212 days dead after 343 days	still viable after 212 days dead after 343 days	still viable after 212 days dead after 343 days

Based on information Crown Copyright, reproduced with the permission of the controller of Her Majesty's Stationery Office.

Left: Always check the floor around a WC pan. This is a source of water and any leak will result in the deterioration of the boards.
Below left: Mould on the wall behind pictures. Is this mildew due to moist air in the building settling on a cold wall—or is there damp penetration?
Below: Dampness in this room?

TIMBER DECAY CAUSED BY FUNGUS

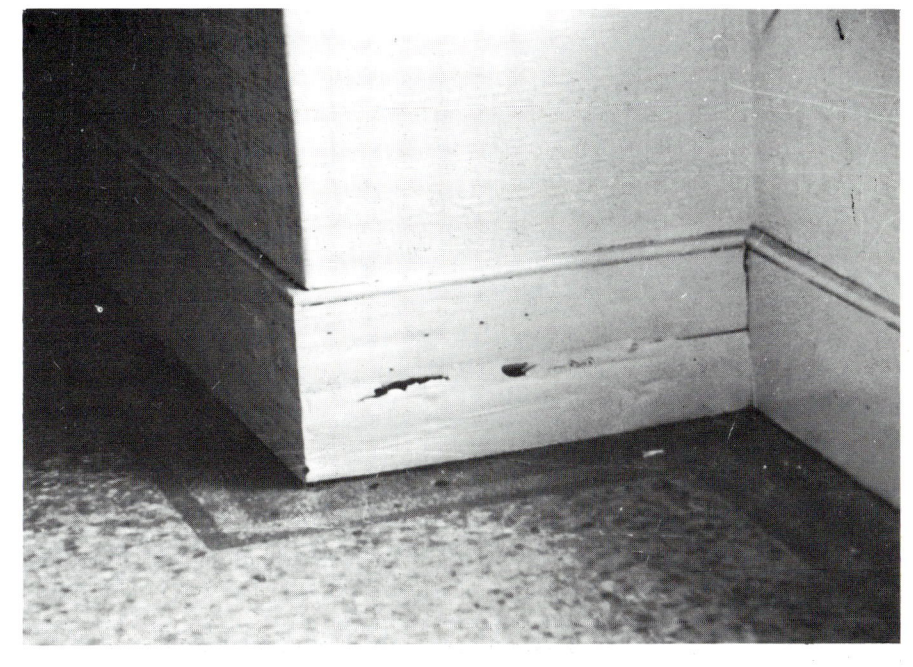

Top and centre: Floorboards adjacent to the external walls of the rear wing or back extension of the older type of terraced house should be lifted wherever possible (provided consent is given). The woodwork in these rooms is vulnerable to damp and worm and must be examined carefully.
Bottom: The first signs of a failure are frequently seen as an irregular surface on a skirting. The testing of the walls for moisture content may indicate that the failure is due to rising damp. The checking of the moisture content of the timbers will indicate if they are wet enough to allow the development of wet or dry rot.

215

TIMBER DECAY CAUSED BY FUNGUS

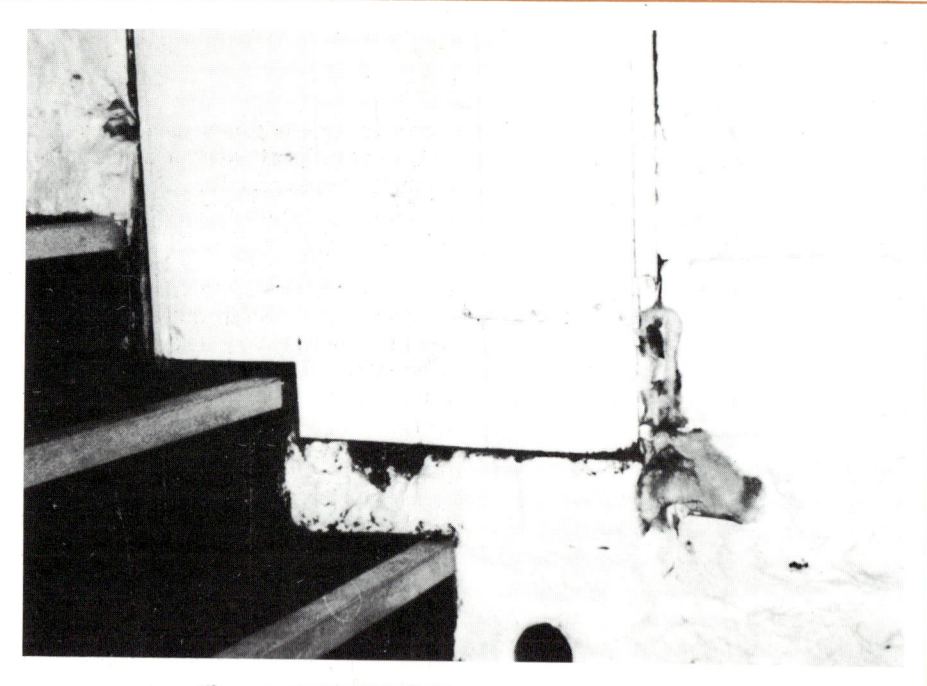

Open all doorways or access panels. That small area of mould showing at the bottom of the door frame is the first sign of dry rot. The cottonwool-like growth on the rubbish in a cellar occurs because there is dampness and no ventilation. In this case there was no other timber than the door and frame.

TIMBER DECAY CAUSED BY FUNGUS

Serpula lacrymans

Dry rot

This rot was originally named after the yellow beak of a blackbird and its ability to produce moisture. The original name *Merulius lacrymans* has changed to the more accurate scientific name of *Serpula lacrymans*. The serpula is a marine worm which inhabits beautifully coloured tortuous calcerous tubes which are often massed together (not a very relevant piece of information). The name comes from the Latin *Serpere:* to creep. and *lacrimans:* tears.

Dry rot falls into a brown rot category and decayed wood tends to develop a cuboidal cracking formation, to the surface of the timber. The moisture content required for propagation is in excess of 20% but a likely optimum is in the range of 35%. Rapid growth can take place with temperatures as low as 5°C but growth stops at 0°C and above 26°C. It is remarkably sensitive to heat and will die on exposure to temperatures of 40°C for a very short period.

The temperature range restricts the fungus to the temperate climates and as a result it is more commonly found inside a property in the United Kingdom.

Its rate of growth, given reasonable conditions, is about 1 metre per year. Most active growth occurs in conditions of bad ventilation and high humidity. The fungus mycelium develops on the surface of infested timber and may take the form of a soft white cushion with a cotton wool texture. Under dryer conditions a skin of silver grey silky mycelium, sometimes with yellow patches and tinges of lilac, may be formed on the wood surface. Droplets of water are produced at the tips of the hyphae and lie on the surface of the mycelium. Feeder strands develop within the mycelium and supply water and nutrients to the growing area. These strands may extend for several feet over a material such as brick or steel and may penetrate behind plaster and through brickwork. One of the major difficulties in eradicating the fungus is to ensure that the feeder strands are killed or rendered ineffective.

The fruit bodies or sporophores are fleshy growths resembling a flat plate or bracket. They are initially pale grey, tinged with yellow, but as the spores develop the spore bearing surface becomes rusty red in colour. This surface may be corrugated into irregular folds (pores) and has a characteristic mushroom like smell. The pores are an irregular overlapping network of intertwined holes. The fruiting body discharges its spores at a phenomenal rate (up to 800 million a day for up to two weeks). The sight of a cobweb or other

Remember dry rot can spread along timber. Here a dado rail provides the food source for the fungus to leapfrog sound and dry construction.

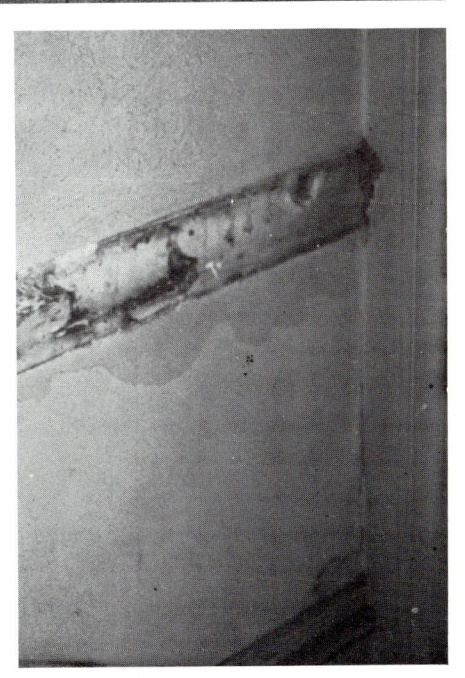

Surface damage
Cuboidal cracking and darkening of the wood. Note deep cross cracking.

Identification
Strands grey, become brittle when dried. Silver grey growth similar to cotton wool, slight yellow tinges to edges. Fruiting body, soft fleshy labyrinthine, white edged, rusty middle.

Rate of growth
1 metre a year for the strands, although flash growth of the mycelium can be greater.

Light
Reactive to light. Required for growth of fruiting body.

Dampness
20%–35% (i.e. damp, not wet).

Temperature
0° to 26°C. Optimum growth at 22°C. Will die if exposed for short periods to temperature of 40°C.

TIMBER DECAY CAUSED BY FUNGUS

surface below the point of discharge gives one some impression of the volume of spores that are pumped out of the body. After this final act of ejaculation the fruiting body shrivels to a dark brown colour and dies.

The presence of light is important in the development of the sporophore (fruiting body). It also is able to promote the rapid cotton wool growth. An inspection of a dark area which is infected can produce a visible microscopic explosion of growth on the timber surface due either to the heat the light produces or to the light itself. The growth whilst fascinating is also quite frightening.

Certain aspects of the growth of dry rot are not fully understood but it is now recognised that the general tendency is for it to attack only damp timbers. In most cases the fungus is restricted to areas where timbers are sufficiently damp.

It is very rare for the fungus to transport water to previously dry areas to allow further spread. Its ability to spread depends on the tiny hyphae and not the larger tentacles, the tentacles are the food wagon for the extension of the attack.

The fruiting body of dry rot is not usually found on the outside of a building. Where this does occur it will be in a shaded position such as under a porch or on a north facing wall. The heat generated by direct sunlight is above its tolerance and it would not be able to survive.

Age of attack

It is not possible to accurately date an outbreak from a visual inspection of the timber. However, the age of an infection can be guessed from an examination of the depth of cracking. In new, open texture timber dry rot can penetrate 18 mm a year, whilst in older, denser timbers the penetration is only about a quarter of this depth in the same period.

The circumstances of the outbreak must be identified. In most cases the start will coincide with the date water penetration occurred.

It is common to find outbreaks which have been in existence for many years, and have been dormant (have not grown) during the past few years. This occurs where the source of dampness was eliminated and insufficient moisture remained for the rot to continue its development.

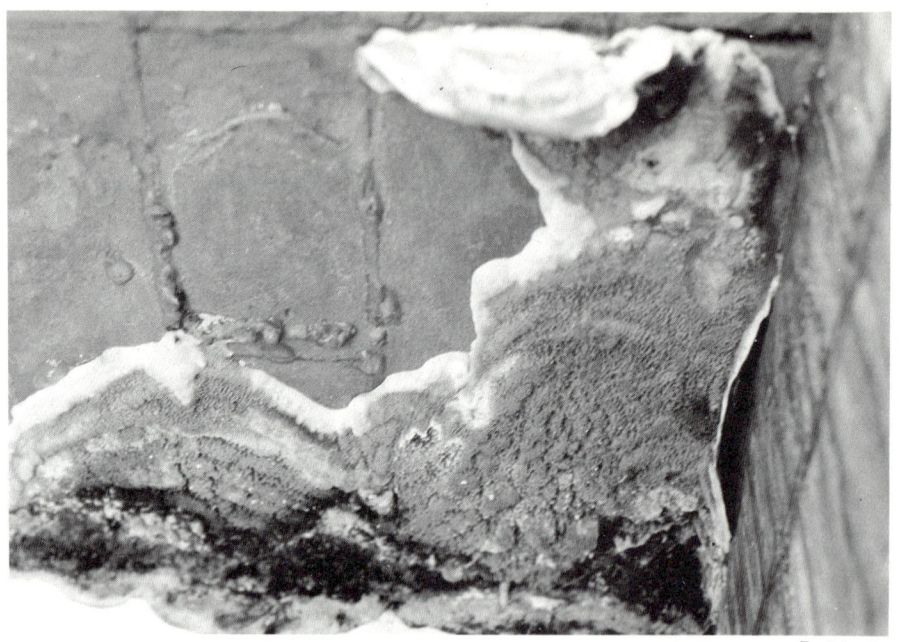

Fruit body.

Close up of fruit body.

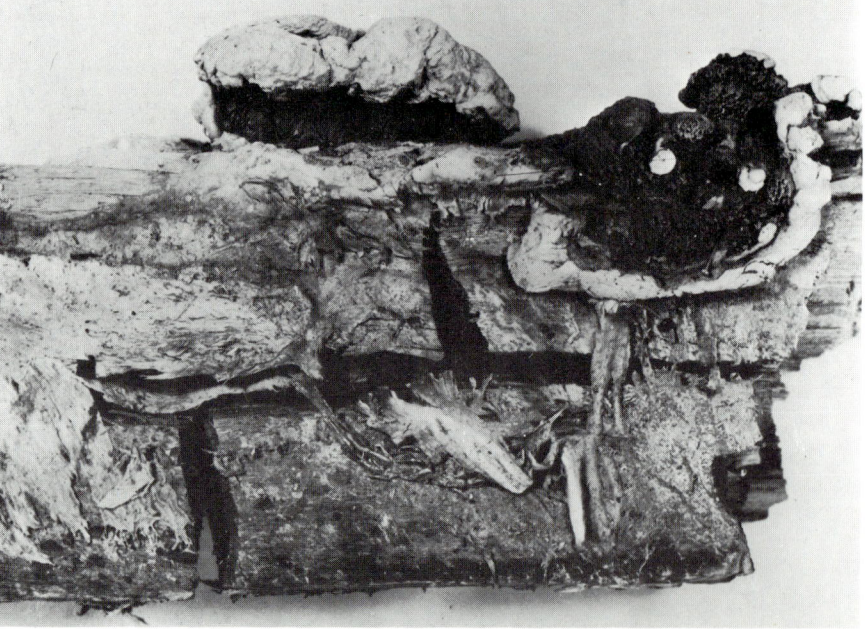

Damage to timber.

Mycelial growth and damage to timber.

TIMBER DECAY CAUSED BY FUNGUS

Coniophora puteana
Wet rot

Also known as Cellar Fungus or Cellar Rot it is the major wet rot in the United Kingdom and thrives in wetter conditions than those which support dry rot. It can also be found in worked timbers and logs in the open air. *Coniophora puteana* is highly vulnerable to fluctuations in moisture content and achieves optimum growth in timbers with a moisture content of between 50 and 60%. However, the very wet timber required by the fungus is also its main vulnerability. It is unable to survive if the timber's moisture content declines below 43%. Infestation results from the fungus spore landing on the damp timber, the spore then germinates and extends its hyphae. It is rare to see a surface mycelium other than in enclosed areas away from daylight. The hyphae extends in individual strands which are initially yellow brown and become brown or dark brown, almost black when reaching maturity. The fruiting body is rarely found inside a building although it is quite common out of doors. It consists of a thin olive brown plate of irregular shape and has a series of irregular lumps. The sterile margin is creamy white and the spore bearing surface is yellow to olive brown.

The wood which has been affected becomes very dark brown. The main cracks which occur tend to run along the grain of the timber with smaller cross-cracks. It is important that one does not over rely on the type of cracking being an indication of the particular fungus. Both wet and dry rots are capable of producing almost identical failures particularly in joists or other large timbers. The smaller timbers of window joinery are more likely to reveal the longitudinal cracking which has become the trade mark of wet rot.

Surface damage
Cracks follow the line of the grain. Minor cracks only across the grain.

Identification
Thread-like strands, yellowish becoming darker brown with age. Fruiting body rarely found. Thin sheetlike shape with spore surface, olive brown. The surface of the fruiting body is knobbly.

Rate of growth
Can cause 40% weight loss in four months.

Light
No recorded reaction, usually found out of light source.

Dampness
Death after 94 days at 49%, so the lower levels show it can survive for a limited period. 50–60% (very wet).

Temperature
−30°C to +40°C. Optimum 23°C.

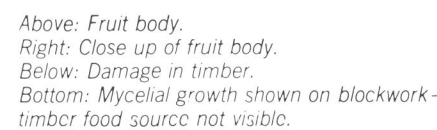

Above: Fruit body.
Right: Close up of fruit body.
Below: Damage in timber.
Bottom: Mycelial growth shown on blockwork – timber food source not visible.

TIMBER DECAY CAUSED BY FUNGUS

Fibrioporia vaillantii — Mine fungus

(Also named *Poria vaporaria*)

The common names for this fungus include white pore or mine fungus. Although this is a common fungus in mines it sometimes does occur in buildings where the timbers have become very damp. A moisture content between 45 and 60% is required, and temperatures up to 36°C.

The characteristics of this rot can be confused with those of dry rot.

It has a white mycellium but does not show the traditional lilac or yellowish edges which is common in the mycelium of *Serpula lacrymans*. When freshly growing the surface of the mycelium is fern like. The strands are smaller than those of dry rot but remain flexible when dry.

It is less likely to have penetrated loose brickwork or lime mortar than dry rot. The fruiting body does not usually occur in buildings. It consists of an irregular white plate which adheres to the timber. The fruiting body is covered with pores, the tubes of which vary in depths between 2 and 12mm. When examined closely it looks like a honeycomb. The regular surface undulation of the fruiting body is very different from that of *Sepula lacrymans*. It produces cracking in the timber which is not quite so pronounced as that in dry rot, although the cuboid shape can be seen.

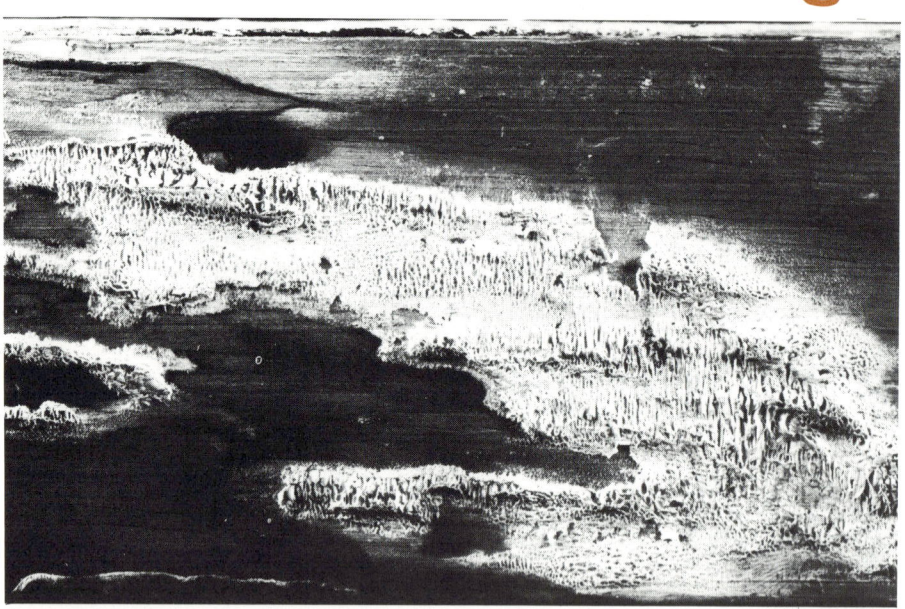

Top: Fruit body.
Centre: Close up of fruit body.
Bottom: Mycelial growth and damaged timber.

Surface damage
Cuboid cracking in the timber when dry. Darkens the wood.

Identification
Strands whitish and seldom thicker than strong twine, remain flexible when dried. Fruiting body plate shaped with honeycomb surface, white in colour. Associated with water leaks. Will not spread to drier part of building.

Rate of growth
1.4 metres a year.

Light
No information available. Not usually found in light.

Dampness
45–60%.

Temperature
Up to 36°C. Optimum 27°C.

Paxillus panuoides

This occurs in softwood timbers which have been allowed to become wet and is also common in damp underground conditions such as mines. It is able to propagate out of doors in soft wood tree stumps or sawdust.

The fruiting body is a brownish or dingy yellow ear-shaped growth between 25–75mm (1"–3") in diameter. It is soft and fleshy and may hang like a bell from the point of attachment or may lie along the wood like a shell. The gills to the underside radiate.

The mycellium is yellow with occasional violet patches, the strands have a hairy appearance (like wool) and do not darken with age. The optimum temperature for its growth is 23°C. The optimum moisture content is between 50 and 70%. The fungus also requires ample aeration so that it tends to be confined to surface layers of decayed timber.

The decayed timber becomes stained to a vivid yellow where the mycelium has made contact. Later this becomes a golden red-brown with fine transverse cracks.

The drying of the timber rapidly kills the fungus which is unable to survive at moisture contents below 45%.

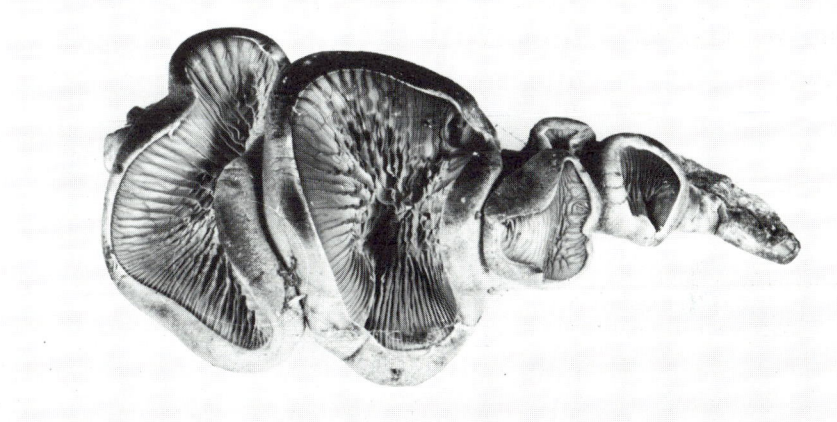

Top: Fruit body.
Centre: Close up of fruit body.
Bottom: Mycelial growth and damaged timber.

Surface damage
Wood becomes soft and cheesy and on drying out has deep cross cracks. Surface damage occurs in softwood.

Identification
Brownish or dingy yellow shell or bell shaped fruiting body. Mycelium yellow with occasional violet patches. It causes a yellow discolouration of wood.

Rate of growth
Up to 0.5m per year.

Light
Will propagate in daylight.

Dampness
50–70%.

Temperature
5°C up to 29°C.

TIMBER DECAY CAUSED BY FUNGUS

Donkioporia expansa

(Also known as *Phellinus megaloporus, Phellinus cryptarum, Poria megalopora*)

In this country it is largely restricted to attacks on oak and chestnut inside buildings and tends only to propagate in warm areas where there is an absence of light. The fruiting body is variable in shape but is usually plate-like up to 250mm (10") and is 25mm (1") thick. It is hard and woody and consists of several layers of pores placed one on top of the other. The colour of the fruiting body varies from pale brown to dark brown.

The fungus causes the wood to be reduced to a loose white fibrous mass and the damage is often concealed from the outside. Damage may occur in a beam where it is embedded in masonry, which has provided the moisture content required for the propagation of the fungus. The optimum temperature for growth is 27°C and the maximum is 35°C.

Donkioporia expansa does not spread over inert material so that attack is more or less confined to the damp timber area.

The decay in oak is more rapid than any other fungus. The decayed wood is particularly suitable for the development of death watch beetle. The control of death watch beetle in a section of timber would depend on the eradication of all traces of the fungus.

Poria medulla-panis is indistinguishable from *Donkioporia*. It is found in oak half-timbering of houses and is less vulnerable to light.

Top: Fruit body.
Centre: Close up of fruit body.
Bottom: Damage in timber.

Surface damage
Reduces wood to fibrous strands.

Identification
Fairly common on damp large oak timbers. Large plate shaped fruiting body up to 200mm long and 25mm thick, often in several layers. The fruiting body is hard and woody, warm buff to brown in colour. Causes an active white stumpy rot. Hyphae are yellow and are plentiful in wood. Can cause decay. 35% weight loss in 4½ months. Linked with death watch beetle.

Rate of growth
Rapid, but will not extend beyond timber.

Light
Absence of light required.

Dampness
20–36%.

Temperature
27°C optimum. 35°C maximum.

TIMBER DECAY CAUSED BY FUNGUS

Lentinus lepideus
Stag's horn fungus

Not a common fungus within the United Kingdom, it is found occasionally in the stump of conifer trees.

The fruiting body can be of two types, the more common type has a fleshy mushroom cap set upside down on an extended stalk. The cap may be up to 4" in diameter and may vary from pale brown to a more purplish-brown colour.

The less common form of fruiting body has a thinner twisted stem which is contorted and the mushroom cap is much smaller or totally absent. These stems resemble entwined roots or the antlers of a stag.

Mycelium appears on the surface of the wood when it is in an advanced state of decay and forms a soft whitish cover. The mycelium is sometimes tinged with brown and frequently has long needle shaped crystals within its surface.

The wood becomes dark in colour and cuboidal cracking occurs. The wood becomes sticky and there is a characteristic balsam like smell.

Right: Fruit body.
Below: Mycelial growth and damaged timber.

Surface damage
Leaves wood in brown cubical shape.

Identification
Occurs in telegraph poles, railway sleepers, paving blocks, mines and occasionally in flat roofs. Distinctive fruiting body of either antler-like development of thin mushroom with extended stalk. Colour from pale brown to a purplish. Tolerant to creosote except in very high quantities. Strong smell like balsam.

Rate of growth
Can be up to 1.5m per year.

Light
Not affected by the presence of light.

Dampness
26 – 44%.

Temperature
27°C. Maximum 37°C.

TIMBER DECAY CAUSED BY FUNGUS

Daedalia quercina

In this country it is found in oak and sometimes attacks chestnut, beech and some other species.

In buildings it is most likely to attack timber window sills or door sills where rainwater has not drained away quickly. It tends to propagate more easily on the warmer south westerly aspect in the southern part of the United Kingdom.

The fruiting body is a thick bracket which may take up various forms. It is a greyish buff colour, the elongated pores are sometimes pale and have rounded edges. The pore surface is tough and corky. Mycelium is usually a yellowish colour and may be seen in the splits and fissures of the decaying timber.

It causes a reddish-brown cuboidal break in the wood, individual cubes can be quite large. The optimum temperature for growth is 23°C and the maximum is 29°C. The optimum moisture content is about 40%.

Right: Fruit bodies on oak.
Below: Various examples of damaged timber.

Surface damage
Reddish-brown cuboidal rot. The individual cubes are often quite large.

Identification
Thin sheets of mycelium in spores in wood. The decayed wood is friable. Attacks hardwoods usually oak. Smells of apples.

Rate of growth
Up to 0.8m per year.

Light
Cultures tend to be more silky in the dark.

Dampness
Optimum 40%.

Temperature
Optimum 23°C. Maximum 30°C.

DECAYS IN TIMBER CAUSED BY FUNGUS

Myxomycetes

Slime moulds

Most moulds fall into one of the following groups, *Penicillium, Trichoderma, Aspergillus and Pullularia*. They grow on damp surfaces such as plaster, brick, wallpaper, fabrics and clothing. They are commonly associated with areas of high condensation, such as occur in restricted areas of air circulation, or in areas of moist humid air. Shower rooms and sports changing rooms frequently suffer from the problem as do bakeries, breweries, and many factories manufacturing goods where a large water discharge occurs. The mould will give off spores very easily. It is the spores that give the colour to the mould growth. These colours change from black to green with some greys, whitish coloured moulds and even yellow and pink occasionally. The *Aspergillus* group which grow on damp patches usually on houses, can cause *Aspergilloma* and *Bronchopulmonary Aspergillosis*. These diseases of the lungs are frequently confused with other bronchitic diseases. Young children and old people are both vulnerable to this infection.

The slime moulds or *Myxomycetes* are formed of delicate globular structures about 20mm in diameter. These are usually black in colour and will release black dust particles if broken. If you wipe them with the back of your hand they seem to leave a larger dirty mark than was capable for the mould on the surface. This is because of the discharge from the slime mould contains.

These moulds do not damage timber.

Top: Fruit body showing pore surface.
Right: Fruit bodies.

TIMBER DECAY CAUSED BY FUNGUS

Phellinus contiguus

This fungus occurs on wood which has been worked. It is more common on window frames and sills. It is found in a wide range of timbers. In New Zealand it is one of the major causes of decay in timbers used in buildings.

The fruiting body is upside down, with very thin layers of the mycelium below the pore zone. It is yellow-ochre in colour, sometimes darker or a dark brown colour.

The fungus causes a stringy white rot in oak and other hardwoods. It has also been found in coniferous wood.

Although it has long been regarded as a minor rot in the United Kingdom, it is becoming a more common cause of decay in window frames and external joinery.

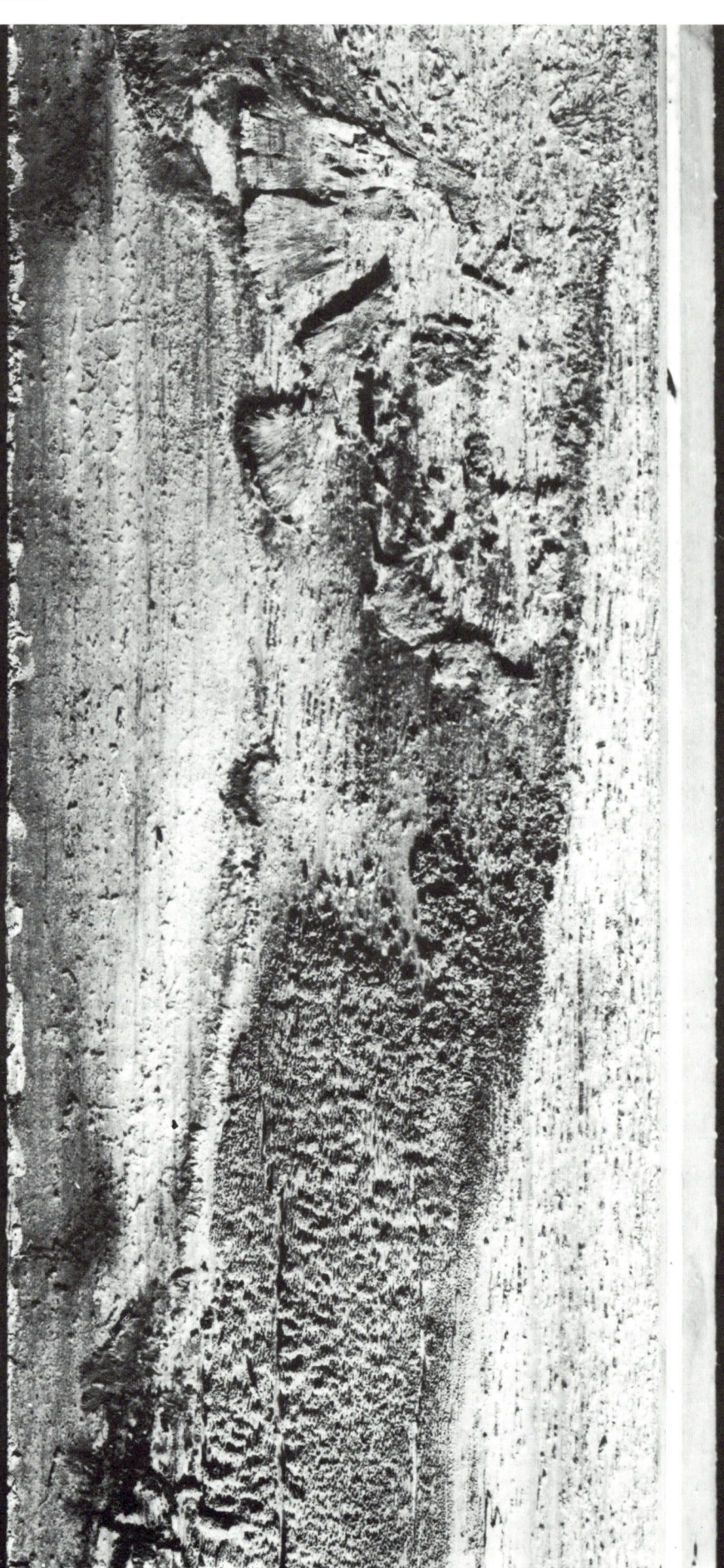

Right: Fruit body and timber damage.

Surface Damage
Softening of the timber, leaving it fibrous and stringy without substance.

Identification
The fruiting body is very thin and tends to mould itself to the surface of the wood, often occupying the gap between casement and frame. The fruiting body is a dull brown in colour. Microscopic identification is easy due to presence of pointed brown form of hyphae.

Rate of Growth
Believed to be about 300 mm per year.

Light Reactive
No information available.

Dampness
22%—upper limit not recorded.

Temperature
0°—31°C.

TIMBER DECAY CAUSED BY FUNGUS

Poria placenta

(Also known as *Poria monticola*)

The fungus frequently causes damage to softwoods, usually where they have been imported from British Columbia. It does not occur naturally in Great Britain. If the timber is used in a dry, well ventilated location the fungus will die out. If the location is suitable for propagation the fungus will spread and can cause a brown rot.

The rot is more prevalent in older trees, usually Sitka spruce and Douglas fir.

The fruiting body is a thin membrane which grows annually and tends to break up after a short period of time, becoming brittle when it dries.

Wood does not show a marked change in the early stages of an attack, and its presence may easily be overlooked. The first signs of the infection are streaks of yellowish-brown or pinkish discolouration. The wood, where damaged, becomes brittle. Even slight attack will have a serious effect upon the wood strength. In Douglas fir the first sign of attack may be a bluish-purple discolouration.

In its more advanced stage, pockets of brown cuboidal rot are formed. Eventually the wood will break up into cuboidal shapes, which easily crumble.

Growth is moderately rapid, being capable of spreading up to two metres a year in ideal circumstances. It will also cause rapid weight loss. The optimum temperature for growth is 28°C (35°C max).

Top: Fruit body.
Right: Mycelial growth.

Surface Damage

Cuboidal cracking and darkening of the wood, deep fractures with and across the grain. Wood crumbles easily.

Identification

Strands become brittle when they dry. The hyphae have numerous clamps. Discolouration or staining of the wood—usually yellowish-brown or pinkish-brown.

Rate of Growth

Can be up to 2 metres a year, rarely exceeds 800 mm in UK.

Light Reactive

Not specifically light reactive.

Dampness

40%—65% requires wet rather than damp conditions.

Temperature

0°C—35°C (optimum 28°C).

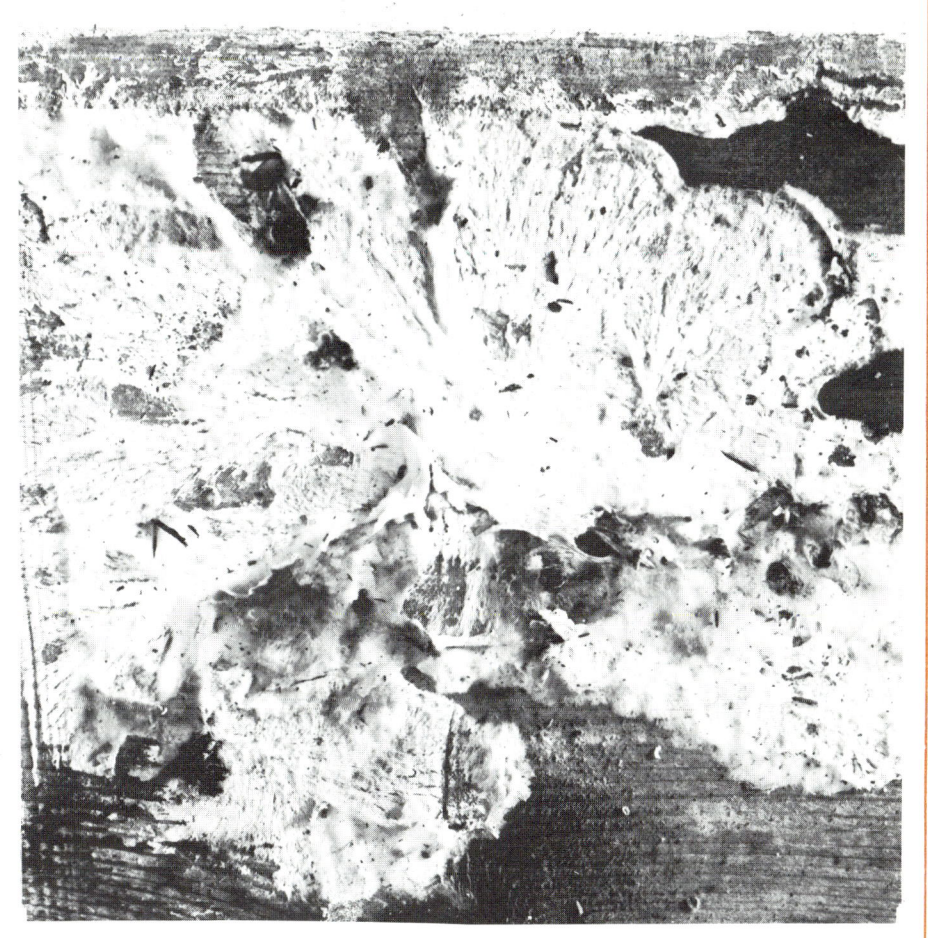

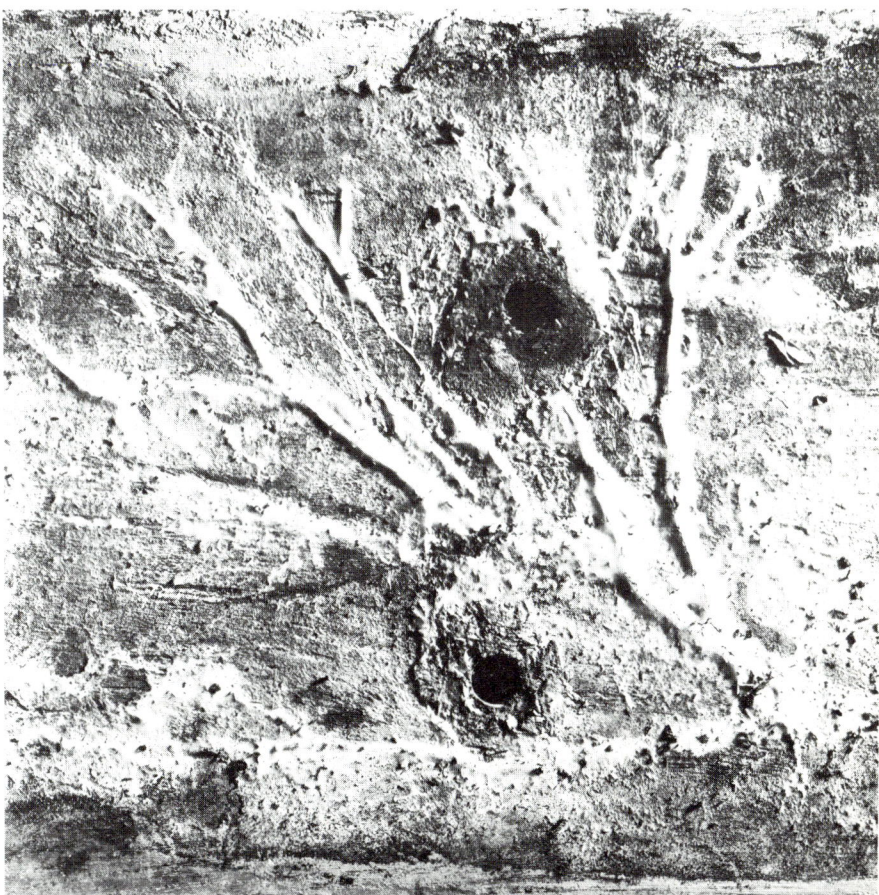

Non wood-rotting fungi

Peziza or 'Elf-cup'

The fruiting bodies of this fungus may be seen in damp plaster or on brick walls where they are saturated. They can be up to a couple of inches wide and are buff or pale yellowy orange and they are fleshy but very fragile and break easily when handled. When they dry they become hard and the colour deepens to a stronger redish orange.

This fungus is not a wood rotting fungus but merely an indication of extreme moisture content within the supporting materials

Corprinus domesticus or 'Ink-cap'

The mycelium of this fungus has an appearance of coconut matting with slightly orangy coloured tufts or may appear as a yellowy orange sheet. The fruiting body is a slender toadstool up to 70mm (3") high. The head of the toadstool is white and it eventually produces a black ink liquid in which the spores are freed. The fruiting body is short-lived and shrivels up on to the material which supported it. This fungus does not damage timber.

Mould

Four of the most common moulds known in residential property in this country are *Penicillium, Trichoderma, Aspergillus* and *Pullularia*. Mould usually appears as unsightly growths in various colours, green, yellow, pink, black, grey or white. It produces spores very readily on microscopic fruiting bodies and it is usually the spores which give the colours that enable one to see where the mould grows. These spores are often rapidly spread by tiny insects, and grow on any damp surface. Moulds require very little nutrient, but moisture is essential for their propagation and they tend to occur where there is condensation. Materials such as wallpaper, clothes, bedlinen and shoes may be damaged by fungus but timber is not affected.

The green mould is usually penicillium and the yellow, brown or black mould is usually aspergillus. Moulds are fungi which germinate from spores carried in the air on to almost any damp surface.

The treating of the affected area with fungicide will restrict the mould growth but the only permanent cure is the elimination of the dampness on the surface.

Arching of woodblock floors

Woodblock flooring is vulnerable to increases in the moisture of the blocks. The expansion of timber sets up forces which if not checked will result in an upward movement in the block to release the pressure. Such failures are usually an indication of a failure in the damp proof membrane in the floor or of water carrying appliances.

In failures in new construction the condition of the damp proof membrane will be suspect. Sometimes they are laid directly upon a bitumen surface which is supposed to perform two functions. The sticking down of the block, and the prevention of rising damp. This is usually unsatisfactory.

Failures have been noted where the blocks are fixed on to standard cement screed laid on top of a damp proof course, and there is a pipe failure somewhere on the same level. The damp proof course provides a horizontal impervious barrier and the water from the leak soaks into the screed to the depth of the d.p.c. Water then spreads horizontally in the form of a lake causing damage some distance away. Both the blocks and the floor surface require drying out before they can be replaced.

Inadequate expansion joints can result in wood block floor failures. The traditional cork compression joint laid around the perimeter may have been too small. Some floor cleaners and polishes fill the pores in the cork and reduce their ability to take up expansion in the wood block. These materials should be checked at intervals of two or three years.

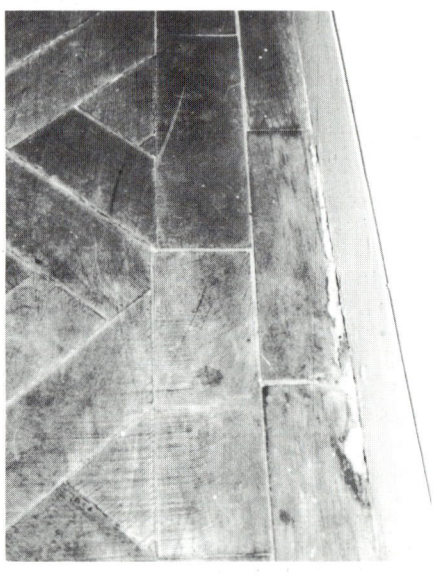

Moisture from a wet wall could cause the woodblocks to expand. Is there room for this?

Electrical services

The rules relating to electrical installations in this country are the Institute of Electrical Engineers Regulations. These are frequently referred to as the IEE Regulations. These Regulations are advisory and it is regretted that they are not yet mandatory. The latest version of the Regulation is the 15th edition.

The Regulations cover requirements for the protection of householders or occupants of buildings against shock, and protection against fire risk stemming from overloads or faults in the system. They also set out the basic requirements for installation practice and the testing and inspection of installations.

These Regulations require that an electrical installation be inspected and tested at least once every 5 years. It would be reasonable in the inspection of a property to ask for the latest test certificate for the electrical installation; if one is not to hand or the last was carried out over 5 years prior to the inspection, an electrical test ought to be recommended in order to comply with the IEE Regulations.

Where the property being inspected is one where there would be a great inherent risk in the event of a failure in the electrical installation, such as sports complexes, swimming pools, caravan sites, areas for the storage of explosives materials, petrol stations or construction sites, inspections should be carried out at much closer intervals and possibly on an annual basis.

At the present time over half the fires which take place in the United Kingdom in residential property are associated with electricity. 15% of these fires have been blamed on defects in the electrical installation and wiring of the properties involved. If the 5-year cycle of re-inspections had been mandatory many of these fires would have been prevented, with a consequence saving of lives.

The first electrical installation to a residential property took place in the 1920s. Up to the 1950s most houses were wired in vulcanised india rubber. The surveyor should be capable of identifying the type of insulation provided to the wiring which has been used in the electrical installation to the property. The most common types of wire insulation are set out below with the dates when the material was in use:

Vulcanised rubber with cotton braid 1900-1920
Usually surface installed behind a ribbed wooden cover. This system will now be faulty and should be condemned.

Lead sheathed cable 1910-1939
Lead cables run on the surface. The wiring is now likely to be faulty and will need replacement.

Tough rubber insulated cable 1925-1935
Usually two wires in round insulation fixed on the surface.

The absence of an earth system in each of these three installations means they fail to meet modern standards.

Vulcanised India rubber 1928-1939
This was often run in screwed conduit on the surface of the walls. Any movement will usually cause failure in the insulation and the absence of an earth wire on the lighting circuits means they do not meet modern requirements.

Tough rubber sheathed cables 1925-1960
These cables were usually run through walls and floors. The majority of the systems using this wire will have used round pin sockets.

PVC insulated cables 1950 onwards
The introduction of square pin sockets coincided with the introduction of pvc insulated cables. For this reason they are easier to recognise before the wire is inspected. The pvc is more durable but will break down on contact with heat, or certain chemicals often found in loft insulation or cavity wall fill.

The deterioration in the insulation will lead to short circuits and earth faults with a consequent risk of fire or electrical shock.

In the testing and reporting upon electrical installations consideration needs to be given to the fuse system or circuit breakers which have been installed. Overloads and short circuits are protected by the use of fuses or miniature circuit breakers. The normal method of protecting against earth leakage is to connect the metal casing of a piece of electrical equipment directly to the earth. This enables fault current to flow through this part to the earth which will provide an overcurrent. Protection devices are installed within the installation to terminate the supply when this takes place. Opinions differ over the minimum amount of electrical current which can prove fatal to people (it can be as low as 0·05 of an amp!) and the small amount of electrical which is necessary to start a fire.

In the older installations the fuse system uses re-wirable fuses. This type of fuse has been banned in nearly every other country in the world because the fuses are unable to detect variations in current of less than one amp. They are also too easy to incorrectly wire using over-strength fuse wire with the result that the protection which they should afford is severely reduced.

Miniature circuit breakers (MCBs)

These operate on the basis that the overload breaks the circuit by "popping" a switch.

The most sensitive type of circuit breaker is the residual current breaker (RCB). This type of unit is required by the latest IEE wiring Regulations in order to protect socket outlets supplying electrical gardening equipment or DIY equipment such as lawn-mowers, hedge cutters, garden lighting, water circulation pumps, power tools, drills and saws which would be used outside the house.

In the event that these circuit breakers are required to operate they break the circuit frequently by pushing out a red marker or switch. To re-connect it is a simple case of pushing that switch back in. There is no need to replace fuses or re-wire

ELECTRICAL SERVICES

the fuse and in this way the human contact is reduced to a minimum, thereby reducing the risk of human error. As the fault is still in existence the circuit breaker will re-open. It is not possible to hold it closed against a fault.

The visual inspection of an electrical installation

Whilst it is recognised that the testing of an electrical installation may well be a specialist job, there are a number of aspects which can be determined by the surveyor so that he can report upon the condition of the installation based upon a surface examination of the wiring:

1. Feel the face of every single socket outlet within the property. Note any covers which are warm or hot to the touch.

2. Where the property is occupied, feel the outside of any plug which is plugged into a socket. Once again note any which are warm or hot.

Turn off the electrical installation at the mains before proceeding further.

3. Carefully examine each socket outlet, particularly looking for burn or scorch marks around the pin positions on the face of the socket.

4. Select a socket in the warmest position in the house. This may be next to a fireplace, backing onto a boiler position, or located above or around the boiler as a supply. Remove the cover in order that you can examine the condition of the cable insulation near the terminal. Being located in the warmest position this will probably be in the poorest condition. If it falls away from the wires then one will make the appropriate comments and advise upon re-wiring of that section of the property.

5. Look in the roof space and examine the location of wires if the roof space has been insulated:

(a) are the wires located beneath or within insulation? This can lead to a build-up in heart within the wiring and the rapid deterioration of the insulation.

(b) is the insulation of the polystyrene granule type? This type reacts with the plastic coating to the wiring and results in what is called migration. This leads to a deterioration in the quality of the insulation.

(c) If junction boxes have been used check that the cover of the insulation runs into the box and is not cut short. Check for any bare wires around junctions or connections within the roof space. Feel the surface of cover plates of junction boxes within the roof space and report upon any which are warm.

General items

1. If you find more than one adaptor has been provided within the property this would be an indication of an inadequate installation with the possible risk of overloading of the circuit taking place.

2. Check the size of the mains fuse in the main entry. In many cases this is not adequate for the number of sockets which are provided within the building.

3. Check the condition of the wiring of the tails on the meter board. Many houses are re-wired from the consumer unit through the test of the building but the connection between the mains head and supplier's fuse and the consumer unit is often of poor quality wiring.

4. Check on the length of cable which has been provided to electrical fittings within a property. Where these cables are in excess of 2.5 m (8′) in length this would be an indication of inadequate electrical sockets or inconvenient location. Excessive wires run beneath carpets or around skirtings are a hazard in themselves and should be avoided wherever possible.

5. Check the bathroom. No power socket should be provided within the bathroom. No switch should be provided within the bathroom. The switch for bathroom lights should be located outside the bathroom or there should be a pull cord.

6. Check the location of electric lights around bathroom fittings. Is it possible to touch a light or light bulb whilst also touching a bath-tap or wash hand basin tap? This should not be possible.

7. Check that any shaver installation is equipped with an isolating transformer. Certain types of shaver socket, particularly those fitted to the end of light fittings with a bathroom are not equipped in this way and should not be installed in a bathroom.

8. Earthing — The 15th edition of the IEE Regulations has increased the requirements for earth bonding in a conventional residential property. The following should be bonded:

(a) A kitchen sink should be earth-bonded to both the hot and cold supplies.

(b) The gas installation should be earth-bonded on the household side of the supply.

(c) Earth bonding should be provided to hot and cold supplies to baths, basins, bidets, as well as the cold supply to the toilet.

(d) Earth bonding should be provided to all the services and connections running into and out of a boiler.

(e) Earth bonding should be provided to water heaters and towel rails.

(f) Every system must have a main earth terminal which must be accessible, reliable and mechanically sound and requiring a tool to be used to disconnect it.

The future

Although the Institute of Electrical Engineers Regulations are not mandatory it is believed that the changes envisaged in the Building Regulations will include them as the statutory requirement for all new installations. This has already taken place in Scotland.

The illustrations show a variety of old, unsightly and dangerous electrical installations.

Appendix 1

Concrete Cube Testing

Albion Concrete Products
Station Road, Llangadog, Dyfed,
Wales SA19 9LT
Tel: Llangadog (0550) 777327

Associated British Consultants (Testing)
Unit 25, Trent South Industrial Park,
Nottingham NG2 4EQ

Borcon Concrete Services
Downhill Quarry, Downhill Lane,
West Boldon, Tyne & Wear NE36 0AX
Tel: Boldon (0783) 360551

Bullne & Partners
Consulting Engineers,
East Hetton Lodge Laboratory, Coxhoe,
County Durham DH6 4JR
Tel: Durham (0385) 770453

C & G Concrete Ltd
Uffington Road, Stamford,
Lincolnshire PE9 2HA
Tel: Stamford (0780) 52161

Calder Testing Laboratories
Coursington Road, Motherwell,
Lanarkshire, Scotland ML1 1NL
Tel: Motherwell (0698) 65171

Chiltern Concrete Testing Ltd
Busgrove Lane, Stoke Row,
Henley on Thames, Oxfordshire RG9 5QB
Tel: Checkendon (0491) 681343. Also
Bristol (0272) 656438

Construction Testing Laboratories Ltd
Unit C, Grafton Road, Croydon, CR0 3RP
Tel: 081-681 2947/3258

Constructional Services Ltd
Nelsons Row, Clapham, London SW4 4JT
Tel: 0/1-622 3481

Contest CMT Ltd
Contest House, 693-5 Uxbridge Road,
Hayes, Middlesex UB4 0RX
Tel: 01-573 2946/7. Also Daventry
(0327) 703828

Francis Testing Services
Francis House, Shopwyke Road,
Chichester, West Sussex PO20 6AD
Tel: Chichester (0243) 780011

Geocrete Services Ltd
Unit 9, St Andrews Way, Deans,
Livingstone, Lothian, Scotland EH54 8QH
Tel: Livingstone (0506) 411785

Hemel Laboratories Ltd
Fosroc Technology Ltd, Bourne Road,
Aston Birmingham B6 7RB
Tel: 021-328 6866

Melbourne Laboratories Ltd
New Melbourne House, Canning Street,
Maidstone, Kent ME14 2RU
Tel: Maidstone (0622) 679951/2

Mid-Sussex Construction Services Ltd
20 Barttelot Road, Horsham, RH12 1DQ
Tel: Horsham (0403) 69276/54279

Oxford Polytechnic
Department of Civil Engineering,
Building and Cartography, Headington,
Oxford OX3 0BP
Tel: Oxford (0865) 64777

Pencrest Ltd
Pendlebury Fold, Bolton,
Greater Manchester BL3 4SF
Tel: Bolton (0204) 62231

Pioneer Testing Services Ltd
Pioneer Concrete (UK) Ltd,
16 Hertford Road, Shoreditch, London N1
Tel: 071-249 0244. Also: Belfast
(0232) 741395/740227; Birmingham
021-328 5340; Leeds (0532) 495620;
Stirling (0786) 64201

Plymouth Polytechnic
Civil Engineering Materials Laboratory,
Drake Circus, Plymouth, Devon PL4 8AA
Tel: Plymouth (0752) 21312 Ext 5552

Probe Technical Services
Tolpits Industrial Estate, Tolpits Lane,
Watford, Hertfordshire WD1 8XA
Tel: Rickmansworth (0923) 778466

Queens University (NIMTS)
Department of Civil Engineering,
Stranmillis Road, Belfast,
Northern Ireland Bt9 5AF
Tel: Belfast (0232) 661111

Royal Military College of Science
Shrivenham, Swindon, Wiltshire SN6 8LA
Tel: Swindon (0793) 782551

STATS Ltd
31-35 High Street, Sandridge,
St Albans, Hertfordshire AL4 9DD
Tel: St Albans (0727) 33261

Frank Saynor & Associates Ltd
60 Dykehead Street, Queenslie, Glasgow,
Scotland G33 4AQ
Tel: 031-556 0383

Site Services (Testing & Drilling) Ltd
12 Hermand Street, Edinburgh, Lothian,
Scotland EH11 1QT
Tel: 031-337 5048/4991

Soils & Materials Testing Ltd
Unit 5, Manor Industrial Estate,
Lower Wash Lane, Latchford, Warrington,
Cheshire WA4 1PL
Tel: Warrington (0925) 37813

Tarmac Construction Holdings Ltd
CED Laboratory, Millfields Road,
Ettingshall, Wolverhampton,
West Midlands WV4 6JP
Tel: Wolverhampton (0902) 41101

Test Houses Ltd
Tarmac Topmix, Swinbourne Road,
Basildon, Essex SS13 1EZ
Tel: Basildon (0268) 727288

Testing Services Ltd
Staines Lane, Chertsey, Surrey KT6 8PP
Tel: Chertsey (093 28) 60211. Also:
Birmingham 021-359 6845; Sheffield
(0742) 431583; Widnes 051-424 7102

Tilcon Testing Services
Cleckheaton Road, Low Moor, Bradford,
West Yorkshire BD12 0QH
Tel: Bradford (0274) 672031. Also:
Blackburn (0245) 62211; Chester-le-
Street, Tyneside (091) 4103180; Glasgow
041-554 1818; Stafford (0785) 59431;
Stockport 061-477 2700

University of Ulster
Materials Testing Laboratory,
Faculty of Technology, Shore Rd
Newton Abbey, County Antrim,
Northern Ireland BT37 0QB
Tel: Whiteabbey (0231) 65131

Wimpey Laboratories Ltd
Concrete Section, Beaconsfield Road,
Hayes, Middlesex UB4 0LS
Tel: 081-573 7744

Appendix 2

Testing Laboratories

Aberdeen Concrete Co Ltd
Greenbank Road, Tullos, Aberdeen,
Scotland AB1 4BQ
Tel: Aberdeen (0224) 871444

Albury Laboratories Ltd
The Old Mill, Albury, Guildford,
Surrey GU5 9AZ
Tel: Shere (048 641) 2041

Bedfordshire County Council
Engineering Laboratory,
Austin Canons Depot, Bedford Road,
Kempston, Bedfordshire MK42 8AA
Tel: 0234 45493

British Ceramic Research Associaion Ltd
Queens Road, Penkhull, Stoke-on-Trent,
Staffordshire ST4 7LQ
Tel: Stoke-on-Trent (0782) 45431

British Gypsum Ltd
Research & Development Department,
East Leake, Loughborough, Leicestershire
LE2 6JQ
Tel: Nottingham (0602) 214321

British Standards Institution
Test House, Marylands Avenue,
Hemel Hempstead, Hertfordshire HP2 4SQ
Tel: Hemel Hempstead (0442) 3111

**The Chatfield Applied Research
Laboratories Ltd**
13 Stafford Road, Croydon,
Surrey CR0 4NG
Tel: 081-688 5689

Fairclough Civil Engineering Ltd
Geotechnical Services (Northern Div),
Chapel Street, Adlington,
Lancashire PR7 4JP
Tel: Adlington (0257) 480264

**Fire Insurers Research & Testing
Organisation**
Melrose Avenue, Borehamwood,
Hertfordshire WD6 2BJ
Tel: 081-207 2345

F H Gilman Ltd
Bolton Hill, Tiers Cross, Haverfordwest,
Dyfed, Wales SA6 3ER
Tel: Johnston (0437) 890481

**Hampshire County Public Analyst &
Scientific Adviser**
Hyde Park Road, Southsea,
Hampshire PO5 4LL
Tel: Portsmouth (0705) 828965

Hepworth Pipe Co Ltd
Hazlehead, Stocksbridge, Sheffield,
South Yorkshire S30 5HG
Tel: Barnsley (0226) 763561

Ibstock Building Products Ltd
Technical Services Department Laboratory,
Ibstock, Leicestershire LE6 1HS
Tel: Ibstock (0530) 60531

Industrial Science Division
Dept of Economic Development,
Antrim Road, Lisburn, County Antrim,
Northern Ireland BT28 3AL
Tel: Lisburn (084 62) 5161

Laing Design & Development Centre
Page Street, London NW7 2ER
Tel: (081) 959 3636

Sir Alfred McAlpine & Son (Southern) Ltd
Kingswood Laboratory, Holyhead Road,
Albrighton, Wolverhampton,
Staffordshire WV7 3AR
Tel: (081) 573 2637

Minton, Treharne & Davies Ltd
Merton House, Bute Crescent, Cardiff,
South Glamorgan, Wales CF1 6NB
Tel: Cardiff (0222) 489002

Nicholes Colton & Partners
7-11 Harding Street, Leicester, LE1 4BH
Tel: Leicester (0533) 536333

Harry Stanger Ltd
The Laboratories, Fortune Lane, Elstree,
Hertfordshire WD6 3HQ
Tel: (081) 207 3191

Taylor Woodrow Construction Ltd
Research Laboratories, Taywood
Engineering Ltd, 345 Ruislip Road,
Southall, Middlesex UD1 2QX
Tel: (081) 575 4509/4849

Timber Research & Development Association
Stocking Lane, Hughenden Valley,
High Wycombe, Bucks HP14 4ND
Tel: Naphill (024 024) 3091

Warrington Research Centre
Holmesfield Road, Warrington,
Cheshire WA1 2DS
Tel: Warrington (0925) 55116

Wimpey Laboratories Ltd
Metallurgy Laboratory and Structures
Laboratory, Beaconsfield Road, Hayes,
Middlesex UB4 0LS
Tel: (081) 573 7744

Yarsley Technical Centre Ltd
Trowers Way, Redhill, Surrey RH1 2JN
Tel: Redhill (0737) 65070/9

Appendix 3
Directory of Equipment Suppliers

Access equipment

Ladders
H C Slingsby Ltd
85-97 Kingsway, London WC2B 6SB
Tel: (071) 405 2551
Fax: (071) 405 7125

Suppliers of Equipment for Hire
EPL International Ltd
Platform Centre, Eaton Green Road, Luton,
Beds LU2 9HD
Tel: (0582) 412715

Damp diagnosis

Channel Electronics
Unit 23A, Cradle Hill Industrial Estate,
Seaford BN25 3JE
Tel: (0323) 894961
Fax: (0323) 893149

Michell Instruments Ltd
Unit 9, Nuffield Close, Nuffield Road,
Cambridge CB4 1SS
Tel: (0223) 312427
Fax: (0223) 66557

Protimeter Ltd
Meter House, Fieldhouse Lane, Marlow,
Bucks SL7 1LX
Tel: (06284) 72722

Sovereign Chemicals Industries Ltd
Park Road, Barrow-in-Furness, Cumbria
LA14 4QU
Tel: (0229) 25045
Fax: (0229) 37290

Analytical Laboratories
Amtac Laboratories Ltd
Norman Road, Broadheath, Altrincham
WA14 4EP
Tel: (061) 928 8924
Fax: (061) 928 7359

BNF Metals Technology Centre
Wantage Business Park, Denchworth Road,
Wantage OX12 9BJ
Tel: (0235) 772992
Fax: (0235) 771144

Building Research Establishment
Garston, Watford WD2 7JR
Tel: (0923) 894040
Fax: (0923) 664010

Fraser-Nash Pendar Ltd
Unit 13 Hamp Industrial Estate, Old Taunton
Road, Bridgewater, Somerset TA6 3NT
Tel: (0278) 456888
Fax: (0278) 453123

International Consulting & Laboratory
Services
Clayton Bostock Hill & Rigby
288 Windsor Street, Birmingham B7 4DW
Tel: (021) 359 5951
Fax: (021) 359 7606

Johnson Matthey PLC
New Garden House, 78 Hatton Garden,
London EC1N 8JP
Tel: (071) 269 8000
Fax: (071) 269 8133

Pore size, area, density
Bremortest
Test for cement or lime content of mortar

Colebrand Instrumentation
Colebrand House, 20 Warwick Street, Regent
Street, London W1R 6BE
Tel: (071) 439 1000
Fax: (071) 734 3358

Salts analysis for identification of rising damp
Dampness Analysis
73 Stratton Road, Princes Risborough,
Aylesbury, Bucks HP17 9AX
Tel: (08444) 5445

Humidity measurement

Acal Auriema Ltd
442 Bath Road, Slough, Berks SL1 6BB
Tel: (0628) 604353
Fax: (0628) 603730

CP Instrument Co. Ltd
PO Box 22, Bishop's Stortford, Herts
Tel: (0279) 757711
Fax: (0279) 755785

F Darton & Co. Ltd
Mercury House, Vale Road, Bushey, Watford,
Herts WD2 2HG
Tel: (0923) 226019
Fax: (0923) 52533

Grant Instruments (Cambridge) Ltd
Barrington, Cambridge CB2 5QZ
Tel: (0763) 60811
Fax: (0763) 62410

Jenway Ltd
Gransmore Green, Felstead, Essex CM6 3LB
Tel: (0371) 820122
Fax: (0371) 821083

Kane-May Ltd
Swallowfield, Welwyn Garden City AL7 1JP
Tel: (0707) 331051
Fax: (0707) 331202

PP Controls Ltd
Cross Lances Road, Hounslow, Middx
TW3 2AF
Tel: (081) 572 3331
Fax: (081) 572 6219

Protimeter Ltd
Meter House, Fieldhouse Lane, Marlow,
Bucks SL7 1LX
Tel: (06284) 72722

Psika Pressure Systems Ltd
Lambert House, Brook Street, Glossop,
Derbyshire SK13 8BG
Tel: (04574) 3358
Fax: (04574) 69364

Wallac (Newbury) Ltd
Crown House, Kings Road West, Newbury,
Berks RG14 5BY
Tel: (0635) 49429
Fax: (0635) 37335

APPENDICES

Relative humidity probe
Protimeter Ltd
Meter House, Fieldhouse Lane, Marlow,
Bucks SL7 1LX
Tel: (06284) 72722

Condensation Locators
Dampness Analysis
73 Stratton Road, Princes Risborough,
Aylesbury, Bucks HP17 9AX
Tel: (08444) 5445

Protimeter Ltd
Meter House, Fieldhouse Lane, Marlow,
Bucks SL7 1LX
Tel: (06284) 72722

Leak detection

Ultrasonic equipment
Medata Systems Ltd
The Parade, Pagham, West Sussex
PO21 4TW
Tel: (0243) 265528

Megger Instruments Ltd
Archcliffe Road, Dover, Kent CT17 9EN
Tel: (0304) 202620
Fax: (0304) 207342

Airflow Measurement
Dantec Electronics Ltd
Techno House, Redcliffe Way, Bristol
BS1 6NU
Tel: (0272) 291436
Fax: (0272) 213532

F Darton & Co. Ltd
Mercury House, Vale Road, Bushey, Watford,
Herts WD2 2HG
Tel: (0923) 226019
Fax: (0923) 52533

Michell Instruments Ltd
Unit 9, Nuffield Close, Nuffield Road,
Cambridge CB4 1SS
Tel: (0223) 312427
Fax: (0223) 66557

PP Controls Ltd
Cross Lances Road, Hounslow, Middx
TW3 2AF
Tel: (081) 572 3331
Fax: (081) 572 6219

Wallac (Newbury Ltd)
Crown House, Kings Road West, Newbury,
Berks RG14 5BY
Tel: (0635) 49429
Fax: (0635) 37335

Concrete reinforcement location

Impulse Radar Services
Harry Stanger
The Laboratories, Fortune Lane, Elstree,
Herts WD6 3HQ
Tel: (081) 207 3191

Kolectric Ltd
Dean International House, Thames Industrial
Estate, Marlow, Bucks SL7 1TB
Tel: (06284) 77266
Fax: (06284) 890178

Ultrasonic testing

Acal Auriema Ltd
442 Bath Road, Slough, Berks SL1 6BB
Tel: (0628) 604353
Fax: (0628) 603730

Electrical supply testing

Megohm Meter/Insulation Testers
Acal Auriema Ltd
442 Bath Road, Slough, Berks SL1 6BB
Tel: (0628) 604353
Fax: (0628) 603730

Clare Instruments Ltd
Woods Way, Goring-by-Sea, Worthing, West
Sussex BN12 4QY
Tel: (0903) 502551
Fax: (0903) 44258

Megger Instruments Ltd
Archcliffe Road, Dover, Kent CT17 9EN
Tel: (0304) 202620
Fax: (0304) 207342

RE Instruments Ltd
Sherwood House, High Street, Crowthorne,
Berks RG11 7AT
Tel: (0344) 772369
Fax: (0344) 778809

Earth Testing (Electrical)
Megger Instruments Ltd
Archcliffe Road, Dover, Kent CT17 9EN
Tel: (0304) 202620
Fax: (0304) 207342

John Minister Instruments Ltd
Unit 15/16, Highfield Industrial Estate,
Bradley Road, Folkestone CT19 6DD
Tel: (0303) 41598
Fax: (0303) 44884

MPE Ltd
Brunswick Road, Cobb's Wood, Ashford,
Kent TN23 1EB
Tel: (0233) 623404
Fax: (0233) 41777

Northern Design (Electronics) Ltd
228 Bolton Road, Bradford BD3 0QW
Tel: (0274) 729533
Fax: (0274) 721074

Clip-on Meters
Channel Electronics
Unit 23A, Cradle Hill Industrial Estate,
Seaford, East Sussex BN25 3JE
Tel: (0323) 894961
Fax: (0323) 893149

Megger Instruments Ltd
Archcliffe Road, Dover, Kent CT17 9EN
Tel: (0304) 202620
Fax: (0304) 207342

John Minister Instruments Ltd
Unit 15/16, Highfield Industrial Estate,
Bradley Road, Folkestone CT19 6DD
Tel: (0303) 41598
Fax: (0303) 44884

Robin Electronics Ltd
GEC Centre, East Lane, Wembley HA9 7YA
Tel: (081) 908 4335
Fax: (081) 908 4101

Laser measurement

Nevill Long & Co. (Boards) Ltd
North Hyde Wharf, Hayes Road, Southall,
Middx UB2 5NL
Tel: (081) 574 6151
Fax: (081) 571 9739

Sonic measurement

Anton Industrial Services
17 The Water Gate, Carpenders Park,
Watford, Herts WD1 5AS
Tel: (081) 428 6418
Fax: (081) 428 5204

C Z Scientific Instruments Ltd
Zeiss England House, PO Box 43, 1 Elstree
Way, Borehamwood, Herts WD6 1NH
Tel: (081) 953 1688
Fax: (081) 953 9456

Ultra Sonic Tape Measure

Solex International
95 Main Street, Broughton Astley, Leics
LE9 6RE
Tel: (0455) 283486
Fax: (0455) 283912

Infra-red photography

Acal Auriema Ltd
442 Bath Road, Slough, Berks SL1 6BB
Tel: (0628) 604353
Fax: (0628) 603730

Agema (Thermovision)
Arden House, West Street, Leighton Buzzard,
Beds LU7 7DD
Tel: (0525) 375660
Fax: (0525) 379271

Thermographic Surveys
Tremco Ltd
86-88 Bestobell Road, Slough, Berks
SL1 4SZ
Tel: (0753) 691696
Fax: (0753) 822640

Kodak Film such as Ektachrome Infrared film
2236 or High-speed film 2481

APPENDICES

Video

Acal Auriema Ltd
442 Bath Road, Slough, Berks SL1 6BB
Tel: (0628) 604353
Fax: (0628) 603730

Metal detection

Dale Electronics Ltd
Dale House, Wharf Road, Frimley Green,
Camberley, Surrey GU16 6LF
Tel: (0252) 835094
Fax: (0252) 837010

Littlemore Scientific Engineering Co.
Railway Lane, Littlemore, Oxford OX4 4PZ
Tel: (0865) 747437
Fax: (0865) 747780

Protovale (Oxford) Ltd
Rectory Lane Trading Estate, Kingston
Bagpuize, Abingdon, Oxon OX13 5AS
Tel: (0865) 821277
Fax: (0865) 820573

Wells Krautkramer Ltd
Blackhorse Road, Letchworth, Herts
SG6 1HF
Tel: (0462) 678151
Fax: (0462) 679599

Woodbridge Electronics Ltd
Deben Way, Melton, Woodbridge, Suffolk
IP12 1RB
Tel: (03943) 6887
Fax: (03943) 380081

Endoscopes

Acal Auriema Ltd
442 Bath Road, Slough, Berks SL1 6BB
Tel: (0628) 604353
Fax: (0628) 603730

Interscope
P W Allen & Company
25 Swan Lane, Evesham, Worcs WR11 4PE
Tel: (0386) 40148

Fibre Optics Research & Technology Ltd
2 Riverdale Estate, Vale Road, Tonbridge,
Kent TN9 1SS
Tel: (0732) 366266
Fax: (0732) 352710

Edward Fletcher & Partners
25 West Park Road, Kew, Richmond, Surrey
TW9 4DB
Tel: (081) 876 2204

Introvision Ltd
Spirella Building, Letchworth, Herts SG6 4ET
Tel: (0462) 673113

Light measurement

Digital Lux Meter

Solex International
95 Main Street, Broughton Astley, Leics
LE9 6RE
Tel: (0455) 283486
Fax: (0455) 283912

Channel Electronics
Unit 23A, Cradle Hill Industrial Estate,
Seaford, East Sussex BN25 3JE
Tel: (0323) 894961
Fax: (0323) 893149

Littlemore Scientific Engineering Co.
Railway Lane, Littlemore, Oxford OX4 4PZ
Tel: (0865) 747437
Fax: (0865) 747780

Macam Photometrics Ltd
10 Kelvin Square, Livingston EH54 5PF
Tel: (0506) 37391
Fax: (0506) 38543

Megatron Ltd
165 Marlborough Road, London N19 4NE
Tel: (071) 272 3739
Fax: (071) 272 5975

Megger Instruments Ltd
Archcliffe Road, Dover, Kent CT17 9EN
Tel: (0304) 202620
Fax: (0304) 207342

Manometer

Air Neotronics Ltd
Monument Industrial Park, Chalgrove, Oxford
OX9 7RW
Tel: (0865) 891190
Fax: (0865) 891541

PP Controls Ltd
Cross Lances Road, Hounslow, Middx
TW3 2AF
Tel: (081) 572 3331
Fax: (081) 572 6219

Vibration excited testing

Vibration Excited Structure Testing
IRD Mechanalysis (UK) Ltd
Bumpers Lane, Sealand Industrial Estate,
Chester CH1 4LT
Tel: (0244) 374914
Fax: (0244) 379870

Non-destructive testing
Diagnostic Sonar Ltd
Kirkton Campus, Livingston EH54 7BX
Tel: (0506) 411877
Fax: (0506) 412410

Pantatron Radiation Engineering Ltd
H & M Engineering Works, Castlemaine
Avenue, Gillingham, Kent ME7 2QG
Tel: (0634) 52359
Fax: (0634) 280335

Wells Krautkramer Ltd
Blackhorse Road, Letchworth, Herts
SG6 1HF
Tel: (0462) 678151
Fax: (0462) 679599

Wells Krautkramer Ltd
Castle Vale Industrial Estate, Minworth,
Sutton Coldfield, Warwicks B76 8AY
Tel: (021) 351 5661
Fax: (021) 351 4837

NDT Services On Site/Hire
Amtac Laboratories Ltd
Norman Road, Broadheath, Altrincham,
Cheshire WA14 4EP
Tel: (061) 928 8924
Fax: (061) 928 7359

Energy measurement

Energy Management Systems
Northern Design (Electronics) Ltd
228 Bolton Road, Bradford BD3 0QW
Tel: (0274) 729533
Fax: (0274) 721074

Burner/Combustion Efficiency Testers
Acal Auriema Ltd
442 Bath Road, Slough, Berks SL1 6BB
Tel: (0628) 604353
Fax: (0628) 603730

Kane-May Ltd
Swallowfield, Welwyn Garden City, Herts
AL7 1JP
Tel: (0707) 331051
Fax: (0707) 331202

PP Controls Ltd
Cross Lances Road, Hounslow, Middx
TW3 2AF
Tel: (081) 572 3331
Fax: (081) 572 6219

Heat/BTU Meters
Hydril UK Ltd
AOT Systems Division, Central Way,
Walworth, Andover, Hants SP10 5BY

Environmental Warmth
Michell Instruments Ltd
Unit 9, Nuffield Close, Nuffield Road,
Cambridge CB4 1SS
Tel: (0223) 312427
Fax: (0223) 66557

Radon detectors

Degassing Unit
Laboratory Impex Ltd
Lion Road, Twickenham, Middx
Tel: (081) 891 4881

Digital Level
Solex International
95 Main Street, Broughton Astley, Leics
LE9 6RE
Tel: (0455) 283486
Fax: (0455) 283912

APPENDICES

Flowmeters

Portable – Measure liquid flow from outside pipe

UFP – 1000

Panametrics
Justin Manor, 341 London Road, Mitcham,
Surrey CR4 4BE
Tel: (081) 640 2252
Fax: (081) 640 5859

Pipe, sewer and cable tracers

UNILEC

Moisture Control & Measurement Ltd
Thorp Arch Trading Estate, Wetherby, Yorks
LS23 7BJ
Tel: (0937) 843927
Fax: (0937) 842524

Bibliography
and Suggestions for Further Reading

Ashurst, J. & N., *Practical Building Conservation*, Vols. 1-5, Gower, 1988
English Heritage Technical Handbook
Five volumes dealing with: Stone Masonry; Brick, Terracotta and Earth; Mortars, Plaster and Renders; Metals; and Wood, Glass and Resins

British Gas, *Gas in Housing*, A technical guide, 1988

Building Research Establishment (BRE), Defects Action Sheet (site) DAS 138, Domestic chimneys: rebuilding or lining existing chimneys
Digest 298, The influence of trees on house foundations in clay soils, June 1985
Digest 299, Dry rot: its recognition and control, July 1985
Digest 312, Flat roof design: the technical options, August 1986
Digest 318, Site investigation for low-rise building: desk studies, February 1987
Digest 327, Insecticidal treatments against wood-boring insects, December 1987
Digest 329, Installing wall ties in existing construction, February 1988
Digests 343 and 344, Simple measurement monitoring of movement in low-rise buildings: Part 1 Cracks; Part 2 Settlement, heave and out of plumb, April and May 1989
Digest 345, Wet rots: recognition and control, June 1989
Information Paper IP 26/81, Solar reflective paints
Information Paper IP 06/86, The spacing of wall ties in cavity walls
Information Paper IP 19/86, Controlling death watch beetle
Information Paper IP 13/87, Ventilating cold deck flat roofs
Information Paper IP 16/88, Ties for cavity walls: new developments
Information Paper IP 17/88, Ties for masonry cladding
Information Paper IP 19/88, House inspections for dampness

BRE reports on the structural condition of most of the individual types of non-traditional house referred to within the Housing Defects Act 1984

Cartwright, K. St G., & Findlay, W. P. K., *Decay of Timber and its Prevention*, 2nd edition, HMSO, 1958

Clifton-Taylor, A., *The Pattern of English Building*, Faber & Faber, 1972
Good guide to materials used in construction

Coggins, C. R., *Decay of Timber in Buildings*, The Rentokil Library, Rentokil, 1980

Concrete Society, *Cracks in Concrete*, Technical Report No. 22, Concrete Society, 1982

Curwell, S. R., Fox, R. C., & March, C. G., *Use of CFCs in Building*, Friends of the Earth, 1988 (071-490 1555)

Curwell, S. R., & March, C. G., *Hazardous Building Materials: A Guide to the Selection of Alternatives*, Spon, 1986

Eldridge, H. J., *Common Defects in Buildings*, HMSO, 1976
Illustrated cause and effect of common building features

Hickin, N., *The Woodworm Problem*, 2nd edition, Rentokil Ltd (The Rentokil Library), 1982

Jaggar, R. W., & Drury, F. E., *Architectural Building Construction*, Vols. 1-3, 1916
Out of print, but a useful guide to construction techniques for the late Victorian house

Jenkins, B., Commentary on the 15th edition of the IEE wiring regulations, Institute of Electrical Engineers, 1981

McKay, W. B., *Building Construction*, Vol. 1, Longman, 1970
Vol 2, 1970
Vol 3, 1974
Vol 4, revised by J. K. McKay, 1988
A useful guide to the construction of Edwardian buildings

Meehan, A. P., *Rats and Mice*, The Rentokil Library, Rentokil, 1984

Petterson, B., & Axen, B., *Testing of the Thermal Insulation and Airtightness of Building*, Swedish Council for Building Research, Stockholm, Sweden, 1980

Richardson, C., A J Guide to structural surveys, *The Architects Journal*, 26 June, 3 July, 7 July, 10 July & 24 July 1985
Reprint of a marvellous series of articles published in 1985

Royal Institution of Chartered Surveyors, *Structural Surveys of Residential Property – A Guidance Note*, Surveyors Publications, 1985
House Buyers Report and Valuation Explanatory Notes to Surveyors, 3rd edition, Surveyors Publications, 1987

Timber Research and Development Association (TRADA),
Section O, Sheet No. 3; Introduction to timber framed housing, 1986
Section O, Sheet No. 5; Timber framed housing – specification notes, 1987
Section O, Sheet No. 10; Structural surveys of timber framed houses, 1984

The Village London Atlas, the Changing Face of London 1822-1903, The Alderman Press, 1986

A Guide to Non-traditional Housing in Scotland 1923-1955, HMSO, 1987
Illustrated examples of non-traditional buildings together with sections of the construction

Non-Traditional Housing in Northern England, Northern Consortium of Housing Authorities, 1988
Information booklet available from NCHA Civic Centre, (091) 3885211 ext. 215

The Housing Defects (Prefabricated Reinforced Dwellings), (England and Wales) Designations 1984, Supplementary Information, HMSO, 1984
Photographs of the standard appearance of the building types referred to within the Act

Index

Index

Access equipment 49, 232
Acer psuedoplatanus (sycamore) *179, 187*
Acrylic sealants 123, 125
Adhesive sealant failures 123
Aesculus hippocastanum (horse chestnut) *179, 186*
Air bricks 20
Air circulation 26
Air conditioning 26
Algae 112
Alterations 4
Ambrosia beetles 199
Anchor bolts 114
Anns v Merton London Borough Council [1978] 30, 36
Anobium punctatum (furniture beetle) *201*
Anthrenus verbasci (carpet beetle) *196–197*
Appointment to inspect 12
Armadillidium vulgare (woodlice) *196–197*
Asbestos 147–149
Asbestos cement roofs 87
Aspect 21
Asphalt 74–76
Assets of company 3

Balconies 97
Bark borer 207
Baxter v F. W. Gapp & Co Ltd [1939] 43, 44
Beech 179, 189
Beetles 194–210
Bell Hotels (1935) Ltd v Motion [1952] 41, 44
Betula pubescens (birch) *179, 191*
Binoculars 46
Birch 179, 191
Biscuit beetle 196–197
Bitumen sealants 123, 125
Borescope 61–62
Bostrich beetle 208
Booklice 196–197
Boundaries 3
Bourne v McEvoy Timber Preservations Ltd [1976] 30, 36
Brickwork 20, 94–102
Building Regulations 8, 26
Building Research Establishment Digest 99
Buckland v Watts [1958] 38, 41, 44
Burglars 26
Burglar alarms 26
Business liability 7
Butyl sealants 123, 125

Calcium silicate bricks 102
Carbide moisture meter 50–51
Caretaker 24
Carpet beetle 196–197
Causation 38
Cavity brickwork 99
Cavity insulation 100
Ceilings 21
Cement 114, 118
Central heating 21
Checklist in preparation for a survey 14
Chimneys 20, 82, 83, 87
Chloride attack 26
Clay 34, 168, 169, 170, 172, 173, 174
Clayton v Woodman & Son (Builders) Ltd [1962] 31, 36
Cleaning buildings 112, 113
Client, what does he want? 4
Coefficient of linear expansion 78
Cohesive sealant failures 124

Collard v Saunders [1972] 43, 44
Combed wheat thatch 84
Communal heating 24
Company sales, survey for 3
Compensation 38
Conclusions 17
Concrete 54, 55, 117–122
Concrete cancer 122
Concrete, lightweight pre-cast houses 69–71
Concrete reinforcement detection 54–55
Condensation 68, 77, 139–140
Coniphora puteana *219*
Conn v Munday [1955] 33, 36
Consent, occupiers' and vendors 12
Contract 5, 7, 8
Contractual relationship 5
Copper 98
Coprinus domesticus (Ink-cap) *228*
Corisand Investments Ltd v Druce & Co [1978] 43, 44
Corrosion of metals 54–55
Cost in use 26, 27, 28
Cracks, classification 65, 167–168

Daedalia quercina *224*
Daisley v B. S. Hall & Co. [1973] 33, 36
Damages 38
Damages, head of 43
Damp, diagnosis of 50–53
Damp proof course 20, 77, 101
Dampness, interpretation from moisture content 52
Death watch beetle 202–203
Decay, resistance to in hardwoods 213
Decorations 21, 151–164
Decorative failures 151–164
Denning MR, Lord Foreword, 30, 40
Devon reed, 84
Dew point 140
Dewpoint indicator 53
Dodd Properties (Kent) Ltd v Canterbury City Council [1980] 38, 44
Donkioporia expansa *222*
Donoghue v Stevenson [1932] 30, 36
Drainage 26, 110
Drinnan v C. W. Ingram & Sons [1967] 33, 36, 43, 44
Dry rot 216–218
Dutch elm disease 179
Dutton v Bognor Regis UDC [1972] 31, 36, 40, 44
Duty, breach of 29–36
Duty of the surveyor 29, 30
Duty, principle of 30

Eagle Star Insurance Co Ltd v Gale & Power [1955] 43, 44
Electric moisture meter 46, 51–53
Electricity 28, 229–230
Electricity testing 63–64, 229–230
Electrics 21, 26, 228
Electrostatic meters 59
Elf-cup 228
Elm, English 179, 185
Endoscopes 61–62
Epoxy resin 144
Equipment 45–65
Ernobius mollis (bark borer beetle) *207*
Escalators 26
Esso Petroleum Co Ltd v Mardon [1976] 30, 36
Europryum confine (wood boring weevils) *206*
Equipment suppliers 232–235

Index

Expanded polystyrene 100, 132
Experience, lack of 33

Factories 26
Fagus sylvatica (beech) 179, 189
Fees 4, 38
Felt 82
Fibre optic probes 61–62, 99, 101
Fibrioporia vaillantii 220
Fire alarms 26
Fire Precautions Act 27
Flashings 82
Flat roofs 73–79
Flats 23, 24
Floors 21
Ford v White & Co [1964] 41, 44
Foundations 21, 170, 174
Frost action 150
Fryer v Bunney [1982] 33, 36
Fungus, timber decay caused by 211–228
Furniture Beetle 201

Gap gauges 65
Gas 25
GLC (Greater London Council) 28
Gribble 209
Ground conditions 165–176
Grove v Jackman & Masters [1950] 33, 36
Guarantees 12
Guttering 20, 78, 105–109

Hammer 46
Hardy v Wamsley Lewis [1967] 34, 36, 41, 44
Harling (see render) 96
Health and Safety at Work Act 8, 27
Health, in the home 68
Heating services 26
Heave 173–174, 179
Hedley Byrne & Co Ltd v Heller & Partners Ltd [1964] 30, 36
High alumina cement (HAC) 26, 118–121
Hill v Debenham, Tewson & Chinnocks [1958] 33, 36
Hoists 26
Hood v Shaw [1960] 34, 36, 43, 44
Horse chestnut 179, 186
House longhorn beetle 198
Humidity 53
Hydraulic lime 114
Hygrometer, whirling 53
Hygroscopicity 52
Hylotrupes bajulus (house longhorn beetle) 198

Identity card 65
IEE Regulations 63, 229–230
Infra red photography 58
Infra red thermography 57
Ink-cap 228
Inspection, inadequate 34
Instructions, confirmation 8, 15
    failure to carry out 33
    letter to confirm 10, 11
    specimen instruction sheet 9
    taking 3
Insurance 28, 173, 174
Interest 44
Intruder alarm system 26

Joint failures 121, 123–128

Kenney v Hall Pain & Foster [1976] 38, 44

Laboratories, testing 231–232
Ladders 49
Law Reform (miscellaneous provisions) Act 1934 44
Laser measurement 56
Lead 98
Lead in paint 164
Leaks, detection 53
Lease 4, 12, 23
Legislative and other controls 27
Lentinus lepideus (stag's horn fungus) 223
Lepisma saccharina (silverfish) 196–197
Liability, protection against 34
Licences 8, 26
Lichen 87
Lifts 23
Light measurement 60
Lime 179, 188
Limitation 44
Limnoria lignorum (gribble) 209
London & South of England Building Society v Stone [1982] 43, 44
Long straw thatch 84–85
Lyctus brunneus (powder post beetle) 200

Maintenance charge 23
Maintenance liability 26
Manometers 60
Mathematical tiling 98
Means of escape in case of fire 8, 24, 27
Mental defectives (see also vandalism) 5
Metal detectors 62, 99
Mine fungus 220
Minors 5
Misrepresentation Act 1967 5, 7
Moisture 50–51, 68, 77, 139–140
Moisture meter 50–51
Morgan v Perry [1973] 41, 44
Mortgagees 43
Movement monitoring 58–59
Mould growth 68, 225
Myxomycetes (slime moulds) 225

Nacerdes melanura (wharf borer beetle) 204
Nail sick 80–82
Nibs (on tiles) 80
Negligence 41
Negligence, contributary 44
Norfolk reed 84, 85

Oak 179, 184
Offices 26
Offices Shops and Railway Premises Act 26
Oniscus asellus (woodlice) 196–197
Openings 20, 133–138
Outbuildings 4
Overseas Tankship (UK) v Morts Docks & Engineering Co Ltd [1961] 40, 44

Parapets 76
Parsons v Way & Waller Ltd [1952] 41, 44
Party walls 82
Paxillus panuoides 221
Peat 170, 175
Pebbledash (see also render) 96
Perry v Sidney Phillips & Son [1982] 40, 41, 44
Pezisa (Elf-cup) 228

Index

Phellinus contiguus   226
Philips v Ward [1956]   40, 41, 43, 44
Photogrammetry   56
Pinhole borer beetles   199
Pirelli General Cable Works Ltd v Oscar Faber & Partners [1982]   44
Pitched roofs   80–92
Plane   179, 190
Plastics   129, 132
Platanus acerifolia (plane)   179, 190
Platypodidae (pinhole borer beetles)   199
Plumb line   47
Plumbing   26
Pointing   94, 95
Polysulphide sealants   123–125
Polyurethane   100
Polyurethane sealants   123, 125
Ponding on roofs   74
Poplar, Lombardy   179, 183
Populus nigra variety italica (Lombardy poplar)   179, 183
Porcellio scaber (woodlice)   196–197
Poria placenta   227
Powder post beetle   200
Proformas   18
Psocids (booklice)   196–197

Quercus robur (oak)   179, 184

Radiation, risk of   68
Radon detectors   63
Rainwater downpipe capacity   108
Redressing stone   114
Relative humidity probe   53
Remoteness   38, 40
Render   96
Repairs, cost of   3, 15
Report, the   15, 20
          inadequate   34
RICS   31, 46, 47
RICS Guidance Note "Structural Surveys of Residential Property"   17, 46
Roof   18, 19, 20, 73–92
Roof pitches   80–87
Roofing felt   21, 76, 82
Root damage   178

Salix fragilis (crack willow)   179, 182
Sand   170, 175, 176
Scolytidae (pinhole borer beetle)   199
Sealants   123–128
Service charges   12, 23
Settlement   166–173
Serpula lacrymans   218
Shingles   86
Shipworm   210
Shops   26
Silicone sealants   123, 125
Silverfish   196–197
Simple Simon Catering Ltd v Binstock Miller & Co [1973]   41, 44
Singer & Friedlander Ltd v John D Wood & Co [1977]   31, 36
Sirex noctillo (wood wasp)   205
Slates   82, 83
Slime moulds   225
Soils   175–176
Solar radiation   164
Solar reflective paint   76
Sonic measurement   56
Sound measurement   60
South Western General Property v Marton [1983]   7

Smith v Eric S Bush D.C. [1989]   7, 31
Sparham-Souter v Town & Country Developments (Essex) Ltd [1976]   41, 44
Special relationship   30
Stag's horn fungus   223
Stairs   21
Standard letters, confirm instructions   10, 11
          confirm access to property   13
Standard paragraphs   18, 20, 21
Stegobium paniceum (biscuit beetle)   196–197
Stewart v H. D. Brechin & Co. [1959]   33, 36
Stone   111–116
Subsidence   173–174
Sulphate attack   26
Supply of Goods and Services Act 1982   7
Surveys, list of types   3
Sycamore   179, 187

Telephone systems   25
Telltales   58
Teredo navale (shipworm)   210
Thatch   84–85
Tiles   80–82, 86, 98
Tilia europaea (lime)   179, 188
Timber frame buildings   141–142
Torch   46
Town and Country Planning Acts   8, 26
Tree, development of the   212
Trees   177–193
Trees, estimation of height   192–193

Ulmus procera (English elm)   179, 185
Ultrasonic leak detector   55–56
Ultrasonic testing   55–56
Ultraviolet radiation   75
Unfair Contract Terms Act 1977   7, 8, 34, 36
Urea-formaldehyde foam   100, 129–132

Valuations   3
Value Added Tax   3, 8
Vandalism   28
Ventilation   78
Verge treatment   77
Video   59

Wall ties   99
Wall tie failure   99
Walls   20, 92–110
Warehouses   26
Water supplies   21
Weather – state during survey   17
Weevils   206
Wet rot   219
Wharf borer   204
Willow, crack   179, 182
Wimpey Construction UK Ltd v Poole   33, 36
Window cleaning equipment   25
Windows   133, 138
Wood wasp   205
Woodblock floors   228
Woodlice   196–197
Wright v British Railways Board   44

Xestobium rufovillosum (death watch beetle)   202–203

Yianni v Edwin Evans & Sons [1981]   31, 33, 36, 43, 44

Zinc   98